KU-572-738

WITHDRAWN

Introduction to LINEAR OPTIMIZATION and EXTENSIONS with MATLAB®

The Operations Research Series

Series Editor: A. Ravi Ravindran

Professor, Department of Industrial and Manufacturing Engineering
The Pennsylvania State University – University Park, PA

Published Titles:

Introduction to Linear Optimization and Extensions with MATLAB®
Roy H. Kwon

Supply Chain Engineering: Models and Applications
A. Ravi Ravindran & Donald Paul Warsing

Analysis of Queues: Methods and Applications
Natarajan Gautam

Integer Programming: Theory and Practice
John K. Karlof

Operations Research and Management Science Handbook
A. Ravi Ravindran

Operations Research Applications
A. Ravi Ravindran

Operations Research: A Practical Introduction
Michael W. Carter & Camille C. Price

Operations Research Calculations Handbook, Second Edition
Dennis Blumenfeld

Operations Research Methodologies
A. Ravi Ravindran

Probability Models in Operations Research
C. Richard Cassady & Joel A. Nachlas

Introduction to LINEAR OPTIMIZATION and EXTENSIONS with MATLAB®

ROY H. KWON

CRC Press is an imprint of the
Taylor & Francis Group, an **informa** business

MATLAB® is a trademark of The MathWorks, Inc. and is used with permission. The MathWorks does not warrant the accuracy of the text or exercises in this book. This book's use or discussion of MATLAB® software or related products does not constitute endorsement or sponsorship by The MathWorks of a particular pedagogical approach or particular use of the MATLAB® software.

CRC Press
Taylor & Francis Group
6000 Broken Sound Parkway NW, Suite 300
Boca Raton, FL 33487-2742

Printed on acid-free paper
Version Date: 20130725

International Standard Book Number-13: 978-1-4398-6263-6 (Hardback)

Visit the Taylor & Francis Web site at
http://www.taylorandfrancis.com

and the CRC Press Web site at
http://www.crcpress.com

This work is dedicated to my wife Gina and son Noah and my parents Donald and Ester.

Preface

This book is an outgrowth of lecture notes used for teaching linear programming to graduate students at the University of Toronto (U of T). There have been hundreds of graduate students over the last decade from various parts of the U of T that have taken my courses, most notably from industrial engineering/operations research, electrical, civil, mechanical, and chemical engineering. This group also includes students in the Masters of Mathematical Finance (MMF) Program at the U of T, where I have been teaching a special course in operations research for which the bulk of topics relate to linear and quadratic programming with applications in finance, e.g., portfolio optimization.

Providing concrete examples and illustrations before more general theory seems to work well for most students, and this book aims to take that path. In fact, the book can be used without the need to go through all of the proofs in the book. Students that plan on specializing in optimization would be encouraged to understand all proofs in the book as well as tackle the more theory-oriented exercises. Thus, the material in this book is designed to be of interest and to be accessible to a wide range of people who may be interested in the serious study of linear optimization. This book may be of special interest to those that are interested in financial optimization and logistics and supply chain management. Many of the students regard the computational aspects as an essential learning experience. This has been reflected in this book in that MATLAB® is integrated along with the learning of the conceptual and theoretical aspects of the material.

A unique feature of this book is the inclusion of material concerning linear programming under uncertainty. Both stochastic programming and robust optimization are introduced as frameworks to deal with parameter uncertainty. It is novel to develop these topics in an introductory book on linear optimization, and important, as most applications require decisions to be made in the face of uncertainty and therefore these topics should be introduced as early as possible.

Furthermore, this book is not encyclopedic and is intended to be used in a one-semester course. The main topics were chosen based on a set of core topics that would be needed as well as additional topics that round out and illustrate the modern development of linear optimization and extensions. For example, this book discusses interior point methods but only develops primal-dual path-following methods and not the myriad other interior point methods for linear programming. To this end, we chose the primal-dual path-following method based on its good theoretical and practical properties and yet at the same time illustrates the key issues involved in designing any interior point method.

This book avoids the use of tableaus in the development of the simplex method. Tableaus have been a mainstay for most presentations of the simplex method-based algorithms for linear programming. However, this books takes the view that the underlying geometry of linear programming is such that the algorithms (not just the simplex method) have a natural geometrical analog in the matrix algebra representation version, which is lost in using tableaus. In particular, simplex method-based algorithms are iterative and are viewed naturally as finding a direction of improvement and step length from a current iterate to get to an improved point and so on until optimality is reached or the problem is discovered to be unbounded. A consequence is that it becomes even more natural for MATLAB to facilitate algorithmic understanding by leaving the elementary row operations in performing inversions to MATLAB instead of requiring a student to do the equivalent by doing a pivot on the tableau.

The prerequisites for this book are courses in linear algebra, multi-variate calculus, and basic proficiency in MATLAB. Well-prepared advanced undergraduates could find the book accessible as well. In fact, only several concepts from linear algebra and multi-variate calculus are needed. The appendix contains those concepts from linear algebra that are especially relevant in this book. The multi-variate calculus is reviewed at those points in the book that require it. What I mean by basic MATLAB proficiency is that one knows how to perform standard matrix algebra operations in MATLAB, e.g., multiplying two matrices together and solving a system of linear equations. In any case, MATLAB is a very convenient and powerful platform for optimization and it is relatively easy to get started. A starting point for review are the excellent on-line tutorials and resources available from MathWorks at the website

http://www.mathworks.com/academia/student_center/tutorials
/launchpad.html

Chapter 1 introduces the linear programming problem and gives many examples starting from the well-known diet problem to more complex network optimization models. Various transformation techniques are given so that one can transform an arbitrary linear program in standard form. The MATLAB function linprog is introduced showing how one can solve linear programs on a computer. A computational (case study) project requires the construction and solution of a larger-sized (compared to examples in the chapter) linear program using real financial data in MATLAB.

Chapter 2 develops the geometry of linear programming. First, the geometry of the feasible set of an LP is considered. The geometry of LP gives insight on the nature of optimal and unbounded solutions in terms of corner points, extreme points, and directions of unboundedness. A key development in this chapter is the corresponding algebraic notions of a basic feasible solution and extreme directions. The chapter culminates with the Fundamental Theorem of Linear Programming, which states that the optimal solution for a finite LP can be attained at a basic feasible solution.

Chapter 3 develops the simplex method. As mentioned, the development eschews the tableau construct and the simplex method is developed within the matrix algebraic representation given by the partition of the problem into basic and non-basic variables. Detailed examples are given that illustrate the various possibilities in executing the simplex method including cycling. The revised simplex method is then developed, which brings to light the importance of numerical linear algebra in solving linear programs. MATLAB code is given that implements the simplex method. The MATLAB code is not claimed to be the most efficient or robust, but serves as an example of how the simplex method, as described in the chapter, may be implemented.

Chapter 4 considers duality theory of linear programming. Duality theory enables the development of another variant of the simplex method called the dual simplex method. Economic interpretations of dual variables are discussed and then sensitivity analysis is developed.

Chapter 5 develops the Dantzig-Wolfe decomposition method and illustrates the very important strategy of exploiting structure in a linear programming problem. MATLAB code is given to the illustrate the implementation of the decomposition.

Chapter 6 considers an interior point strategy to solve linear programs. In particular, the class of primal-dual path following methods are developed and then a variant from this class called the predictor-corrector method is considered and implemented in MATLAB.

Chapter 7 develops quadratic programming theory and develops optimality conditions for both unconstrained and constrained versions of the problem. The mean-variance portfolio optimization problem is used as an example of a quadratic program and is featured in several of the numerical examples in the chapter. The MATLAB function quadprog is illustrated, which enables quadratic programs to be solved on computer. An application in generating the efficient frontier of a mean-variance portfolio problem is given. Quadratic programming is discussed in the context of convex optimization. A predictor-corrector interior point method for convex quadratic programming is given.

Chapter 8 considers linear programming under uncertainty. The stochastic programming with recourse framework is developed first. The L-Shaped method is developed to solve two-stage stochastic programs with recourse. Then, robust optimization is developed. Examples of developing robust counterparts are considered and illustrated through a robust portfolio problem. A key theme here is the emphasis on tractable robust formulations.

This book was designed to be used in a semester-long course. Chapters 1 through 4 would be considered as the core part of a course based on this book. The remaining chapters do not have to be considered in a linear order. A course emphasizing interior point methods can cover both Chapters 6 and 7. Parts of Chapter 8 depend on Chapter 5, e.g., the proof of convergence of the L-Shaped method needs the development of the Dantzig-Wolfe decomposition.

This book has nearly 100 exercises and a complete solutions manual is available to instructors. Several of the problems require the use of MATLAB. PowerPoint slides are also available for each chapter. Additional material is available from the CRC Web site: http://www.crcpress.com/product/isbn/9781439862636.

Acknowledgments

This book could not have been completed without the help of several key people. First, I would like to thank Dr. Jonathan Li for putting the book together in LaTex and for also giving some useful feedback concerning various parts of the book. In addition, he also combed through the book with me to clean up and improve the presentation. Second, Dexter Wu helped greatly with the coding of the various algorithms in the book in MATLAB® and provided solutions to the computational problems in the book. His diligence and enthusiasm in helping with the book was immensely helpful and greatly appreciated. Gina Kim provided help with many of the figures in the book and it was a joy to work with her on this project. Finally, I would like to also thank my editor Cindy Carelli for giving me the opportunity to write this book and for her patience and understanding of the several delays that occurred in the development of this work.

About the Author

Roy H. Kwon is currently associate professor in the Department of Mechanical and Industrial Engineering at the University of Toronto, St. George Campus. Also, he is a member of the faculty in the Masters of Mathematical Finance (MMF) Program at the University of Toronto. He received his Ph.D. from the University of Pennsylvania in operations research from the Department of Electrical and Systems Engineering in 2002. Dr. Kwon has published articles in such journals as *Management Science, Naval Research Logistics*, the *European Journal of Operational Research*, and *Operations Research Letters* among others. In addition, he has worked and consulted in the use of operations research (optimization) for the military, financial, and service sectors.

Contents

1

Linear Programming

1.1 Introduction

Linear programming (LP) is the problem of optimizing (maximizing or minimizing) a linear function subject to linear constraints. A wide variety of practical problems, from nutrition, transportation, production planning, finance, and many more areas can be modeled as linear programs. We begin by introducing one of the earliest examples, the diet problem, and then give some additional applications in the areas of production management, transportation, finance, and personnel scheduling. Some of these examples are not initially linear programs, but are amenable to being transformed into LPs and techniques for conversion are illustrated. The embedded assumptions behind linear optimization problems are discussed. A definition of a standard form of a linear optimization problem is given and techniques for converting an LP into standard form are illustrated.

1.1.1 The Diet Problem

Due to a limited budget you would like to find the most economical mix of food items subject to providing sufficient daily nutrition. The available food items are peanut butter, bananas, and chocolate, and the cost per serving is 20 cents, 10 cents, and 15 cents, respectively. The amount of nutrients of fat, carbohydrates, and protein per serving of peanut butter is 130 grams, 51.6 grams, and 64.7 grams. For bananas, the amounts of these nutrients per serving are 1 gram, 51 grams, and 2 grams, whereas for chocolate the amounts per serving are 12 grams, 22 grams, and 2 grams. Suppose it is decided by a dietician that you need at least 35 grams of fat, at least 130 grams of carbohydrate, and at least 76 grams of protein daily. Then, a combination of these three food items should provide you with sufficient amounts of each nutrient for the day. The problem of determining the least cost combination of the food items that provide a sufficient amount of nutrients can be represented as a problem of the following form:

minimize	cost of food items used
subject to	food items must provide enough fat
	food items provide must enough carbohydrates
	food items provide must enough protein

This form of the problem reveals two major components, i.e., (1) there is a goal or objective (minimize cost of food items used) and (2) a set of constraints that represent the requirements for the problem (food items must provide enough nutrition). A linear programming problem will exhibit the same form, but with mathematical representations of the components of the problem. In general, a linear program will consist of decision variables, an objective, and a set of constraints.

To illustrate the formulation of the diet problem as a linear program, let

$$\begin{aligned} x_{pb} &= \text{servings of peanut butter} \\ x_b &= \text{servings of bananas} \\ x_c &= \text{servings of chocolate.} \end{aligned}$$

These variables represent the quantities of each food item to be used and should be non-negative, i.e., $x_{pb} \geq 0$, $x_b \geq 0$, and $x_c \geq 0$, and are called decision variables since they must be determined.

The total cost associated with any particular combination of food items x_{pb}, x_b, and x_c is

$$.20x_{pb} + .10x_b + .15x_c, \tag{1.1}$$

which says that the total cost of food items used is the sum of the costs incurred from the use of each food item.

To ensure that the mix of the three foods will have enough fat one can impose the following linear inequality constraint

$$130x_{pb} + 1x_b + 12x_c \geq 35, \tag{1.2}$$

which expresses that the total amount of fat from the three food items should be at least the required minimum of 35 grams. Similarly, one can impose the inequality

$$51.6x_{pb} + 51x_b + 22x_c \geq 130 \tag{1.3}$$

to ensure that a combination of food items will have enough carbohydrates and finally impose the constraint

$$64.7x_{pb} + 2x_b + 2x_c \geq 76 \tag{1.4}$$

to ensure enough protein.

Then, the diet problem linear program can be expressed as

$$\begin{aligned}
\textit{minimize} \quad & .20x_{pb} + .10x_b + .15x_c \\
\textit{subject to} \quad & 130x_{pb} + 1x_b + 12x_c \geq 35 \\
& 51.6x_{pb} + 51x_b + 22x_c \geq 130 \\
& 64.7x_{pb} + 2x_b + 2x_c \geq 76 \\
& x_{pb}, x_b, x_c \geq 0.
\end{aligned} \tag{1.5}$$

The problem can be interpreted as the problem of determining non-negative values of x_{pb}, x_b, and x_c so as to minimize the (cost) function $.20x_{pb}+.10x_b+.15x_c$ subject to meeting the nutritional requirements as embodied in the constraints (1.2)–(1.4). The function $.20x_{pb} + .10x_b + .15x_c$ is called the objective function, and the objective is the minimization of this function. It is important to observe that the objective function and the left-hand sides of the constraints are all linear, i.e., all variables are taken to the power of 1. The diet problem highlights the major components of a linear programming problem where an LP problem consists of decision variables, an objective (optimize a linear objective function), linear constraints, and possibly some sign restrictions on variables. The diet problem can be generalized to where there are n types of food items and m nutritional requirements, where x_i represents the number of servings of food item i $(i = 1, ..., n)$, c_i is the cost of one serving of food item i $(i = 1, , , .n)$, a_{ij} is the amount of nutrient i in one serving of food item j $(j = 1, ..., m)$, and b_j is the minimum amount of nutrient j required $(j = 1, ..., m)$. Then, the general formulation of the diet problem can be written as

$$\begin{aligned}
\text{minimize} \quad & c_1x_1 + c_2x_2 + \cdots + c_nx_n \\
\text{subject to} \quad & a_{11}x_1 + a_{12}x_2 + \cdots + a_{1n}x_n \geq b_1 \\
& a_{21}x_1 + a_{22}x_2 + \cdots + a_{2n}x_n \geq b_2 \\
& \qquad \vdots \\
& a_{m1}x_1 + a_{m2}x_2 + \cdots + a_{mn}x_n \geq b_m \\
& x_1, x_2, ..., x_n \geq 0.
\end{aligned}$$

In matrix form, the general diet problem can be represented as

$$\begin{aligned}
\text{minimize} \quad & c^T x \\
\text{subject to} \quad & Ax \geq b \\
& x \geq 0
\end{aligned}$$

where

$$x = \begin{bmatrix} x_1 \\ x_2 \\ \vdots \\ x_n \end{bmatrix} \text{ is the vector of food items,}$$

$$A = [a_{ij}] = \begin{bmatrix} a_{11} & a_{12} & \cdots & a_{1n} \\ a_{21} & a_{22} & \cdots & a_{2n} \\ \cdot & & \ddots & \\ a_{m1} & a_{m2} & & a_{mn} \end{bmatrix}$$

$$= \begin{bmatrix} a_1^T \\ \vdots \\ a_m^T \end{bmatrix}$$

is a matrix of dimension $m \times n$ and a_j^T is an n-dimensional row vector that represents the jth row of A i.e. the total amount of nutrient j that would be obtained from the amount of food items given by x,

$$c = \begin{bmatrix} c_1 \\ c_2 \\ \vdots \\ c_n \end{bmatrix} \text{ is the vector of costs of food items,}$$

and

$$b = \begin{bmatrix} b_1 \\ b_2 \\ \vdots \\ b_n \end{bmatrix} \text{ is the vector of nutrition requirements.}$$

For example, for the diet problem (1.5) we have $n = 3$ and $m = 3$ with

$$A = \begin{bmatrix} 130 & 1 & 12 \\ 51.6 & 51 & 22 \\ 64.7 & 2 & 2 \end{bmatrix}, c = \begin{bmatrix} .20 \\ .10 \\ .15 \end{bmatrix}, x = \begin{bmatrix} x_{pb} \\ x_b \\ x_c \end{bmatrix}, \text{ and } b = \begin{bmatrix} 35 \\ 130 \\ 76 \end{bmatrix}.$$

The objective function is $c^T x = \begin{bmatrix} .20 & .10 & .15 \end{bmatrix} \cdot \begin{bmatrix} x_{pb} \\ x_b \\ x_c \end{bmatrix} = .20x_{pb} +$ $.10x_b + .15x_c$.

(Note: The $\cdot$ indicates the dot product.)

The first constraint is

$$a_1^T x = \begin{bmatrix} 130 & 1 & 12 \end{bmatrix} \cdot \begin{bmatrix} x_{pb} \\ x_b \\ x_c \end{bmatrix} = 130x_{pb} + 1x_b + 12x_c \geq b_1 = 35.$$

The second constraint is

$$a_2^T x = \begin{bmatrix} 51.6 & 51 & 22 \end{bmatrix} \cdot \begin{bmatrix} x_{pb} \\ x_b \\ x_c \end{bmatrix} = 51.6x_{pb} + 51x_b + 22x_c \geq b_2 = 130.$$

The third constraint is

$$a_3^T x = \begin{bmatrix} 64.7 & 2 & 2 \end{bmatrix} \cdot \begin{bmatrix} x_{pb} \\ x_b \\ x_c \end{bmatrix} = 64.7x_{pb} + 2x_b + 2x_c \geq b_3 = 76.$$

1.1.2 Embedded Assumptions

It is important to realize the assumptions behind a linear programming problem. In particular, an LP model is characterized by proportionality, divisibility, additivity, and certainty.

1. *Proportionality* means that the contribution toward the objective function and constraints of decisions are directly proportional to its values, i.e., there are no economies of scale such as quantity-based discounts. For example, in the diet problem, every unit (serving) of peanut butter x_{pb} will contribute .20 towards the overall cost and 130 grams of fat toward the fat nutrient requirement. In particular, this means that decision variables in an LP are raised to the first power only. So terms of the form $c_i x_i^{1/2}$ or $c_i x_i^2$ (where c_i is a coefficient and x_i a decision variable) are not permitted in an LP.

2. *Divisibility* means that the decision variables can take on any real number. For example, the amount of peanut butter in the diet problem can be less than one serving, like 35% of a serving, $x_{pb} = .35$, or in general, some fractional amount like $x_{pb} = 1.7$ servings.

3. *Additivity* means that the contribution of a decision variable toward the objective or constraints does not depend on other decision variables. In other words, the total contribution is the sum of individual contributions of each decision variable. For example, the contribution of every serving of peanut butter in the diet problem toward the overall fat requirement of 35 grams is 130 grams, independent of the amount of servings of the other food items.

4. *Certainty* means that the data used as coefficients for a linear programming model, such as the objective coefficients c and constraint coefficients A and b, are known with certainty. For example, it is assumed in the diet model that the cost of one serving of bananas is 10 cents and that one serving of peanut butter provides 130 grams of fat, and these values are assumed as correct or valid for the model.

The embedded assumptions may seem to be overly restrictive, but in fact a wide range of problems can be modeled as linear programs. Recent advances in optimization technology have considered the relaxation of the certainty assumption. In particular, developments in stochastic programming and robust optimization have enabled the incorporation of uncertainty of coefficients in linear programming models. These exciting topics will be covered in Chapter 8.

1.2 General Linear Programming Problems

The diet problem is only one possible form of a linear optimization problem. It is possible to have an LP problem that maximizes an objective function instead of minimizing and a constraint may be of the form $a^T x \geq b$ (greater than or equal inequality), $a^T x \leq b$ (less than or equal inequality), or $a^T x = b$ (equality). Furthermore, variables may be non-negative, non-positive, or unrestricted (i.e., the value can be negative, positive, or zero). For example, the following is a linear programming problem with equality and inequality constraints as well as variables that are non-negative, non-positive, and unrestricted.

$$\begin{array}{ll} \textit{maximize} & 5x_1 + x_2 - 3x_3 \\ \textit{subject to} & x_1 + x_2 \leq 6 \\ & x_2 + x_3 \geq 7 \\ & x_1 - x_3 = 2 \\ & x_1 \geq 0 \\ & x_2 \leq 0 \\ & x_3 \textit{ unrestricted.} \end{array} \tag{1.6}$$

In general, a linear programming problem can be represented in the form

$$\begin{array}{lll} \textit{minimize or maximize} & c^T x & \\ \textit{subject to} & a_i^T x \leq b_i & i \in L \\ & a_i^T x \geq b_i & i \in G \\ & a_i^T x = b_i & i \in E \\ & x_j \geq 0 & j \in NN \\ & x_j \leq 0 & j \in NP. \end{array}$$

where L, G, and E are index sets of constraints that are of the less than or equal to type, greater than or equal to type, and of equality type, respectively. NN (NP) is an index set of variables that are non-negative (non-positive). All other variables are assumed to be unrestricted.

Furthermore, an equality constraint $a^T x = b$ can be represented as the two inequalities $a^T x \leq b$ and $a^T x \geq b$, and an inequality constraint of the form $a^T x \geq b$ can be written as a less than or equal type by multiplying both sides by -1 to get $-a^T x \leq -b$. Thus, the LP (1.6) is equivalent to the following

$$
\begin{array}{ll}
\textit{maximize} & 5x_1 + x_2 - 2x_3 \\
\textit{subject to} & x_1 + x_2 \leq 6 \\
& -x_2 - x_3 \leq -7 \\
& x_1 - x_3 \leq 2 \\
& -x_1 + x_3 \leq -2 \\
& x_1 \geq 0 \\
& x_2 \leq 0 \\
& x_3 \text{ unrestricted.}
\end{array}
$$

1.2.1 Standard Form of a Linear Program

In this section, we define the *standard form* of a linear program. A linear programming problem is said to be in standard form if (1) the objective is to minimize, (2) all constraints are of the equality type, and (3) all variables are non-negative. The following LP is in standard form:

$$
\begin{array}{ll}
\text{minimize} & c^T x \\
\text{subject to} & Ax = b \\
& x \geq 0.
\end{array}
$$

The standard form of a linear programming problem is important because some important algorithms that solve linear programs e.g. the simplex method require linear programs to be in standard form. This requirement is not too restrictive since any linear program can be converted into an equivalent linear program in standard form. The following conversion rules can be used to convert an LP in standard form.

1. *Converting unrestricted variables*

If a decision variable x is initially defined to be unrestricted, i.e., the variable can take on any real number regardless of sign, then x can be expressed as the difference between two non-negative numbers $x^+ \geq 0$ and $x^- \geq 0$ so that $x = x^+ - x^-$. For example, if $x = -5$, then $x^+ = 0$ and $x^- = 5$.

2. *Converting inequality constraints*

If a constraint i is initially of the form $a_{i1}x_1 + a_{i2}x_2 + \cdots + a_{in}x_n \leq b_i$, then a non-negative slack variable s_i can be added to the left-hand side of the constraint to get $a_{i1}x_1 + a_{i2}x_2 + \cdots + a_{in}x_n + s_i = b_i$ where $s_i \geq 0$.

If a constraint i is initially of the form $a_{i1}x_1 + a_{i2}x_2 + \cdots + a_{in}x_n \geq b_i$, then a non-negative surplus variable s_i can be subtracted from the left-hand side of the constraint to get $g_i(x) = a_{i1}x_1 + a_{i2}x_2 + \cdots + a_{in}x_n - s_i = b_i$ where $s_i \geq 0$.

3. *Converting maximization to minimization*
It follows that since *maximize* $c^T x = -$*minimize* $-c^T x$, any maximization problem can be converted to an equivalent minimization problem by minimizing the negated terms in the original objective function. It is common to omit the outer negation in formulations since it will not affect the optimization.

Example 1.1
Convert the LP

$$\begin{array}{ll} \textit{maximize} & 5x_1 - 4x_2 + 6x_3 \\ \textit{subject to} & -x_1 + x_2 \leq -7 \\ & 2x_2 - x_3 \geq 2 \\ & x_1 + 2x_3 = 7 \\ & x_1 \geq 0, x_2 \geq 0, x_3 \text{ unrestricted} \end{array}$$

into standard form.

Solution: The variable x_3 is unrestricted, so let $x_3 = x_3^+ - x_3^-$ where $x_3^+ \geq 0$ and $x_3^- \geq 0$. Then, for every occurrence of x_3 in the objective function and constraints, replace with $x_3^+ - x_3^-$. Next, negate the terms of the objective function to get $-5x_1 + 4x_2 - 6x_3^+ + 6x_3^-$ and then the new objective is to *minimize* $-5x_1 + 4x_2 - 6x_3^+ + 6x_3^-$. Add a slack variable s_1 to the left-hand side of the first constraint to get $-x_1 + x_2 + s_1 = -7$ and a surplus variable s_2 to the second to get $2x_2 - x_3^+ + x_3^- - s_2 = 2$. The third constraint is already in equality form. Then, the standard form of the LP is

$$\begin{array}{llllllll} \textit{minimize} & -5x_1 + & 4x_2 - 6x_3^+ + 6x_3^- & & & & \\ \textit{subject to} & -x_1 + & x_2 + & & s_1 & & = -7 \\ & & 2x_2 - x_3^+ + x_3^- - & & & s_2 & = 2 \\ & x_1 + & & 2x_3^+ - 2x_3^- & & & = 7 \\ & \multicolumn{6}{l}{x_1 \geq 0, x_2 \geq 0, x_3^+ \geq 0, x_3^- \geq 0, s_1 \geq 0, s_2 \geq 0.} \end{array}$$

1.2.2 Linear Programming Terminology

We discuss some terminology for linear programming problems, and without loss of generality, assume that a LP is in standard form, i.e.,

$$\begin{array}{ll} \text{minimize} & c^T x \\ \text{subject to} & Ax = b \\ & x \geq 0. \end{array}$$

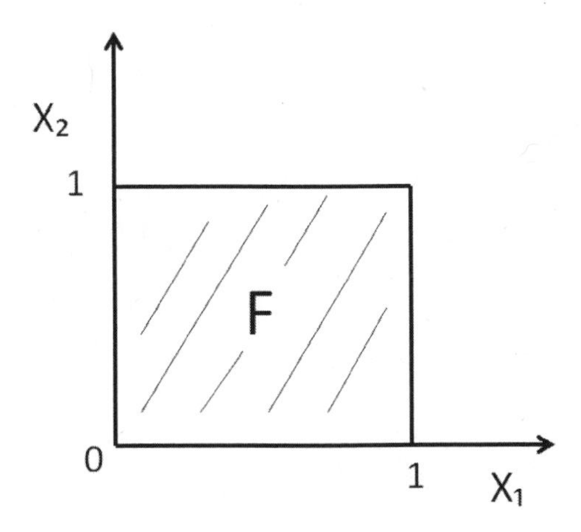

FIGURE 1.1
Bounded feasible set.

The matrix A is called the *constraint matrix* and has dimension $m \times n$. The vector c is called the *cost vector* with dimension $n \times 1$, and b is called the *right-hand side vector* which is a column vector of dimension $m \times 1$. The vector x is called the *decision vector* and has dimension $n \times 1$.

The set $F = \{x \in R^n | Ax = b, x \geq 0\}$ is called the *feasible set* of the linear programming problem. A vector $x \in R^n$ is said to be *feasible* for a linear program if $x \in F$, otherwise x is said to be *infeasible*. A linear program is said to be *consistent* if $F \neq \varnothing$, otherwise the linear program is *inconsistent.* The feasible set is *bounded* if $\|x\| \leq M$ for all $x \in F$ for some positive constant M ($\|\cdot\|$ is a norm on x). Intuitively, the feasible set is bounded if there is a sphere or rectangle that can completely contain the feasible set.

A vector x^* is an optimal solution for a linear programming problem if $x^* \in F$ and $c^T x^* \leq c^T x$ for all $x \in F$, else x^* is said to be *sub-optimal.* Also, a linear programming problem is *bounded* if $L \leq c^T x$ for all $x \in F$ for some constant L, else the LP is said to be unbounded. Clearly, if F is bounded, then the linear program is bounded.

Example 1.2
Consider the following linear program (P):

$$\begin{array}{lll} \textit{minimize} & -x_1 - x_2 & \\ \textit{subject to} & x_1 & \leq 1 \\ & \qquad x_2 & \leq 1 \\ & x_1 \geq 0, x_2 \geq 0. & \end{array} \tag{1.7}$$

The feasible set is $F = \{x = (x_1, x_2)^T \in R^2 | \ x_1 \leq 1, x_2 \leq 1, x_1 \geq 0, x_2 \geq 0\}$. Converting to standard form by adding slack variables, the linear program becomes

$$\begin{array}{lllllll} \textit{minimize} & -x_1 - x_2 & & & & \\ \textit{subject to} & x_1 & & + & x_3 & & = 1 \\ & & x_2 & + & & x_4 & = 1 \\ & \multicolumn{6}{l}{x_1 \geq 0, x_2 \geq 0, x_3 \geq 0, x_4 \geq 0.} \end{array} \quad (1.8)$$

The feasible set for (1.8) is

$$\bar{F} = \{x = (x_1, x_2, x_3, x_4)^T \in R^4 |\ x_1 + x_3 = 1, x_2 + x_4 = 1, x_1 \geq 0, x_2 \geq 0, x_3 \geq 0, x_4 \geq 0\}.$$

Observe that the linear program (1.7) is consistent since there is at least one x in F, e.g., $x = \begin{bmatrix} 0.5 \\ 0.5 \end{bmatrix}$ and the feasible set is bounded since for any vector $x \in F$ we have $\|x\| \leq 1$ and so (1.7) is bounded. Therefore, the linear program (1.8) is also consistent and bounded with $\bar{F}$ bounded. A graph of the feasible set F is given in Figure 1.1.

The optimal solution in this example is $x^* = \begin{bmatrix} 1 \\ 1 \end{bmatrix} \in F$, and the corresponding optimal objective function value is -2. Chapter 3 will discuss methods for generation and verification of optimal solutions.

Example 1.3
Consider the linear program

$$\begin{array}{llllll} \textit{minimize} & -x_1 - x_2 & & & & \\ \textit{subject to} & & & & & \\ & -x_1 & & +x_3 & & = 1 \\ & & x_2 & & +x_4 & = 1 \\ & \multicolumn{5}{l}{x_1 \geq 0, x_2 \geq 0, x_3 \geq 0, x_4 \geq 0.} \end{array}$$

The feasible region in two dimensions is equivalent to $F = \{x = (x_1, x_2)^T \in R^2 |\ -x_1 \leq 1, x_2 \leq 1, x_1 \geq 0, x_2 \geq 0\}$ see Figure 1.2. In this case, the feasible region is not bounded as the graph extends infinitely to the right on the x-axis. In fact, for the sequence of vectors $x^{(k)} = \begin{bmatrix} k \\ 1 \end{bmatrix}$, $k = 1, 2, 3, \ldots$, the objective function value $-k - 1 \to -\infty$ as $k \to \infty$ and so the LP is unbounded.

1.3 More Linear Programming Examples

This section covers some additional examples to highlight the broad range of problems that can be formulated as linear programs. Modeling a problem as a

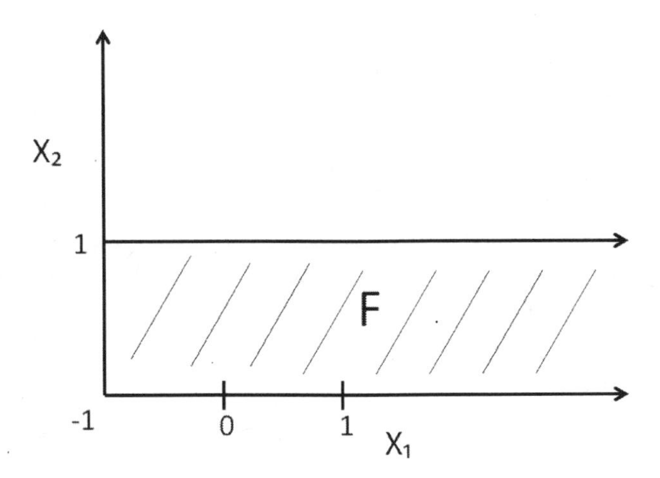

FIGURE 1.2
Unbounded feasible set.

linear program is an art and there is no unique way to formulate problems, but the basic requirements are the need to define decision variables, the objective, and constraints.

Example 1.4 (Production Planning)

Consider a company that produces n different products. Each product uses m different resources. Suppose that resources are limited and the company has only b_i units of resource i available for each $i = 1, ..., m$. Further, each product j requires a_{ij} units of resource i for production. Each unit of product j made generates a revenue of p_j dollars. The company wishes to find a production plan, i.e., the quantity of each product to produce, that maximizes revenue.

The problem can be formulated as a linear program. The first step is to define the decision variables. Let x_j be a decision variable that represents the amount of product j produced by the company. Suppose that fractional amounts of a product are allowed and this amount should be non-negative since a negative value for x_j is meaningless. Then, a production plan is represented by the vector $x = \begin{bmatrix} x_1 & \cdots & x_n \end{bmatrix}^T$. The contribution toward revenue from the production of x_j units of product j is $p_j x_j$ and so the total revenue from a production plan x is then $p_1x_1 + p_2x_2 + \cdots + p_nx = \sum_{j=1}^{n} p_jx_j$. The contribution toward using resource i from the production of x_j units of product j is $a_{ij}x_j$, and so the total consumption of resource i by production plan x is $a_{i1}x_1 + a_{i2}x_2 + \cdots + a_{in}x_j = \sum_{j=1}^{n} a_{ij}x_j$ and this quantity can not exceed b_i. Since total revenue is to be maximized and resources limitations must be observed over all resources, then the LP is

$$\begin{array}{ll} \text{maximize} & \sum_{j=1}^{n} p_j x_j \\ \text{subject to} & \sum_{j=1}^{n} a_{ij} x_j \leq b_i \quad i = 1, ..., m \\ & x_j \geq 0 \quad j = 1, ..., n. \end{array}$$

Example 1.5 (Multi-period Production Planning)

In the previous example, the planning horizon was assumed to be one period and it was implicitly assumed that the optimal production plan generated from the model would be able to be entirely sold for the single period in order to achieve the maximum revenue. That is, the model assumed that production would occur just once and did not incorporate any consideration of demand levels and future production. We now consider a multi-period production model where production decisions are made for more than one period. Consider a single product that has demand in number of units over the next year as follows:

Fall	Winter	Spring	Summer
30	40	10	20

We wish to meet all demand for each period and allow excess production for a period so that it may be carried over to meet demand at future time periods. However, there will be a unit holding cost of $10 for inventory at the end of each period. Assume that there is 5 units of inventory at the start of Fall, and there is to be no inventory at the end of the Summer period.

To formulate the multi-period production model, there are some additional constructs that must be developed to capture the dynamics of production in a multiple period setting. First, in addition to decision variables that give the amount of production for each time period, there needs to be another quantity that links production from one time period to the next. These are inventory variables that indicate the amount of product in excess of the demand at the end of a time period. Then, the dynamics from one time period to the next can be captured through the following inventory balance constraint

current inventory + production for current period = amount of product used to meet current demand + inventory for next period.

Now, let x_t = number of units of product produced in period t, i_t = units of inventory at the end of period t where Fall, Winter, Spring, and Summer correspond to periods $t = 1, 2, 3$, and, 4, respectively. Also $t = 0$ will refer to the start of Fall and d_t = demand for period t.

Thus, the inventory balance constraints take the form $i_{t-1} + x_t = d_t + i_t$ for $t = 1, 2, 3, 4$. The objective function is to minimize total inventory costs $10i_1 + 10i_2 + 10i_3 + 10i_4$. Then, the multi-period production model is

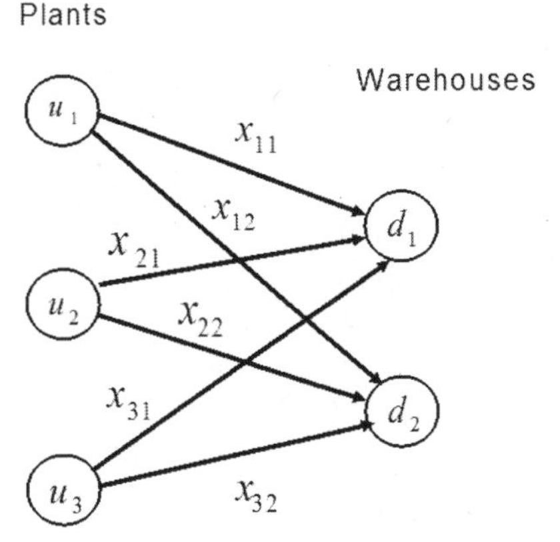

FIGURE 1.3
Transportation problem.

$$\begin{array}{ll} \textit{minimize} & 10i_1 + 10i_2 + 10i_3 + 10i_4 \\ \textit{subject to} & i_0 + x_1 = 30 + i_1 \\ & i_1 + x_2 = 40 + i_2 \\ & i_2 + x_3 = 10 + i_3 \\ & i_3 + x_4 = 20 + i_4 \\ & i_0 = 5 \\ & i_4 = 0 \\ & i_0, i_1, i_2, i_3, i_4, x_1, x_2, x_3, x_4 \geq 0. \end{array}$$

Example 1.6 (Transportation Problem)
Reactiveeno Inc. is a company that produces a special type of facial cream that aids in reducing acne. This product is manufactured in n plants across North America. Every month the facial cream product is shipped from the n plants to m warehouses. Plant i has a supply of u_i units of the product. Warehouse j has a demand of d_j units. The cost of shipping one unit from plant i to warehouse j is c_{ij}. The problem of finding the least-cost shipping pattern from plants to warehouses can be formulated as an LP.

Let x_{ij} be the number of units of product shipped from plant i to warehouse j. So the objective is to minimize the cost of shipping over all possible plant warehouse pairs (see Figure 1.3). There are two classes of constraints. One class of constraints must ensure that the total amount shipped from a plant i to warehouses, i.e., $\sum_{j=1}^{m} x_{ij}$ does not exceed the supply u_i of plant i, whereas the other class of constraints ensures that the demand d_j of a warehouse j is met from the total amount $\sum_{i=1}^{n} x_{ij}$ shipped by plants. Then, the transportation problem can be formulated as the following LP.

$$\begin{array}{lll} \textit{minimize} & \sum_{i=1}^{n}\sum_{j=1}^{m} c_{ij}x_{ij} & \\ \textit{subject to} & \sum_{j=1}^{m} x_{ij} \leq u_i & i = 1, ..., n \\ & \sum_{i=1}^{n} x_{ij} \geq d_j & j = 1, ..., m \\ & x_{ij} \geq 0 & i = 1, ..., n, \ j = 1, ..., m \end{array}$$

Example 1.7 (The Assignment Problem)

A special case of the transportation problem above is when there are as many plants as warehouses, i.e., $m = n$, and each warehouse demands exactly one unit of the product, and each plant produces only one unit of the product, i.e., $d_i = u_i = 1$. Then, the model takes the following form

$$\begin{array}{lll} \textit{minimize} & \sum_{i=1}^{n}\sum_{j=1}^{n} c_{ij}x_{ij} & \\ \textit{subject to} & \sum_{j=1}^{n} x_{ij} = 1 & i = 1, ..., n \\ & \sum_{i=1}^{n} x_{ij} = 1 & j = 1, ..., n \\ & x_{ij} \geq 0 & i = 1, ..., n, \ j = 1, ..., n. \end{array}$$

This special case is known as the assignment problem and has the interpretation of matching persons with jobs so that each person gets one job and each job gets one person. c_{ij} in this context represents the cost of assigning person i to job j. This quantity can reflect the different skill levels of workers.

Example 1.8 (Workforce Scheduling)

You are the human resources manager of a company and one of your major duties is to schedule workers for each day of the week. Each day of the week must have a required number of workers as given in the following table.

Day of Week	Required Number of Workers
Monday	25
Tuesday	25
Wednesday	22
Thursday	21
Friday	23
Saturday	20
Sunday	18

Each worker must work 5 consecutive days and then will have the next two days off. You wish to minimize the total number of workers scheduled for the week subject to providing each day with the required number of workers. The challenging aspect of this formulation resides in defining the decision variables. The variables should be defined so that it facilitates the expression of the constraints that impose a 5 consecutive day work schedule for workers. One such possibility is to let x_j = number of workers that start on day j where $j = 1$ corresponds to Monday, $j = 2$ corresponds to Tuesday, etc., then, the model is

$$
\begin{array}{lllr}
\textit{minimize} & x_1 + x_2 + x_3 + x_4 + x_5 + x_6 + x_7 & & \\
\textit{subject to} & x_1 + \qquad\qquad\quad x_4 + x_5 + x_6 + x_7 & \geq 25 & \\
& x_1 + x_2 + \qquad\qquad x_5 + x_6 + x_7 & \geq 25 & \\
& x_1 + x_2 + x_3 \qquad\qquad + x_6 + x_7 & \geq 22 & \\
& x_1 + x_2 + x_3 + x_4 \qquad\qquad + x_7 & \geq 21 & \\
& x_1 + x_2 + x_3 + x_4 + x_5 & \geq 23 & \\
& \quad x_2 + x_3 + x_4 + x_5 + x_6 & \geq 20 & \\
& \qquad x_3 + x_4 + x_5 + x_6 + x_7 & \geq 18 & \\
& x_i \geq 0 \text{ and integer} & i = 1, ..., 7. &
\end{array}
$$

The jth constraint ensures that there will be enough workers for day j by ensuring that there are enough workers that start from those days of the week for which a worker will be working on day j. For example, the first constraint ensures that there are enough workers for Monday by ensuring that there are enough workers that *start* work on Monday, Thursday, Friday, Saturday, and Sunday, so that there will be at least 25 workers on Monday. Observe that the coefficients of the variables x_2 and x_3 are zero in the first constraint and hence these variables do not appear in the first constraint since any worker starting on Tuesday or Wednesday and working 5 consecutive days will not be working on Monday. Note that there is an integer value requirement on the variables in addition to the non-negativity restriction since it is unreasonable to have a fractional value of a worker! So the model above violates the divisibility assumption of linear programming. The model is in fact what is called an *integer program* due to the integrality restriction of the decision variables. Otherwise, the model is very close to a linear program since the objective and constraints are linear.

Example 1.9 (Bond Portfolio Optimization)
Suppose that a bank has the following liability schedule

Year 1	Year 2	Year 3
\$12,000	\$18,000	\$20,000

That is, the bank needs to pay \$12,000 at the end of the first year, \$18,000 at the end of the second year, and \$20,000 at the end of the third year. Bonds are securities that are sold by agencies, such as corporations or governments, that entitle the buyer to periodic interest (coupon) payments and the payment of the principle (face value) at some time in the future (maturity).

The bank wishes to use the three bonds below to form a portfolio (a collection of bonds) today to hold until all bonds have matured and that will generate the required cash to meet the liabilities. All bonds have a face value of a \$100 and the coupons are annual (with one coupon per year). For example, one unit of Bond 2 costs \$99 now and the holder will receive \$3.50 after 1 year and then \$3.50 plus the face value of \$100 at the end of the second year which is the maturity of Bond 2.

Bond	1	2	3
Price	$102	$99	$98
Coupon	$5	$3.5	$3.5
Maturity year	1	2	3

The bank wishes to purchase Bonds 1, 2, and 3 in amounts whose total cash flow will offset the liabilities. Assume that fractional amounts of each bond are permitted. A linear programming model can be formulated to find the lowest-cost set of bonds (i.e., portfolio) consisting of Bonds 1, 2, and 3 above that will meet the liabilities. Let the x_i = amount of bond i purchased. Then, the problem can be modeled as follows

$$\begin{array}{lll} \textit{minimize} & 102x_1 + 99x_2 + 98x_3 & \\ \textit{subject to} & 105x_1 + 3.5x_2 + 3.5x_3 & \geq 12000 \\ & \quad 103.5x_2 + 3.5x_3 & \geq 18000 \\ & \quad\quad 103.5x_3 & \geq 20000 \\ & x_1, x_2, x_3 \geq 0. & \end{array}$$

The objective is to minimize the cost of a portfolio and each constraint ensures that that cash flow generated by the bonds for a given time period is sufficient to match the liability for that period. For example, in the first constraint each unit of Bond 1 purchased will generate $5 from the coupon plus the $100 face value (since bonds of type 1 mature at the end of year 1) so the total cash contribution from Bond 1 is $\$105x_1$, the total cash flow from Bond 2 is only $\$3.5x_2$ since these bonds do not mature until the end of year 2 but only payout $3.5 per unit for the coupon at the end of year 1. The total cash flow from bonds of type 3 is also $\$3.5x_3$ at the end of the first year. Note that in constraint 2 there is no term involving bonds of type 1 since they have already matured after one year and can no longer generate cash flow.

1.3.1 Converting Minimization Problems with Absolute Value

Consider an optimization problem of the following form

$$\begin{array}{lll} \textit{minimize} & c_1|x_1| + c_2|x_2| + \cdots + c_n|x_n| & \\ \textit{subject to} & x_i \text{ unrestricted} & i = 1, ...n \end{array}$$

where $c_i > 0$ for all $i = 1, ...n$. The problem in the form above is not a linear program since the absolute value terms in the objective function are not linear. However, the problem can be converted into an equivalent linear program through the following transformation. Let

$$|x_i| = x_i^+ + x_i^- \text{ and } x_i = x_i^+ - x_i^- \text{ where } x_i^+, x_i^- \geq 0.$$

Since the objective is to minimize and $c_i > 0$, then $x_i^+ \times x_i^- = 0$ will hold at optimality, this will ensure that the transformation is equivalent to the absolute value of x_i where if $x_i \geq 0$, then $|x_i| = x_i^+ = x_i$ and $x_i^- = 0$ else, $|x_i| = x_i^- = -x_i$ and $x_i^+ = 0$.

Example 1.10

The optimization problem above can be transformed to an LP by replacing each occurrence of $|x_i|$ with $x_i^+ + x_i^-$ and by adding $x_i = x_i^+ - x_i^-$ as a constraint along with the restrictions $x_i^+ \geq 0$ and $x_i^- \geq 0$. The model then becomes

$$\begin{array}{ll}
\textit{minimize} & c_1(x_1^+ + x_1^-) + c_2(x_2^+ + x_2^-) + \cdots + c_n(x_n^+ + x_n^-) \\
\textit{subject to} & x_1 = \ x_1^+ - x_1^- \\
 & x_2 = \ x_2^+ - x_2^- \\
 & \qquad \vdots \\
 & x_n = \ x_n^+ - x_n^- \\
 & x_1^+ \geq 0, x_1^- \geq 0, x_2^+ \geq 0, x_2^- \geq 0, ..., x_n^+ \geq 0, x_n^- \geq 0.
\end{array}$$

Example 1.11 (Application: Portfolio Optimization)

Consider the problem of investing money in n stocks where each stock i has a random rate of return r_i with an expected return of μ_i. r_i models the price uncertainty for stock i and is often assumed to be a normal distribution. In addition, the covariance between the returns of stock i and stock j is σ_{ij}. Let x_i= the proportion of wealth invested in stock i. A portfolio is then represented by the vector $x = (x_1, ..., x_n)^T$. A reasonable model to use to construct a portfolio is the following model developed by Markowitz (1952), which is a one-period model where an investment is made now and held until a future point in time T.

$$\begin{array}{lll}
\textit{minimize} & \sum_{i=1}^{n} \sum_{j=1}^{n} \sigma_{ij} x_i x_j & \\
\textit{subject to} & \sum_{i=1}^{n} \mu_i x_i & = R \\
 & \sum_{i=1}^{n} x_i & = 1 \\
 & \quad x_i \geq 0 & i = 1, ..., n
\end{array}$$

The objective function of the model is the variance of the return of the portfolio x where the variance represents the risk of the portfolio. Then, the objective is to minimize the risk (portfolio variance) subject to meeting an expected return goal of R for the portfolio (first constraint) and ensuring that the budget is exhausted (second constraint). In other words, the goal is to find the portfolio among all of the portfolios that can achieve an expected return of R while exhausting the budget, and that has the smallest variance among them.

However, there are several challenges in using the Markowitz model. First, observe that the model is non-linear since the terms in the objective function are non-linear, i.e., $\sigma_{ij}x_ix_j$. Second, obtaining all of the parameters is a significant endeavor. For example, with n stocks there will be the need to estimate $n(n-1)/2$ covariance terms (for a portfolio of 1000 stocks that means nearly half a million terms to estimate).

Fortunately, it is possible to convert the Markowitz model to a linear programming model. Instead of using portfolio variance or equivalently portfolio standard deviation

$$\sigma = \sqrt{E(\sum_{i=1}^{n}(r_i - \mu_i)x_i)^2} = \sqrt{\sum_{i=1}^{n}\sum_{j=1}^{n}\sigma_{ij}x_ix_j}$$

as the objective function ($E(\cdot)$ denotes the expectation operator), we consider

$$\omega = E|\sum_{i=1}^{n}(r_i - \mu_i)x_i|$$

which measures the absolute deviation of the portfolio return from the expected portfolio return. The following result by Konno and Yamazaki (1991) justifies the use of ω as the objective.

If the vector $(r_1, r_2, ..., r_n)$ is multi-variate normally distributed then, $\omega = \sqrt{\frac{2}{\pi}}\sigma$.

Thus, it will suffice to minimize ω since it is the same as minimizing the variance. The following model is equivalent to the Markowitz model.

$$\begin{array}{lll} \textit{minimize} & E|\sum_{i=1}^{n}(r_i - \mu_i)x_i| & \\ \textit{subject to} & \sum_{i=1}^{n}\mu_i x_i & = R \\ & \sum_{i=1}^{n}x_i & = 1 \\ & x_i \geq 0 & i = 1, ..., n \end{array}$$

The objective function now has absolute value terms, but can be converted by using the technique above. First, the time period is divided into T sub-periods to get

$$u_i = \frac{1}{T}\sum_{t=1}^{T}r_{it}$$

where r_{it} = the *realized* return for asset i for for sub-period t and then

$$E|\sum_{i=1}^{n}(r_i - \mu_i)x_i| = \frac{1}{T}\sum_{t=1}^{T}|\sum_{i=1}^{n}(r_{it} - \mu_i)x_i|.$$

Then, let

$$|\sum_{i=1}^{n}(r_{it} - \mu_i)x_i| = s_t + y_t$$

and

$\sum_{i=1}^{n}(r_{it} - \mu_i)x_i = s_t - y_t$ for $t = 1, ..., T$ where $s_t \geq 0$ and $y_t \geq 0$.

Then, the model transforms to the following LP, which is called the mean-absolute deviation (MAD) portfolio optimization model.

$$\begin{array}{lll} \textit{minimize} & \sum_{t=1}^{T}(s_t + y_t) & \\ \textit{subject to} & \sum_{i=1}^{n}(r_{it} - \mu_i)x_i = s_t - y_t & t = 1, ..., T \\ & \sum_{i=1}^{n} \mu_i x_i = R & \\ & \sum_{i=1}^{n} x_i = 1 & \\ & x_i \geq 0 & i = 1, ..., n \\ & s_t \geq 0, y_t \geq 0 & t = 1, ..., T \end{array}$$

Note that the model does not explicitly require the covariance terms.

1.3.2 Network Optimization Models

Many important applications can be stated as problems over networks. For example, the transportation problem in Example 1.6 involves a transportation network from plants to warehouses where a link from a plant to a warehouse in Figure 1.3 can represent distance (or cost related to distance) between the two entities. Each plant and warehouse represents nodes in the network, which in this case represents the location of each entity. The problem involves the determination of an amount of a product to be shipped from plants to warehouses, so it involves the flow of physical items (i.e., products) through links of the network. Networks can be used to represent many other problems where there is a notion of flow over an entity that can be represented as a network. For example, routing data over the Internet is another problem which involves the flow of data over a network where a node would be a router and links would correspond to fiber- optic cable connections between routers. We now present the formal definition of a network.

Definition 1.12

A network (or graph) G consists of a set of nodes N and a set of edges E, i.e., $G = (N, E)$.

G is an *undirected graph* when each edge $e \in E$ is an unordered pair of distinct nodes i and j in N. An edge between nodes i and j can be denoted in this case as $\{i, j\}$ or $\{j, i\}$.

G is a *directed graph* when each edge $e \in E$ is an ordered pair of distinct nodes i and j in N. An edge will be denoted in this case as (i, j) and indicates that the direction of the edge is from node i to node j.

Example 1.13

Consider the following undirected graph $G = (N, E)$ in Figure 1.4 where $N = \{1, 2, 3, 4, 5\}$ and $E = \{\{1, 2\}, \{1, 3\}, \{2, 3\}, \{2, 4\}, \{3, 4\}, \{3, 5\}, \{4, 5\}\}$.

Example 1.14

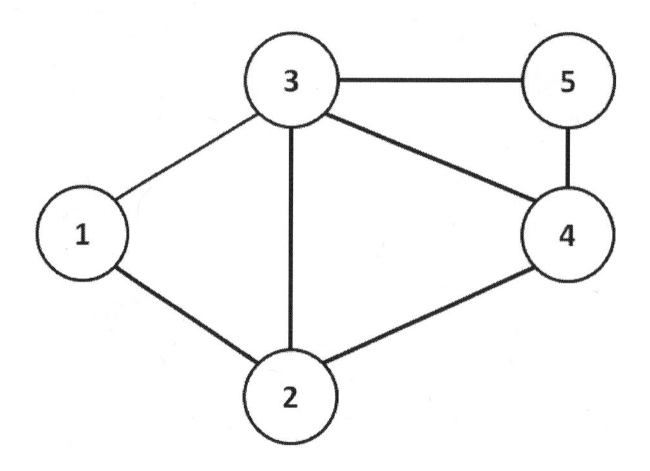

FIGURE 1.4
Undirected graph.

Consider the following directed graph in Figure 1.5 where $N = \{1, 2, 3, 4, 5\}$ and $E = \{(1, 2), (1, 3), (3, 2), (4, 2), (4, 3), (3, 5), (5, 4)\}$.

Definition 1.15
A path between nodes i and j of a graph G is a sequence of nodes and edges starting from i and ending with j, such that no nodes are repeated. It is important to note that all edges specified in a path must be in E.

A path is said to be directed if in addition to being a path, all edges must be of the same orientation (direction).

For the undirected graph in Figure 1.4 there is a path between any two distinct pair of nodes i and j and the direction of edges does not matter. In the directed graph in Figure 1.5 there is a directed path from node 1 to node 5, consisting of the sequence node 1, edge $(1, 3)$, node 3, edge $(3, 5)$, node 5 where edges $(1, 3)$ and $(3, 5)$ are of the same orientation. Observe that there is no directed path from node 5 to node 1.

Definition 1.16
We say that a graph $G = (N, A)$ is connected if there is a path between every pair of distinct nodes i and $j \in N$.

The graphs in Figures 1.4, 1.5, 1.6, and 1.7 are all connected.

1.3.2.1 Minimum Cost Flow Problem

The minimum cost flow problem is to determine a least-cost shipment (flow) of a product (item) through a connected directed network $G = (N, E)$ in order to satisfy demand at various nodes from various supply nodes. It is

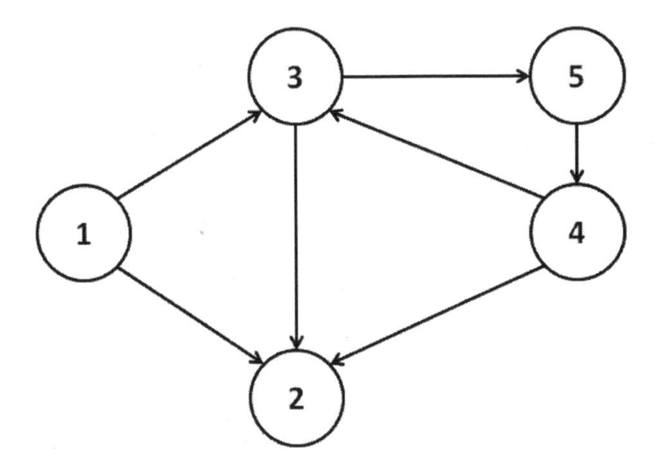

FIGURE 1.5
Directed graph.

a generalization of the transportation problem in Example 1.6 and can be formulated as a linear program. There are three types of nodes. The first type of node is a supply node, and such a node i has a supply $b_i > 0$ units of the product. The second type of node is a demand node, and such a node i requires a net amount $|b_i|$ of the product where $b_i < 0$. The third type of node is called a transshipment node and such a node does not have a supply of products and do not require net any amount of the product and so $b_i = 0$. Transshipment nodes serve as nodes that a flow of products can go through without leaving any amount of the product.

For each $(i,j) \in E$, let c_{ij} be the cost of shipping one unit of product from node i to node j. Define x_{ij} to represent the amount of product to be shipped from node i to node j. The vector x consisting of the decision variables x_{ij} is called a flow and represents the amount of shipment along an edge $(i,j) \in E$. For each $(i,j) \in E$, let l_{ij} be a constant that is a lower bound on the flow on edge (i,j) and u_{ij} a constant that is an upper bound on the flow on edge (i,j). If there is no lower (upper) bound on an edge (i,j), then $l_{ij} = 0$ $(u_{ij} = \infty)$. We assume that $\sum\limits_{i \in N} b_i = 0$ and that all cost and lower and upper-bound values are non-negative and integral. There is one more required assumption; see Exercise 17.

For example, in the graph in Figure 1.6, nodes 1 and 5 are supply nodes with 10 units of a product at node 1 and 15 at node 5. Nodes 3, 4, and 6 are demand nodes with demands of 11, 19, and 4, respectively. Node 2 is the only transshipment node. The value next to an edge (i,j) gives the unit cost c_{ij} of shipping a product along that edge.

Then, the minimum cost flow problem can be formulated as a linear programming problem. The objective is to find a minimum cost flow x. The

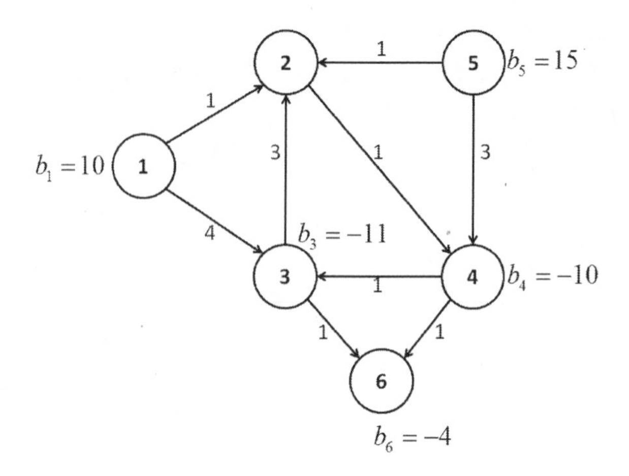

FIGURE 1.6
Minimum cost flow network.

constraints must ensure that demand is met at the appropriated nodes. This can be achieved by requiring that for each node i, the total flow out of i minus the total flow into node i must equal b_i. For example, for a transshipment node j the total flow out of j minus total flow into j must be 0, whereas for a demand node k, this difference must be a negative value whose magnitude is equal to the demand required at node k. These constraints go by the name of flow conservation or mass balance constraints. The problem can then be formulated as follows.

$$\begin{array}{lll} \textit{minimize} & \sum\limits_{(i,j)\in E} c_{ij}x_{ij} & \\ \textit{subject to} & \sum\limits_{\{j:(i,j)\in E\}} x_{ij} - \sum\limits_{\{j:(j,i)\in E\}} x_{ji} = b_i & \text{for all } i \in N \\ & l_{ij} \leq x_{ij} \leq u_{ij} & \text{for all } (i,j) \in E \end{array}$$

Example 1.17

The minimum cost flow linear program corresponding to the network in Figure 1.6 assuming lower bounds of 0 and no upper bounds on the edges is a linear program in standard form

$$\begin{array}{ll} \textit{minimize} & c^T x \\ \textit{subject to} & Ax = b \\ & x \geq 0 \end{array}$$

where

$$x = \begin{bmatrix} x_{12} \\ x_{13} \\ x_{24} \\ x_{32} \\ x_{36} \\ x_{43} \\ x_{46} \\ x_{52} \\ x_{54} \end{bmatrix}, c = \begin{bmatrix} c_{12} \\ c_{13} \\ c_{24} \\ c_{32} \\ c_{36} \\ c_{43} \\ c_{46} \\ c_{52} \\ c_{54} \end{bmatrix} = \begin{bmatrix} 1 \\ 4 \\ 1 \\ 3 \\ 1 \\ 1 \\ 1 \\ 1 \\ 3 \end{bmatrix}, b = \begin{bmatrix} b_1 \\ b_2 \\ b_3 \\ b_4 \\ b_5 \\ b_6 \end{bmatrix} = \begin{bmatrix} 10 \\ 0 \\ -11 \\ -10 \\ 15 \\ -4 \end{bmatrix}$$

$$A = \begin{bmatrix} 1 & 1 & 0 & 0 & 0 & 0 & 0 & 0 & 0 \\ -1 & 0 & 1 & -1 & 0 & 0 & 0 & -1 & 0 \\ 0 & -1 & 0 & 1 & 1 & -1 & 0 & 0 & 0 \\ 0 & 0 & -1 & 0 & 0 & 1 & 1 & 0 & -1 \\ 0 & 0 & 0 & 0 & 0 & 0 & 0 & 1 & 1 \\ 0 & 0 & 0 & 0 & -1 & 0 & -1 & 0 & 0 \end{bmatrix}.$$

The optimal solution is to ship 10 units of the product from node 1 to node 2 ($x_{12} = 10$) and ship 15 units from node 5 to node 2 ($x_{52} = 15$), then 25 units are then shipped from node 2 to node 4 ($x_{24} = 25$), and finally 11 units are shipped from node 4 to node 3 ($x_{43} = 11$) and 4 units are shipped from node 4 to node 6 ($x_{46} = 4$). This shipping flow leaves all demand nodes satisfied, and for any transshipment node i in G the optimal solution satisfies the mass balance requirement, e.g., the total flow into transshipment node 2 is $x_{12} + x_{52} = 10 + 15 = 25$, which is equal to the total flow out of node 2, i.e., $x_{24} = 25$.

1.3.2.2 Maximum Flow Problem

Another important class of problems over a network is the maximum (max) flow problem. The max flow problem is to determine the maximum amount of flow that can be sent from a source node s to a sink node t over a directed network. Many steady-state flow problems can be modeled as max flow problems, such as determining the steady state of oil in an oil pipeline network or vehicles on a highway network, or data flow in a communications network, or electricity on a power network (grid). For example, in Figure 1.7, the goal is to send as much flow from node s through the network to node t while not violating the capacity constraints of each edge.

The max flow problem can be formulated as a linear program. Let $G = (N, E)$ be a directed graph and u_{ij} be the capacity of edge $(i, j) \in E$. Similar to the minimum cost flow problem, there will be mass balance constraints that govern the flow through the directed network where all nodes in the graph except the source s and the sink t are transshipment nodes. The net

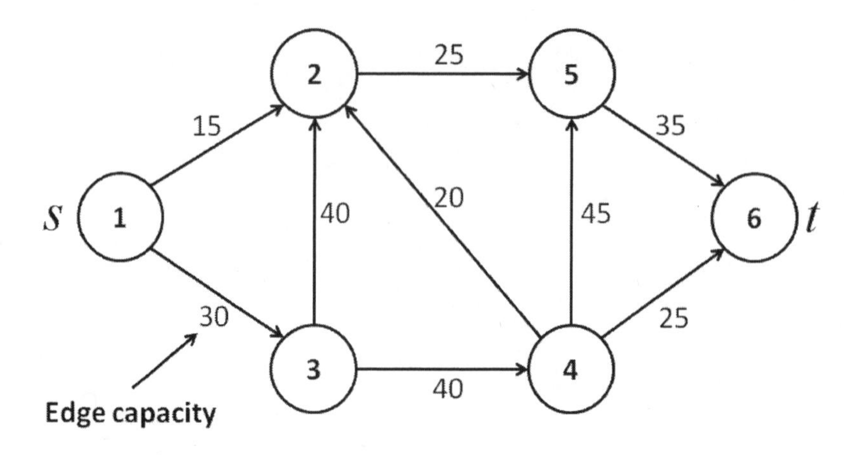

FIGURE 1.7
Directed graph for max flow problem.

flow out of the source will be a decision variable v that represents the total flow out of the source. The sink node t will have a net inflow of v, and so it can be seen as having a b_i value equal to $-v$. The variables x_{ij} are defined as in the minimum cost flow problem and represent the amount of flow on edge $(i,j) \in E$. Then, the linear program is defined as follows

$$\begin{array}{lllll}
\textit{maximize} & v & & & \\
\textit{subject to} & \sum\limits_{\{j:(i,j)\in E\}} x_{ij} - \sum\limits_{\{j:(j,i)\in E\}} x_{ji} & = v & \text{for } i = s \\
 & \sum\limits_{\{j:(i,j)\in E\}} x_{ij} - \sum\limits_{\{j:(j,i)\in E\}} x_{ji} & = 0 & \text{for all } i \in N \text{ except } t \text{ and } s \\
 & \sum\limits_{\{j:(i,j)\in E\}} x_{ij} - \sum\limits_{\{j:(j,i)\in E\}} x_{ji} & = -v & \text{for } i = t \\
 & 0 \leq x_{ij} \leq u_{ij} & & \text{for all } (i,j) \in E
\end{array}$$

The assumptions needed for the max flow problem are:

(1) The edge capacities are non-negative and integral.

(2) There is no directed path from s to t, all of whose edges are uncapacitated (i.e., infinite capacity).

(3) There are no parallel edges (i.e., two or more edges with the same starting node i and ending node j).

(4) For any edge $(i,j) \in E$, the edge $(j,i) \in E$ as well.

Example 1.18

Suppose that TorontoCo Oil company generates crude oil from its refineries in the oil sands of Alberta and wishes to ship the maximum amount of the crude oil per day from its Alberta refinery to a storage facility in Winnipeg, Manitoba, through a network of pipelines depicted in the directed graph G of Figure 1.7. The Alberta refinery is the source s (node 1) and the storage facility is the sink t (node 6), and the flow of oil may go through several oil pipeline transshipment points which are represented by nodes 2 thru 5. There are capacities on each edge $(i, j) \in E$ indicating the maximum amount of oil (in units of 1000 barrels) that can be shipped from node i to node j. Then, the linear program that represents the problem of determining the maximum amount of crude oil that can be shipped daily from Alberta to Winnipeg can be written as

$$\begin{array}{lll}
\textit{maximize} & v & \\
\textit{subject to} & x_{12} + x_{13} & = v \\
& x_{25} - x_{12} - x_{32} - x_{42} & = 0 \\
& x_{32} + x_{34} - x_{13} & = 0 \\
& x_{42} + x_{45} + x_{46} - x_{34} & = 0 \\
& x_{56} - x_{25} - x_{45} & = 0 \\
& -x_{46} - x_{56} & = -v \\
& \quad 0 \le x_{12} \le 15 & \\
& \quad 0 \le x_{13} \le 30 & \\
& \quad 0 \le x_{32} \le 40 & \\
& \quad 0 \le x_{25} \le 25 & \\
& \quad 0 \le x_{34} \le 40 & \\
& \quad 0 \le x_{42} \le 20 & \\
& \quad 0 \le x_{45} \le 45 & \\
& \quad 0 \le x_{56} \le 35 & \\
& \quad 0 \le x_{46} \le 25. &
\end{array}$$

1.3.3 Solving Linear Programs with MATLAB®

In this section, we demonstrate the basic steps in using the Optimization Toolbox of MATLAB to solve linear programming problems. Linear programs can be solved with MATLAB by using the function *linprog*. To use *linprog*, a linear program is specified in the following form

$$\begin{array}{ll}
\textit{minimize} & f^T x \\
\textit{subject to} & Ax \le b \\
& A_{eq} x = b_{eq} \\
& l_b \le x \le u_b,
\end{array}$$

which assumes that constraints are grouped according to inequality constraints, equality constraints, and bounds on the decision variables. The first set of constraints $Ax \leq b$ represents inequality constraints (of the less than or equal to type). Note that any constraint that is originally an inequality constraint that is of the greater than or equal to type ($\leq$) must be converted to a less than or equal to equivalent. The second set of constraints $A_{eq}x = b_{eq}$ represents the equality constraints, and $l_b \leq x \leq u_b$ represents the lower and upper bounds on the decision variables. Then, A, A_{eq} are matrices and b, b_{eq}, l_b, u_b are vectors. f is a vector that represents the cost coefficients of the objective function. These quantities are represented in MATLAB as f, A, b, Aeq, beq, lb, and ub and are used as arguments for the *linprog* function.

For example, the statement

[x, fval] = linprog(f, A, b, Aeq, beq, lb, ub)

returns a vector x that represents the optimal solution and the optimal objective function value $fval$ of the linear program specified by the data.

Example 1.19
Consider the linear program

$$\begin{array}{ll} \textit{minimize} & -3x_1 - 5x_2 - 3x_3 \\ \textit{subject to} & x_1 - x_2 + x_3 \leq 15 \\ & 2x_1 + 3x_2 + 6x_3 \leq 30 \\ & 2x_1 + x_2 = 30 \\ & 0 \leq x_1 \leq 15 \\ & 0 \leq x_2 \leq 10 \\ & 0 \leq x_3 \leq 5. \end{array}$$

Then,

$$f = \begin{pmatrix} -3 \\ -5 \\ -3 \end{pmatrix}, A = \begin{pmatrix} 1 & -1 & 1 \\ 2 & 3 & 6 \end{pmatrix}, b = \begin{pmatrix} 15 \\ 30 \end{pmatrix},$$

$$A_{eq} = \begin{pmatrix} 2 & 1 & 0 \end{pmatrix}, b_{eq} = (30), l_b = \begin{pmatrix} 0 \\ 0 \\ 0 \end{pmatrix}, \text{ and } u_b = \begin{pmatrix} 15 \\ 10 \\ 5 \end{pmatrix}.$$

The vectors and matrices for this LP are created in MATLAB by the following statements

```
>> f=[-3,-5,-3];
>> A=[1,-1,1;2,3,6];
>> b=[15;30];
>> Aeq=[2,1,0];
>> beq=[30];
>> lb=[0;0;0];
```

```
>> ub=[15;10;5];
```

The linear program function can then be called with the following statement

$$[x, fval] = linprog(f, A, b, Aeq, beq, lb, ub)$$

which outputs the following values

```
x =
15.0000
  0.0000
  0.0000

fval =
-45.0000
```

If there are no inequality constraints, then one can set A = [ ] and b = [ ], while if there are no equality constraints, one can set Aeq = [ ] and beq= [ ]. Also, if there are no lower bounds (upper bounds), then lb(ub) can be omitted.

Example 1.20
Consider the following LP:

$$\begin{array}{rl} \textit{minimize} & -3x_1 - 5x_2 - 3x_3 \\ \textit{subject to} & x_1 - x_2 + x_3 \leq 15 \\ & 2x_1 + 3x_2 + 6x_3 \leq 30 \\ & 0 \leq x_1 \\ & 0 \leq x_2 \\ & 0 \leq x_3. \end{array}$$

The data is entered as

```
>> f=[-3,-5,-3];
>> A=[1,-1,1;2,3,6];
>> b=[15;30];
>> Aeq=[ ];
>> beq=[ ];
>> lb=[0;0;0];
```

Then, the LP can be solved by execution of the statement

$$[x, fval] = linprog(f, A, b, Aeq, beq, lb),$$

which outputs the following values

```
x =
  0.0000
10.0000
  0.0000

fval =
-50.0000
```

Alternatively, the statement

[x, fval] = linprog(f, A, b, [ ], [ ], lb)

can be used to obtain the same results.

Example 1.21
Consider the LP

$$
\begin{array}{ll}
\textit{minimize} & -3x_1 - 5x_2 - 3x_3 \\
\textit{subject to} & x_1 - x_2 + x_3 \leq 15 \\
& 2x_1 + 3x_2 + 6x_3 \leq 30 \\
& 0 \leq x_1 \leq 15 \\
& 0 \leq x_2 \\
& 0 \leq x_3 \leq 20.
\end{array}
$$

In this case, the variables all have a lower bound of 0, but x_2 does not have an upper bound. In this case, the vector lb can be defined as $\begin{pmatrix} 15 \\ Inf \\ 20 \end{pmatrix}$ where Inf represents infinity.

Example 1.22
We solve the minimum cost flow in Example 1.17. The data is entered as

```
>> f=[1,4,1,3,1,1,1,1,3];
>> Aeq=[1,1,0,0,0,0,0,0,0; -1,0,1,-1,0,0,0,-1,0; 0,-1,0,1,1,-1,0,0,0; 0,0,-1,0,0,1,1,0,-1; 0,0,0,0,0,0,0,1,1; 0,0,0,0,-1,0,-1,0,0];
>> A=[ ];
>> b=[ ];
>> beq=[10;0;-11;-10;15;-4];
>> lb=zeros(9,1);
>> ub=[ ];
```

Then, the linear program function is called with the following statement [x,fval]=linprog(f,A,b,Aeq,beq,lb,ub),

which outputs the following values

x' =
10.0000 0.0000 25.0000 0.0000 0.0000 11.0000 4.0000 15.0000 0.0000

fval =
65.0000

Alternatively, the statement

[x, fval] = linprog(f, [ ], [ ], Aeq, beq, lb, [ ])

could be used to produce the same results.

Remark MATLAB requires an LP to be in a specific form, which should not be confused with the definition of standard form that is defined in Section 1.3.1. Standard form will be required in the development of the simplex method to solve linear programs in Chapter 3.

1.4 Exercises

Exercise 1.1 (a) Convert the following optimization problem into a linear program in standard form

$$\begin{array}{lll} \textit{maximize} & 2x_1 - 4x_2 - 3|x_3| & \\ \textit{subject to} & x_1 + x_2 - 2x_3 & \geq 1 \\ & \quad x_2 + \ x_3 & \leq 1 \\ & x_1 + x_2 + \ x_3 & = 4 \\ & \quad x_1 \geq 0, x_2 \geq 0. & \end{array}$$

(b) Specify the components A, x, b, and c for the LP you obtained in (a).

Exercise 1.2
An alternative approach to handle absolute value terms is by using the following observation: $|x|$ is the smallest value z that satisfies $x \leq z$ and $-x \leq z$. Using this approach, convert the following problem into a linear program.

$$\begin{array}{lllll} \textit{minimize} & 2x_1 & + & 3|x_2| & \\ \textit{subject to} & x_1 & + & x_2 & \geq 6 \end{array}$$

Exercise 1.3

Can the following optimization problem be transformed to a linear program? If so, write the LP in standard form.

$$\begin{array}{lllll} \textit{minimize} & 2x_1 + x_2^2 + x_3 & & & \\ \textit{subject to} & & x_2^2 - x_3 & & = 0 \\ & 5x_1 & & +3x_3 & \leq 5 \\ & & x_1 \geq 0, x_2 \geq 0, x_3 \geq 0 & & \end{array}$$

Exercise 1.4

Consider the following optimization problem

$$\begin{array}{ll} \textit{maximize} & \min\{c_0^1 + c^1 x, c_0^2 + c^2 x, ..., c_0^n + c^n x\} \\ \textit{subject to} & Ax = b \end{array}$$

where c_0^i and c^i are constants for $i = 1, ..., n$.

(a) Convert the problem to an equivalent linear program.

(b) Give an economic interpretation of the problem.

Exercise 1.5

Sketch each feasible set F below and state whether it is bounded, unbounded, or empty.

$$\begin{array}{ll} & x_1 + x_2 \leq 6 \\ \text{(a)} & 4x_1 - 2x_2 \leq 12 \\ & x_1 \leq 2 \end{array}$$

$$\begin{array}{ll} & 2x_1 + 6x_2 \leq 22 \\ \text{(b)} & x_1 - x_2 \leq 5 \\ & x_1 \geq 0, x_2 \geq 0 \end{array}$$

$$\begin{array}{ll} & x_1 + x_2 \leq 4 \\ \text{(c)} & x_1 + 2x_2 \geq 12 \\ & x_1 \geq 0 \end{array}$$

Exercise 1.6

(a) Prove that *"If the feasible set of a linear program in bounded, then the linear program is bounded"* is a true statement.

(b) Show that the converse *"If a linear program is bounded, then the feasible set is bounded"* is false.

Exercise 1.7

Consider the LP

$$\begin{array}{ll} \textit{minimize} & 2x_1 + x_2 \\ \textit{subject to} & x_2 \geq 2 \\ & x_1 - 3x_2 \leq 5 \end{array}$$

Determine if the LP is unbounded or not. If it is, specify a sequence of vectors $x^{(k)}$ such that the objective function $\longrightarrow -\infty$ as $k \longrightarrow \infty$.

Exercise 1.8

Suppose that there are 4 different projects and 4 workers and each worker must be assigned a project and each project must be assigned a worker. It costs $20 an hour for a worker. The following table gives the time required (in hours) for each worker i to complete a particular project j:

	Project 1	Project 2	Project 3	Project 4
Worker 1	7	3	6	10
Worker 2	5	4	9	9
Worker 3	6	4	7	10
Worker 4	5	5	6	8

(a) Formulate this problem of assigning workers and jobs at minimum cost as a linear program.

(b) Solve the model in (a) using the MATLAB linprog function. If you get a fractional optimal solution, find a feasible integer solution (i.e., all variables are 0 or 1) with same optimal objective function value.

Exercise 1.9

The Ontario Steel Company ONASCO needs to produce a new type of steel that will be created from a mixture of various types of iron, alloy, and steel. The new type of steel should have a chrome content of at least 1% but no more than 1.25%, and should have a silicon content of at least 3%, but no more than 4%. ONASCO has the following materials for creating the new type of steel

	Amount Available	Cost	Chrome % per lb	Silicon % per lb
Iron A	unlimited	$.05/lb	0.01	1.75
Iron B	unlimited	$.07/lb	7.00	17.00
Steel A	unlimited	$.02/lb	0.00	0.00
Steel B	unlimited	$.025/lb	0.00	0.00
Alloy A	unlimited	$.10/lb	0.00	22.00
Alloy B	unlimited	$.12/lb	0.00	31.00

ONASCO would like to make 1500 tons of the new type of steel at minimum cost.

(a) Formulate this problem as a linear program.

(b) Solve the model in (a) using MATLAB.

Exercise 1.10

ONASCO has estimated production requirements of 6,000, 5,000, 7,000, and 1,000 tons of the new type of steel (see Exercise 1.9) for the next four quarters. The current workers at ONASCO are capable of producing 2,000 tons per quarter and the current amount of the new steel in inventory is 1,000 tons. It costs ONASCO $50 per ton with the current workforce. There is a cost of $10 per ton for any inventory at the end of a quarter. At the end of the fourth quarter ONASCO wants to have an inventory of at least 1000 tons. Assume that the maximum production capacity per quarter is 6000 tons.

(a) Formulate the problem above as a linear program and solve using MATLAB.

(b) Suppose now it is possible to increase or decrease the amount of regular workers. Also, additional overtime (non-regular) workers for any quarter can be hired for $100 per ton. It will cost $150 per ton to increase the amount of regular workers from one quarter to the next, and to decrease regular workers from one quarter to the next it will cost $100 per ton. Finally, ONASCO would like there to be a regular workforce amount equivalent to being able to produce 2000 tons at the end of the last quarter. Reformulate this problem as a linear program and solve using MATLAB.

Exercise 1.11

Suppose that an aerospace company builds engine turbines in production facilities in Montreal, Seattle, and Sarnia. The turbines are shipped to warehouses in Atlanta, Calgary, and Los Angeles for ultimate delivery to Tokyo, Shanghai, Toronto, and Paris airports. The costs of shipping a turbine from the various origins to destinations are as follows:

	Atlanta	Calgary	Los Angeles
Montreal	$5000/turbine	$3000/turbine	$6500/turbine
Seattle	$6000/turbine	$3200/turbine	$2500/turbine
Sarnia	$5500/turbine	$2300/turbine	$6200/turbine

	Tokyo	Shanghai	Toronto	Paris
Atlanta	$6000/turbine	$4500/turbine	$5000/turbine	$4000/turbine
Calgary	$5500/turbine	$3200/turbine	$2500/turbine	$4700/turbine
Los Angeles	$5500/turbine	$2300/turbine	$6200/turbine	$8000/turbine

(a) Suppose that the demand in Tokyo is 50 turbines, Shanghai 130 turbines, Toronto 75 turbines, and Paris 90 turbines. Assume that the production capacity at Montreal, Seattle, and Sarnia is unlimited and that the capacity of the warehouses is also unlimited. Formulate a linear program to determine the minimum-cost shipping pattern so that all demand is met and solve using MATLAB.

(b) Suppose that only two of the three production facilities can be used for production of turbines, and that the capacity of the Los Angeles warehouse is now 75 turbines (the capacities of the Atlanta and Calgary warehouses remain unlimited). Formulate a model for this modification of the problem. (Hint: define variables that indicate whether a production facility is selected or not).

Exercise 1.12

An international airport has a security force to help combat crime. The head of the security force has divided a day into two time periods of 12 hours and wishes to staff security robots across the airport for the two time periods. The first time period is from 6AM to 6PM, the second time period is from 6PM to 6AM, and the number of security robots that should work during the first period is 34 and 17 for the second time period. A security robot can work for 12 or 18 consecutive hours. It costs $120 dollars to operate a single robot that works for 12 hours (consecutively) and $252 dollars per robot that works 18 hours (consecutively). Formulate an LP that minimizes the cost of deployment of robots.

Exercise 1.13

A warehouse can store 10,000 units of a certain commodity. A commodities trader has access to the warehouse and currently has 2000 units of the commodity in storage in the warehouse. At the start of each month t, for the next 12 months ($t = 1, ..., 12$), the trader can buy an amount of the commodity at price p_t subject to the capacity of the warehouse at the start of month t or can sell an amount of the commodity at price s_t at the start of the month subject to how much of the commodity is currently in the warehouse. In addition, there is a unit inventory cost i_t for holding the commodity in the warehouse, which is incurred at the start of month t. The trader wishes to have no inventory of the commodity at the end of the year. Formulate an LP that maximizes profit from the buying or selling of the commodity over the 12-month time horizon.

Exercise 1.14

In Example 1.9 (Bond Portfolio) suppose that one is allowed to invest in a portfolio of bonds such that the cash generated for a time period may exceed the amount of the liability for that time period. Any excess cash generated is re-invested at an interest rate of 2% for one time period and can be used toward future liabilities. Assume that fractional investments in bonds are allowed. Formulate this modified version of the Bond Portfolio problem as a linear program and solve in MATLAB. Why would it be advantageous to allow excess cash generation and reinvestment?

Exercise 1.15

Suppose that n projects are available for investment. Each project is unique and not related to other available projects. There is a benefit b_i that is obtained from investment in project i. The cost of investment into project i is denoted by c_i. You have a budget of C dollars for investing in the projects.

(a) Assume that one can invest in a fractional amount of each project and then formulate the problem of selecting a portfolio of projects such that total benefit is maximized subject to not violating the budget as a linear program.

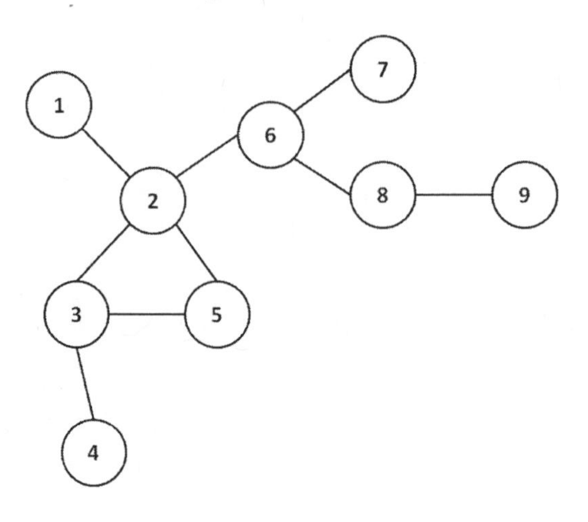

FIGURE 1.8
Power grid for Exercise 1.16.

(b) Now assume for each project one must invest in the entire project or not at all. What modifications to the model you developed in (a) have to be made to accommodate the all-or-none selection requirement? Is the resulting model a linear program?

(c) Suppose that there are 4 projects with cost and benefit data as follows

Project	Cost	Benefit
1	50	150
2	20	50
3	100	233
4	40	88

One simple heuristic to select projects is to use the benefit-to-cost ratio of each project and then invest in the project with the highest ratio and then in the project with the next highest ratio and so on, until no more of the money is left. Let the budget $C = \$100$.

Is this strategy optimal for the linear program from part (a)?

Is this strategy optimal for the optimization problem from part (b)? Why or why not?

Exercise 1.16

Consider a power grid consisting of electricity producers that are connected to consumption points on the grid. The consumption points are affiliated with regional retail power companies that then distribute the power to their end users. The undirected graph in Figure 1.8 gives the structure of the network where nodes 1, 2, and 3 are the location of power generation plants and nodes

4, 5, 7, and 9 are power consumption points. The cost of transmitting power over any edge is $13 per megawatt hour and there is no capacity constraint on an edge. It is possible for power to traverse both orientations on an edge since the graph is undirected. Each power generator has a maximum generation capacity in megawatt hours and a cost per megawatt hour generated as given in the table below.

Generator	1	2	3
Cost ($/Megawatt hour)	18	15	24
Capacity (hundreds of megawatt hours)	80	130	190

The power consumption points have demand as indicated below.

Consumption Point	4	5	7	9
Demand (hundreds of megawatt hours)	70	100	50	120

Formulate the problem of finding the minimum-cost power generation and distribution strategy over the network to satisfy all consumption points as a linear program.

Exercise 1.17

One standard assumption about the minimum-cost network flow problem is that a directed graph $G = (N, E)$ should have an uncapacitated (every edge has no upper limit on the flow it can have on it) directed path between every pair of nodes.

How can one in general modify a graph G if it does not have this property?

Exercise 1.18

Yorkville Airlines has service from Calgary to each of the following cities Winnipeg, Hamilton, Toronto, Montreal, and Quebec City. As an operations research analyst for Yorkville Airlines you have been asked to determine how many daily flights there should be from Calgary to Quebec City with the condition that flights must connect via Winnipeg and then to either Hamilton, Montreal, or Toronto, and then finally to Quebec City. It has already been determined by the government transportation bureau that Yorkville Airlines must have no more than the following flights for each connecting segment as given in the table below.

City pairs	Max # of Daily Flights
Calgary to Winnipeg	5
Winnipeg to Montreal	4
Winnipeg to Hamilton	5
Winnipeg to Toronto	2
Montreal to Quebec City	2
Hamilton to Quebec City	1
Toronto to Quebec City	3

Formulate this problem as a linear program and solve in MATLAB.

Exercise 1.19
How can you solve a max flow problem as a minimum-cost flow problem?

1.5 Computational Project

(Portfolio Investment Problem)

This project considers the mean variance portfolio optimization problem in Example 1.11 and considers the use of linear programming via MATLAB in a non-trivial (larger) application. We wish to compute optimal mean-variance portfolios. That is, we wish to generate portfolios x that result from solving the model

$$
\begin{array}{lll}
\textit{minimize} & \sum_{i=1}^{n}\sum_{j=1}^{n}\sigma_{ij}x_i x_j & \\
\textit{subject to} & \sum_{i=1}^{n}\mu_i x_i & = R \\
 & \sum_{i=1}^{n} x_i & = 1 \\
 & x_i \geq 0 & i = 1, ..., n
\end{array}
$$

where μ_i is the expected return of asset i and σ_i is the standard deviation of random return of asset i. σ_{ij} is the covariance of the returns of assets i and j.

Parameter Estimation

Recall that the MAD model only requires estimates of the expected returns μ_i of each asset i under the assumption of multi-variate normality of asset returns. The following formulas to can be used to compute the parameters:

$r_{it} = \frac{I_{i,t} - I_{i,t-1}}{I_{i,t-1}}$ (*rate* of return for stock i over period t where $I_{i,t}$ is the *total* return for stock i for month t; for example, if the price of stock i is 100 at the start of month t and the price of this stock is 110 at the end of month t, then $I_{i,t} = 10\%$), then the expected return of each asset i over a time horizon $[0, T]$ is $\mu_i = (\Pi_{t=1}^{T}(1 + r_{it}))^{\frac{1}{T}} - 1$.

Note that a period does not have to be one month in duration, e.g., it could be 4 months in duration.

MAD Model 3 Asset Example

We formulate the mean-absolute deviation (MAD) LP model for a 3 asset problem. Recall that the MAD model can be formulated as follows (see Example 1.11).

$$\begin{array}{lll} \textit{minimize} & E|\sum_{i=1}^{n}(r_i - \mu_i)x_i| & \\ \textit{subject to} & \sum_{i=1}^{n}\mu_i x_i & = R \\ & \sum_{i=1}^{n} x_i & = 1 \\ & x_i \geq 0 & i = 1, ..., n \end{array}$$

which as demonstrated earlier can be formulated as the linear program

$$\begin{array}{lll} \textit{minimize} & \sum_{t=1}^{T}(s_t + y_t) & \\ \textit{subject to} & \sum_{i=1}^{n}(r_{it} - \mu_i)x_i = s_t - y_t & t = 1, ..., T \\ & \sum_{i=1}^{n}\mu_i x_i = R & \\ & \sum_{i=1}^{n} x_i = 1 & \\ & x_i \geq 0 & i = 1, ..., n \\ & s_t \geq 0, y_t \geq 0 & t = 1, ..., T \end{array}$$

We illustrate the MAD model on a 3-asset example. The three assets are Apple, TD, and XOM. The historical price of each asset is obtained every four months over a one-year period starting from May 1, 2009 to May 3, 2010. This will define 3 periods each 4 months long, e.g., the first period is from May 1, 2009 ($t = 0$) to Sep 1, 2009 ($t = 1$) and the second period is from Sep 1, 2009 to Jan 4, 2010 ($t = 2$), etc. The price data is in the table below

Month	Apple	TD	XOM
May 1 2009	125.83	74.46	66.67
Sep 1 2009	168.21	67.51	69.15
Jan 4 2010	210.73	74.31	68.19
May 3 2010	261.09	86.63	67.77

Using the formulas above for parameter estimation, we get the following table of rate of returns and expected returns of each asset.

	Apple	TD	XOM
r_{i1}	0.3368	-0.0933	0.0372
r_{i2}	0.2528	0.1007	-0.0139
r_{i3}	0.2390	0.1658	-0.0062
μ_i	0.2755	0.0518	0.0055

then, the MAD model can be formulated as

minimize $1/3(|(0.3368 - 0.2755)x_1 + (-0.0933 - 0.0518)x_2 + (0.0372 - 0.0055)x_3|$
$+ |(0.2528 - 0.2755)x_1 + (0.1007 - 0.0518)x_2 + (-0.0139 - 0.0055)x_3|$
$+ |(0.2390 - 0.2755)x_1 + (0.1658 - 0.0518)x_2 + (-0.0062 - 0.0055)x_3|)$

subject to

$$0.2755x_1 + 0.0518x_2 + 0.0055x_3 = R$$
$$x_1 + x_2 + x_3 = 1$$
$$x_1 \geq 0, x_2 \geq 0, x_3 \geq 0$$

So the MAD model above becomes the linear program

minimize $1/3(s_1 + y_1 + s_2 + y_2 + s_3 + y_3)$

subject to

$$
\begin{gathered}
(0.3368 - 0.2755)x_1 + (-0.0933 - 0.0518)x_2 + (0.0372 - 0.0055)x_3 = s_1 - y_1 \\
(0.2528 - 0.2755)x_1 + (0.1007 - 0.0518)x_2 + (-0.0139 - 0.0055)x_3 = s_2 - y_1 \\
(0.2390 - 0.2755)x_1 + (0.1658 - 0.0518)x_2 + (-0.0062 - 0.0055)x_3 = s_3 - y_3 \\
0.2755x_1 + 0.0518x_2 + 0.0055x_3 = R \\
x_1 + x_2 + x_3 = 1 \\
x_1 \geq 0, x_2 \geq 0, x_3 \geq 0
\end{gathered}
$$

Project Assignment

(1) The project is to construct optimal portfolios by using the mean-absolute deviation (MAD) linear programming model instead of the Markowitz (MVO) model, since the MVO model is a non-linear programming problem (MVO will be explored further in Chapter 6). The portfolios will involve all stocks of the S&P 500 index, which is a set of 500 stocks with large capitalization in the U.S. stock market. The capitalization of a stock is the number of outstanding shares in the market times the price per share. For more details about the stocks in the S&P 500 see http://en.wikipedia.org/wiki/List_of_S%26P_500_companies.

Historical monthly price information from March 31, 2008 to Oct. 31, 2012 is to be used for all stocks that appear in the S&P 500 index to compute the expected returns for use in the MAD linear program. Use the formulas above for estimating parameters.

(2) Compute an optimal portfolio by solving the MAD LP model in MATLAB for each R starting from $R = \min\{\mu_i | \mu_i \geq 0\}$ to $R = \max\{\mu_i\}$ in increments of 0.5%.

(3) Provide a table where you list for each value R the corresponding optimal volatility and optimal proportions invested in the stocks.

(4) Using the optimal variance (volatility) value for each optimized portfolio from (3), construct a plot in MATLAB where the y-axis represents the volatility of the portfolio and the x-axis represents R. What does your plot reveal about the relationship between return and risk?

(Note: Volatility is the objective function value of MAD.)

MAD MATLAB Code

The MATLAB code is given below for the project problem. The code assumes that there is a data file called ProjectDataCh1 that contains the monthly rate of return (in_return) for each stock in the S&P 500.

```
clear
clc
% MATLAB code for Computational Project Chapter 1
load ProjectDataCh1
```

```
% ProjectDataCh1 is a file that contains monthly per period
% returns r_it in in_return
n = 500; % the number of stocks can be chose
T = 24; % number of time periods
% computing geometric means
for i = 1:n
    mu(i) = (prod(1+in_return(:,i)))^(1/T)-1;
end
R=0:.005: max(mu);          % range of expected return goals R
% compute MAD objective coefficients
c = [zeros(n,1); ones(T,1); ones(T,1)];
Aeq = [];
for t=1:T
    Aeq = cat(1, Aeq, in_return(t,:)-mu);
end
Aeq = [Aeq -eye(T) eye(T);  % constraint coefficients for MAD
    mu zeros(1,2*T);
    ones(1,n) zeros(1,2*T);];
lb = zeros(n+T+T,1);        % lower bound on variables

% computing optimal portfolios over range of return goals R
for a = 1:length(R)
    %right hand side coefficients for each R
    beq = [zeros(T,1); R(a); 1];
    [x_MAD(:,a), fval_MAD(a)] = linprog(c, [],[], Aeq,beq, lb,[]);
end

fval_MAD = (1/T)*fval_MAD;  %  minimizing (1/T)w
devi = (pi/2)^.5*fval_MAD;  %  w = sqrt(2/pi)*SD
invest_frac = x_MAD(1:n, :);% optimal portfolio weights for each R

% create figure for optimal portfolios
figure(1)
[xx, yy] = meshgrid(1:n, R);
mesh(xx,yy, invest_frac')
colormap bone
axis([0 500 0 max(mu) 0 1])
xlabel('stocks')
ylabel('expected return R')
zlabel('investment fraction')
title('Portfolio Composition under different R')

% create figure for the efficient frontier of MAD
figure(2)
plot(devi, R, '-k*')
```

```
xlabel('volatility \sigma')
ylabel('expected return R')
title('The efficient frontier of MAD')
```

Notes and References

Linear programming has its origins in the study of linear systems of inequalities, which dates as far back as 200 hundred years ago. More recently, the work of Dantzig (1963) in the development of the simplex method for solving linear programs greatly accelerated the field of linear programming inspiring extensions and new methodological developments as well as the discovery of new applications. Today, linear programming enjoys a prominent role in the applied mathematical and management sciences due to the wide variety of problems that can be modeled as linear programs. As seen in this chapter, linear programs are prevalent in the areas of logistics and supply chain management, finance, and production planning. Williams (1999) considers general model building principles in using mathematical programming and covers a wide range of applications. In addition, linear programming plays an important role in the methodologies of other mathematical optimization problems, such as integer and non-linear programming. Early books on the topic include Dantzig (1963), Simonnard (1966), Gass (1975), Bradley, Hax, and Magnanti (1977), Murty (1983), Chvatal (1983), and Bazaraa, Jarvis, and Sherali (1977). More recent books include those by Bertsimas and Tsitsiklis (1997), Saigal (1995), Winston and Venkataramanan (2003), and Vanderbei (2008). Network flows is an important topic that is covered comprehensively by Murty (1992), and Ahuja, Magnanti, and Orlin (1993).

The diet problem was originally formulated by Stigler (1945) and the portfolio optimization problem in Example 1.11 was developed by Markowitz (1952). The linear programming MAD equivalent was developed by Konno and Yamazaki (1991). The book by Zenios (2008) covers a wide variety of financial optimization models, many of which are linear programs.

2

Geometry of Linear Programming

2.1 Introduction

In this chapter, we explore the geometry of linear programming and gain geometric insight into optimal solutions. Also, the corresponding algebraic representations of the geometry are developed. This leads to the Fundamental Theorem of Linear Programming, which serves as the basis for algorithm development for linear programs.

2.2 Geometry of the Feasible Set

We have seen that the feasible set for a linear program can be bounded, unbounded, or infeasible. In this section, we explore additional geometric properties of feasible sets of general linear programs that are consistent, i.e., whose feasible set is non-empty. Consider the following linear program

$$\begin{array}{ll} \textit{minimize} & -x_1 - 2x_2 \\ \textit{subject to} & x_1 + x_2 \leq 20 \\ & 2x_1 + x_2 \leq 30 \\ & x_1,\ x_2 \geq 0. \end{array} \tag{2.1}$$

The feasible set $P = \{x = \binom{x_1}{x_2} |\ x_1 + x_2 \leq 20, 2x_1 + x_2 \leq 30, x_1,\ x_2 \geq 0\}$. The graph of P is in Figure 2.1. For this linear program, the feasible set P is clearly bounded. Any point $x = \begin{bmatrix} x_1 \\ x_2 \end{bmatrix}$ in P must satisfy each contraint, e.g., $x_1 + x_2 \leq 20$, $2x_1 + x_2 \leq 30$, $x_1 \geq 0$ and $x_2 \geq 0$. The first constraint requires that only points x in R^2 whose sum of components is less than or equal to 20 can be considered. The second constraint requires further that the sum of twice the first component and the second component be less than or equal to 30. Finally, each component must be non-negative. Each constraint of this linear program is an inequality constraint. An inequality constraint is also

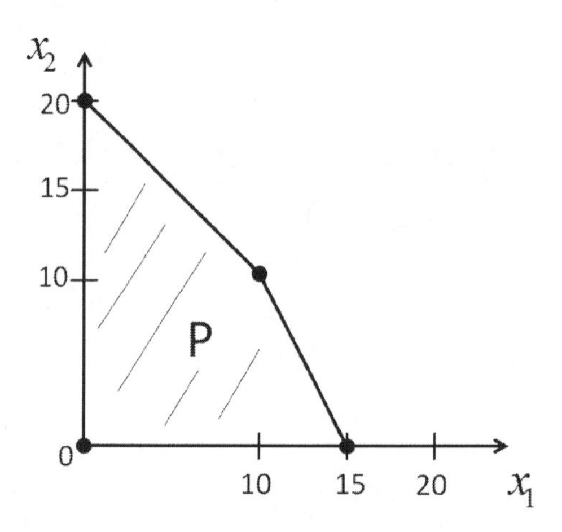

FIGURE 2.1
Graph of feasible set of LP (2.1).

known as a closed halfspace since the vectors that will satisfy the constraint will lie on one side of a straight line (or hyperplane in higher dimensions) where the points on the line also satisfy the constraint. The definition of a closed halfspace is as follows.

Definition 2.1
A *closed halfspace* is a set of the form $H_{\leq} = \{x \in R^n | a^T x \leq \beta\}$ or $H_{\geq} = \{x \in R^n | a^T x \geq \beta\}$.

Example 2.2
The constraint $x_1 + x_2 \leq 20$ is a closed halfspace $H_{\leq}$ where $a = \begin{bmatrix} 1 \\ 1 \end{bmatrix}$ and $\beta = 20$; see Figure 2.2. Likewise the constraint $2x_1 + x_2 \leq 30$ is a closed halfspace $H_{\leq}$ with $a = \begin{bmatrix} 2 \\ 1 \end{bmatrix}$ and $\beta = 30$. The constraint $x_1 \geq 0$ is a closed halfspace $H_{\geq}$ with $a = \begin{bmatrix} 1 \\ 0 \end{bmatrix}$ and $\beta = 0$, and $x_2 \geq 0$ is a closed halfspace $H_{\geq}$ with $a = \begin{bmatrix} 0 \\ 1 \end{bmatrix}$ and $\beta = 0$; see Figure 2.3 for graph of $H_{\geq}$.

Thus, the feasible set P of the linear program can be seen to be the intersection of closed halfspaces that results in Figure 2.1 since any feasible $x \in P$ must lie in all of the closed halfspaces.

In general, a linear program may have some equality constraints as well as inequality constraints, and these equality constraints will correspond to hyperplanes.

Definition 2.3

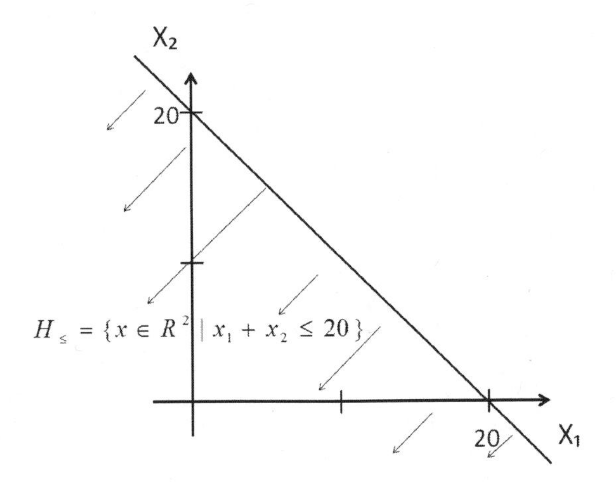

FIGURE 2.2
Closed halfspace $x_1 + x_2 \leq 20$.

A *hyperplane* is a set of the form $H = \{x \in R^n | a^T x = \beta\}$ where a is a non-zero vector, i.e., $a \neq 0$ and $\beta \in R^1$is a scalar.

In particular, an equality constraint $a_i^T x = b_i$ in a linear program is a hyperplane $H = \{x \in R^n | a_i^T x = b_i\}$. Geometrically, a hyperplane H splits R^n into two halves. In R^2 a hyperplane H is a line that splits the plane into two halves; see Figure 2.4. In R^3, a hyperplane is a plane that splits 3-dimensional space into two halves.

Furthermore, the vector a in the definition of the hyperplane H is perpendicular to H. a is called the *normal vector* of H. To show that a is perpendicular to H, let z and y be in H, then $a^T(z - y) = a^T z - a^T y = 0 - 0 = 0$. The vector $z - y$ is parallel to H, thus a is perpendicular to H.

Example 2.4

The hyperplane

$$H = \{x = \begin{bmatrix} x_1 \\ x_2 \end{bmatrix} \in R^2 | x_1 + x_2 = 20\}$$

has vector $a = \begin{bmatrix} 1 \\ 1 \end{bmatrix}$ and this vector is perpendicular to all vectors x in H and all vectors parallel to H. In addition, the vector $-a$ is also perpendicular to vectors in and parallel to a hyperplane H, but in the opposite direction to a; see Figure 2.5.

Each hyperplane $H = \{x \in R^n | a_i^T x = b_i\}$ associated with an equality constraint $a_i^T x = b_i$ can be re-written as two closed halfspaces $H_L = \{x \in$

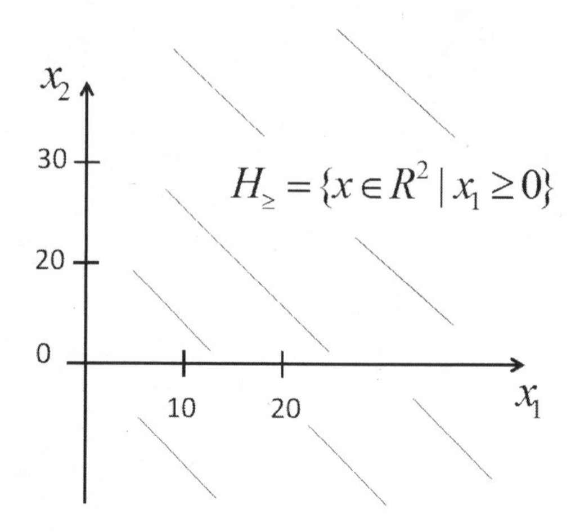

FIGURE 2.3
Closed halfspace $x_1 \geq 0$.

$R^n | a_i^T x \leq b_i\}$ and $H_U = \{x \in R^n | a_i^T x \geq b_i\}$, meaning that x belongs to H if and only if it belongs in the intersection of H_L and H_U.

Thus, we can conclude that any feasible point $x \in P$ lies in the intersection of closed halfspaces. When the number of closed halfspaces is finite, as is the case for linear programs, we have the following definition.

Definition 2.5
The intersection of a finite number of closed halfspaces is called a *polyhedron* (or *polyhedral set*). A bounded polyhedron is called a *polytope*.

So the feasible set P of any linear program is a polyhedral set. In particular, the set $P = \{x = \binom{x_1}{x_2} | \ x_1 + x_2 \leq 20, 2x_1 + x_2 \leq 30, x_1, \ x_2 \geq 0\}$ is a polytope.

The closed halfspaces $H_{\leq}$ and $H_{\geq}$ have the following property. If one takes two points x and y in $H_{\leq}(H_{\geq})$, then the line segment between x and y is also contained in $H_{\leq}(H_{\geq})$; see Figure 2.6. Now the line segment between two points x and y in $C \subseteq R^n$ can be expressed as $\lambda x + (1 - \lambda)y$ for $0 \leq \lambda \leq 1$. More formally, we have the following definition.

Definition 2.6
A set $C \subseteq R^n$ is said to be convex if for any x and y in C, $\lambda x + (1 - \lambda)y \in C$ for all $\lambda \in [0, 1]$.

Figure 2.7 depicts sets that are convex and not convex.

We can now show that a closed halfspace $H_{\leq}$ is a convex set. Let $z = \begin{bmatrix} z_1 \\ z_2 \end{bmatrix}$ and $y = \begin{bmatrix} y_1 \\ y_2 \end{bmatrix}$ be any pair of points in $H_{\leq} = \{x \in R^n | a^T x \leq \beta\}$. Then, consider any point on the line segment between z and y, i.e., $\lambda z + (1 - \lambda)y$ for some $\lambda \in [0, 1]$. We must show that $a^T(\lambda z + (1 - \lambda)y) \leq \beta$. Now

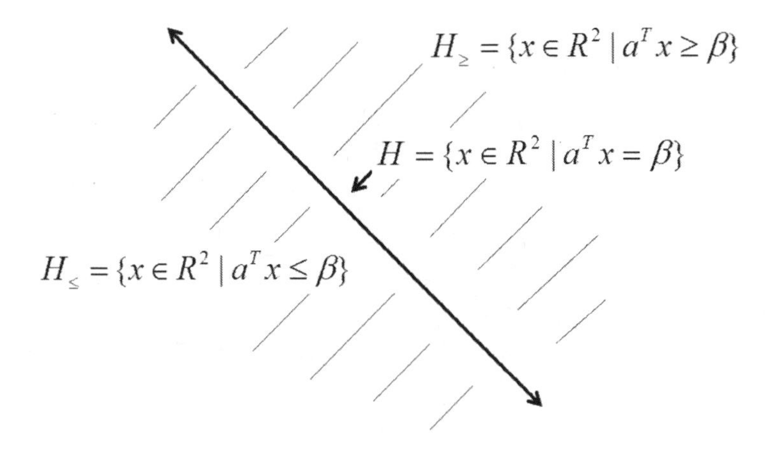

FIGURE 2.4
Hyperplane in R^2.

$a^T(\lambda z + (1-\lambda)y) = \lambda a^T z + (1-\lambda)a^T y \leq \lambda\beta + (1-\lambda)\beta = \beta$ where the inequality holds since λ and $(1-\lambda)$ are both non-negative and z and y are in $H_{\leq}$. Thus, $H_{\leq}$ is convex. A similar argument shows that $H_{\geq}$ is convex. We summarize in the following statement.

Theorem 2.7
The closed halfspaces $H_{\leq}$ and $H_{\geq}$ are convex sets.

We now aim to show that the feasible set P of a linear program is a convex set. First we need the following theorem.

Theorem 2.8
The intersection of convex sets is convex.

Proof: Suppose there is an arbitrary collection of convex sets S_i indexed by the set I. Consider the intersection $\cap_{i \in I} S_i$ and let x and y be elements of this intersection. For any $\lambda \in [0,1]$, then $z = \lambda x + (1-\lambda)y$ is in every set S_i since x and y are in S_i for every $i \in I$ and S_i is a convex set. Thus, $\cap_{i \in I} S_i$ is a convex set. ■

Corollary 2.9
The feasible set of a linear program is a convex set.

Proof: Immediately follows from Theorem 2.8 since a feasible set P of a linear program is a polyhedron (i.e., equivalent to the intersection of closed halfspaces) and that each closed halfspace is a convex set by Theorem 2.7. ■

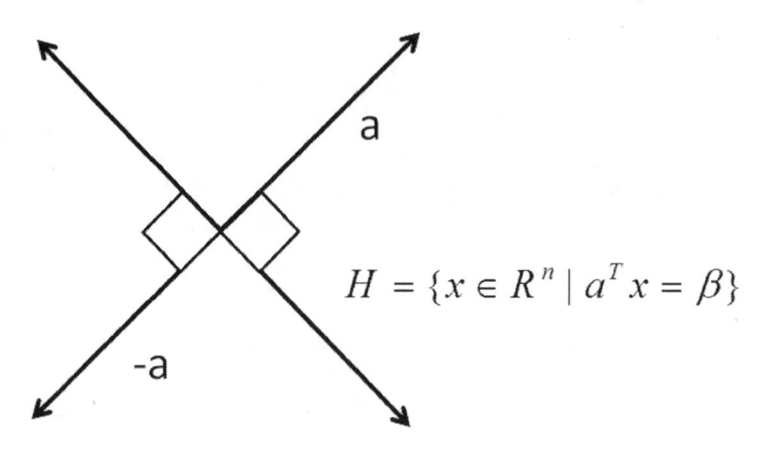

FIGURE 2.5
a and $-a$ are perpendicular to H.

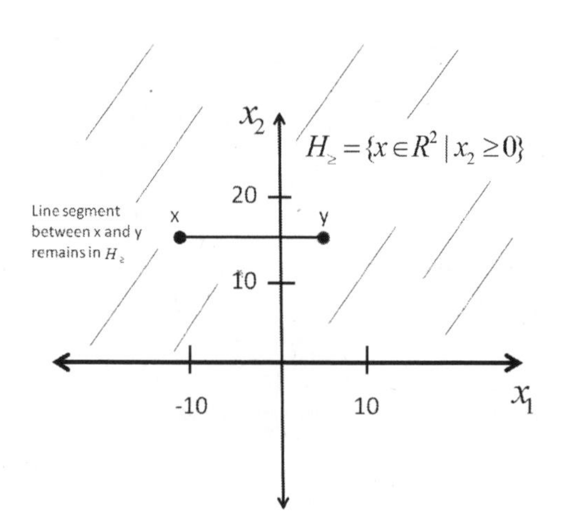

FIGURE 2.6
Convexity of $x_2 \geq 0$.

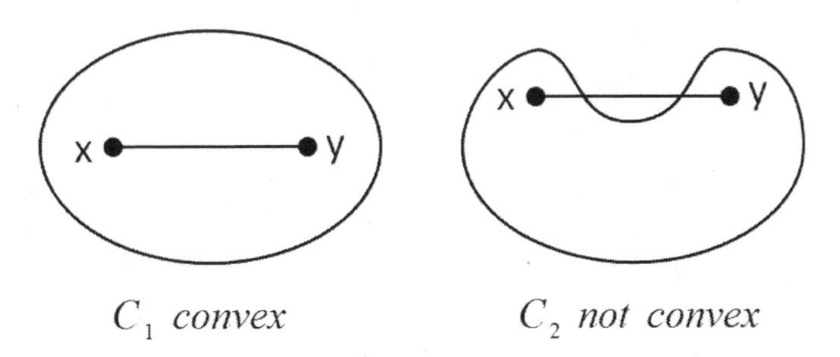

FIGURE 2.7
Convexity and non-convexity.

2.2.1 Geometry of Optimal Solutions

In this section we develop a geometric characterization of optimal solutions for linear programs based on insights from low dimensional problems. Consider the linear program (2.1). The objective function is $-x_1 - 2x_2 = c^T x$. Consider contours of the objective function $H = \{x \in R^2 | -x_1 - 2x_2 = \beta\}$. A key insight is that the contours are hyperplanes and the negative of the gradient of the objective function, i.e., $-c = \begin{bmatrix} 1 \\ 2 \end{bmatrix}$ is perpendicular to all such contours. Thus, to decrease the objective function in the direction of most rapid descent, the contours of the objective should be moved in the direction of $-c$ while remaining perpendicular to $-c$; see Figure 2.8.

However, the contours of the objective function cannot be moved indefinitely in the direction of $-c$ as points on the contours must be feasible for the LP. Thus, one should move the contours in the direction $-c$ as far as possible while ensuring that the contours intersect the feasible region. It is not enough to have a contour of the objective just intersecting the feasible set P. At optimality, the feasible set should be completely contained within one of the closed halfspaces of the contour of the objective function.

For example, consider the feasible set P for the LP (2.1). One can move the contours in the direction of $-c = \begin{bmatrix} 1 \\ 2 \end{bmatrix}$ until moving further would create infeasibility; see Figure 2.8. Observe that the intersection of the contour and the feasible set P in this case is the single point $x^* = \begin{bmatrix} 0 \\ 20 \end{bmatrix}$, which happens to be the optimal solution to the LP. Furthermore, the feasible set P is completely contained in the closed halfspace $H_{\geq} = \{x \in R^2 | -x_1 - x_2 \geq -20\}$ or the equivalent halfspace $H_{\leq} = \{x \in R^2 | x_1 + x_2 \leq 20\}$.

In general, we can characterize optimal solutions to LPs via the geometry of the hyperplane and feasible set intersection.

Geometric Characterization of Optimality

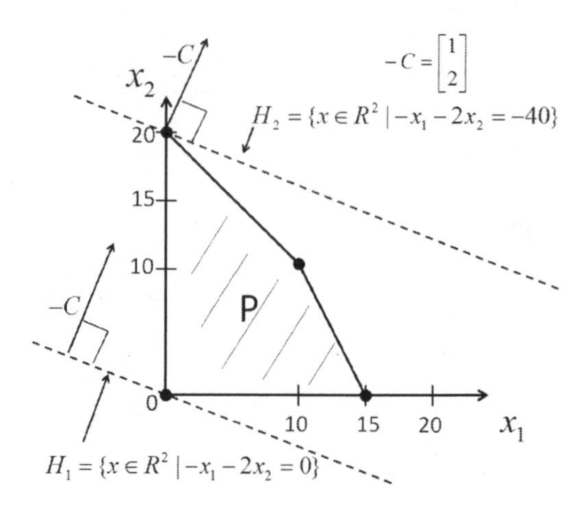

FIGURE 2.8
Hyperplane characterization of optimality for LP (2.1).

Let $P \neq \emptyset$ be the feasible set of a linear program with an objective to minimize $c^T x$ and let $H = \{x \in R^n \mid -c^T x = \beta\}$. If $P \subset H_{\leq} = \{x \in R^2 \mid -c^T x \leq \beta\}$ for some $\beta \in R^1$, then any x in the intersection of P and H is an optimal solution for the linear program.

Unique Intersection

As seen in the previous example for the linear program (2.1) for $\beta = 40$, the feasible set P is completely contained in the halfspace $H_{\leq} = \{x \in R^2 | x_1 + x_2 \leq 20\}$ and $x^* = \begin{bmatrix} 0 \\ 20 \end{bmatrix}$ is in both P and $H = \{x \in R^2 | x_1 + 2x_2 = 40\}$ and is the only such point. An important observation is that the optimal point x^* is a "corner point" which intuitively in two dimensions is a point not in the interior of P and not in the interior of a line segment on a side or "edge" of P; see Figure 2.9. The other three corner points of P are $v_1 = \binom{0}{0}$, $v_2 = \binom{15}{0}$, and $v_3 = \binom{10}{10}$. These "corner" points of the feasible set P play a very special role in algorithms for solving linear programs such as the simplex method in Chapter 3.

Infinite Intersection

It is possible for the set of optimal solutions for an LP to be infinite. Consider the following LP

$$\begin{array}{ll} \textit{minimize} & -x_1 \\ \textit{subject to} & x_1 \leq 1 \\ & x_2 \leq 1 \\ & x_1, x_2 \geq 0. \end{array} \tag{2.2}$$

The feasible set P for this LP and hyperplanes (objective contours) are in

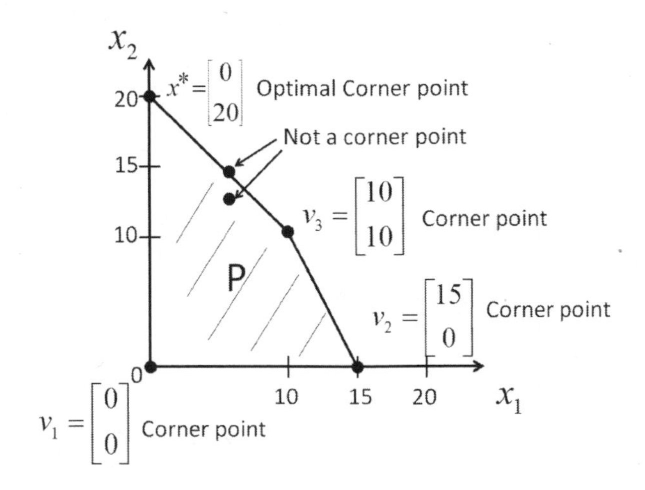

FIGURE 2.9
Corner points of feasible set of LP (2.1).

Figure 2.10. The "corner points" for P are $v_1 = \begin{bmatrix} 0 \\ 0 \end{bmatrix}$, $v_2 = \begin{bmatrix} 1 \\ 0 \end{bmatrix}$, $v_3 = \begin{bmatrix} 0 \\ 1 \end{bmatrix}$, and $v_4 = \begin{bmatrix} 1 \\ 1 \end{bmatrix}$.

It is clear that the contours $H = \{x \in R^2 | \ x_1 = \beta\}$ can be pushed in the direction of $-c = \begin{bmatrix} 1 \\ 0 \end{bmatrix}$ as far as $\beta = 1$, and it is at this point that P is completely contained in $H_{\leq} = \{x \in R^2 | \ x_1 \leq 1\}$ while having a non-empty intersection with H. From Figure 2.10 we see that the line segment in P between the corner points v_3 and v_4 (inclusive of these corner points) intersects with $H^* = \{x \in R^2 | \ x_1 = 1\}$, and thus all points on this line segment are optimal solutions. Of course, there are an infinite number of points on this line segment (in particular an uncountably infinite number of points). It is crucial to observe that either of the endpoints of this line segment are optimal as well, meaning that there is at least one corner point that is optimal.

Unbounded Case

It is possible that an LP is consistent but unbounded. Consider the following LP

$$\begin{aligned} &\textit{minimize} && -x_1 - x_2 \\ &\textit{subject to} && x_1 + x_2 \geq 1 \\ & && x_1 \geq 0, x_2 \geq 0. \end{aligned} \tag{2.3}$$

In this case, the contours of the objective function can be moved indefinitely in the direction of $-c$ while always intersecting the feasible set P since it is

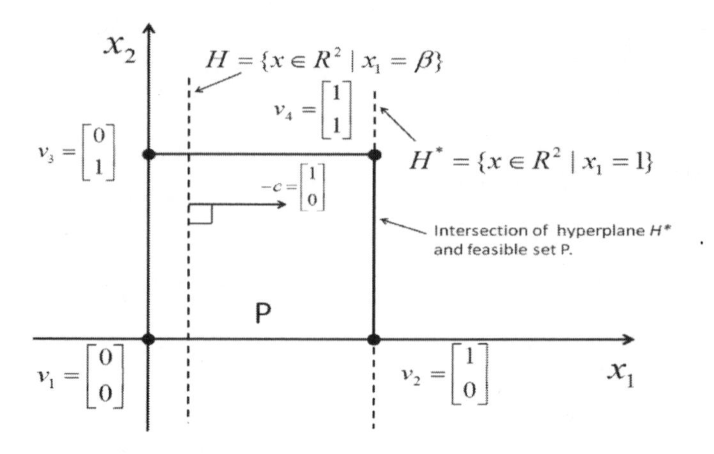

FIGURE 2.10
Hyperplane characterization of infinite optimal solutions for LP (2.2).

unbounded; see Figure 2.11. That is, for any positive value of β, the hyperplane $H = \{x \in R^2 | \; x_1 + x_2 = \beta\}$ will always intersect the feasible set $P = \{x \in R^2 | \; x_1 + x_2 \geq 1, x_1 \geq 0, x_2 \geq 0\}$.

A main insight from the geometry of the two-dimensional examples is that if an LP has a finite optimal solution, then an optimal solution can be attained at a corner point. This observation will turn out to hold true in higher dimensions and is called the Fundamental Theorem of Linear Programming. It will be proved in Section 4 of this chapter. This suggests that one can plot the feasible set and find all of the corner points and then evaluate them to find the optimal solution it exists. Unfortunately, this strategy is effective for only small problems, e.g., LPs with just two variables. Graphing feasible sets in higher dimensions is not a practical endeavor, however, the insights from the geometry of LPs in low dimensions hold for higher dimensional problems.

2.3 Extreme Points and Basic Feasible Solutions

In this section, we develop an algebraic representation of corner points through the corresponding geometric notion of extreme points. Then, an algebraic representation of extreme points is given. The algebraic representation will allow higher-dimensional LP problems to be considered without relying on the geometry of the underlying problem, and will serve as the basis for development of practical algorithms effective for large LP problems.

Extreme Points

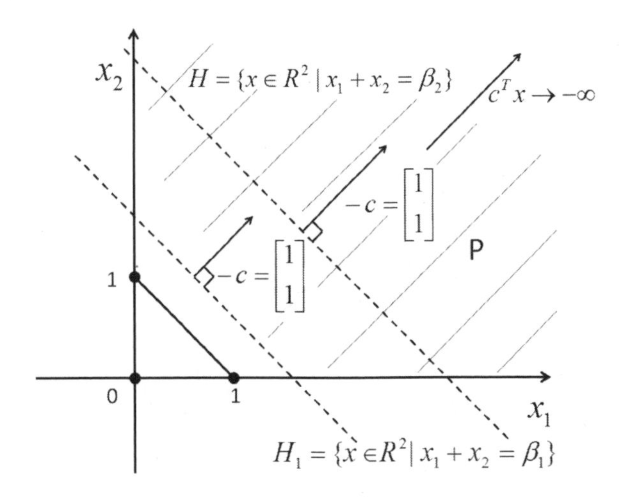

FIGURE 2.11
Unbounded LP (2.3).

We first define a geometric entity that will correspond to the notion of a corner point for any dimension. Recall that in two dimensions, a corner point of a feasible set P of a linear program should not lie in the interior of P and should not lie in the interior of any line segment on the edge of P. A mathematical representation of this requirement is embodied in the definition of an extreme point. First, we need the following.

Definition 2.10
A *convex combination* of vectors $x_1, x_2, ..., x_k \in R^n$ is a linear combination $\sum_{i=1}^{k} \lambda_i x_i$ of these vectors such that $\sum_{i=1}^{k} \lambda_i = 1$ and $\lambda_i \geq 0$ for $i = 1, ..., k$.

For example, the convex combination of two distinct points z and y in R^n is the set $l = \{x \in R^n | x = \lambda z + (1 - \lambda)y \text{ for } \lambda \in [0, 1]\}$, which is the line segment between z and y. We are now ready to give the definition of an extreme point.

Definition 2.11
Let $C \subseteq R^n$ be a convex set and $x \in C$. A point x is an extreme point of C if it can't be expressed as a convex combination of other points in C.

The definition of extreme points will be vital in developing an algebraic representation of corner points.

Example 2.12
For the feasible set $P = \{x = \binom{x_1}{x_2} | \ x_1 + x_2 \leq 20, 2x_1 + x_2 \leq 30, x_1, \ x_2 \geq 0\}$, the corner points $v_1 = \binom{0}{0}$, $v_2 = \binom{15}{0}$, $v_3 = \binom{0}{20}$, and $v_4 = \binom{10}{10}$ are extreme points since they cannot be written as the convex combination of more than one point in P; see Figure 2.12.

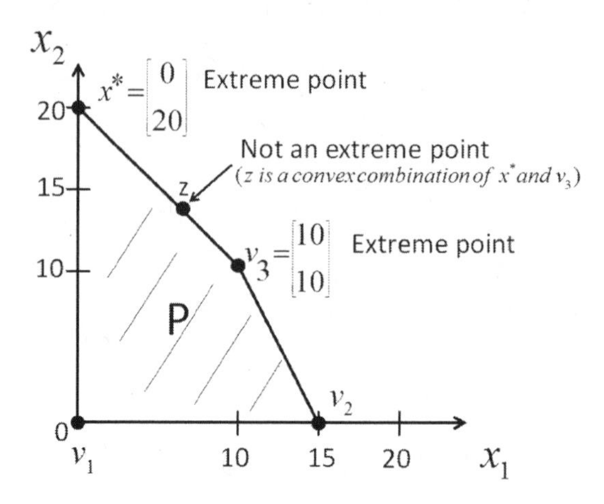

FIGURE 2.12
Extreme points of feasible set of LP (2.1).

At this point it is not evident how extreme points can facilitate an algebraic representation since it is a geometric entity. The missing connection is how an extreme point x is related to the corresponding parameters (data) in the linear program. To see this, convert the LP (2.1) to standard form, i.e., minimize $c^T x$ subject to $Ax = b$ and $x \geq 0$ by adding slack variables x_3 and x_4 to get

$$\begin{array}{ll} \textit{minimize} & -x_1 - 2x_2 \\ \textit{subject to} & x_1 + x_2 + x_3 = 20 \\ & 2x_1 + x_2 + \quad x_4 = 30 \\ & x_1,\ x_2,\ x_3,\ x_4 \geq 0 \end{array}$$

where corresponding matrix entities are

$$A = \begin{bmatrix} 1 & 1 & 1 & 0 \\ 2 & 1 & 0 & 1 \end{bmatrix},\ b = \begin{bmatrix} 20 \\ 30 \end{bmatrix},\ \text{and } c = \begin{bmatrix} -1 \\ -2 \end{bmatrix}.$$

Let A_i be the column in A associated with x_i for $i = 1, ..., 4$ so that

$$A_1 = \begin{bmatrix} 1 \\ 2 \end{bmatrix},\ A_2 = \begin{bmatrix} 1 \\ 1 \end{bmatrix},\ A_3 = \begin{bmatrix} 1 \\ 0 \end{bmatrix},\ \text{and } A_4 = \begin{bmatrix} 0 \\ 1 \end{bmatrix}.$$

Consider the corner point $v_4 = \binom{10}{10} = \binom{x_1}{x_2}$ in the feasible set P of the original LP, then the corresponding feasible solution in standard form must have the slack variables at zero, i.e., $x_3 = x_4 = 0$ and so the corresponding vector is

$$z = \begin{pmatrix} 10 \\ 10 \\ 0 \\ 0 \end{pmatrix} = \begin{pmatrix} x_1 \\ x_2 \\ x_3 \\ x_4 \end{pmatrix}.$$

A important observation here is that if we take the columns in A associated with the variables that are positive in z, i.e., A_1 (associated with $x_1 = 10 > 0$) and A_2 (associated with $x_2 = 10 > 0$) to form the sub-matrix $B = \begin{bmatrix} A_1 & A_2 \end{bmatrix} = \begin{bmatrix} 1 & 1 \\ 2 & 1 \end{bmatrix}$, then the resulting submatrix B is non-singular (i.e., the determinant of B is non-zero).

The other corner points for this LP can all be extended to corresponding feasible solutions for the standard form version of the LP and a corresponding non-singular square submatrix B can be obtained for each corner point; see Table 2.1.

Table 2.1 Feasible solutions in standard form for LP (2.1)

Corner point	Standard form feasible solution	Submatrix B
$v_1 = \binom{x_1}{x_2} = \binom{0}{0}$	$x_1 = 0, x_2 = 0, x_3 = 20, x_4 = 30$	$[A_3\ A_4] = \begin{bmatrix} 1 & 0 \\ 0 & 1 \end{bmatrix}$
$v_1 = \binom{x_1}{x_2} = \binom{15}{0}$	$x_1 = 15, x_2 = 0, x_3 = 5, x_4 = 0$	$[A_1\ A_3] = \begin{bmatrix} 1 & 1 \\ 2 & 0 \end{bmatrix}$
$v_1 = \binom{x_1}{x_2} = \binom{0}{20}$	$x_1 = 0, x_2 = 20, x_3 = 0, x_4 = 10$	$[A_2\ A_4] = \begin{bmatrix} 1 & 0 \\ 1 & 1 \end{bmatrix}$
$v_1 = \binom{x_1}{x_2} = \binom{10}{10}$	$x_1 = 10, x_2 = 10, x_3 = 0, x_4 = 0$	$[A_1\ A_2] = \begin{bmatrix} 1 & 1 \\ 2 & 1 \end{bmatrix}$

The correspondence between the positive components of an extreme point and the invertibility of the corresponding matrix B turns out to algebraically characterize extreme points.

Theorem 2.13

Consider a linear program in standard form where the feasible set $P = \{x \in R^n | Ax = b, x \geq 0\}$ is non-empty. A vector $x \in P$ is an extreme point if and only if the columns of A corresponding to positive components of x are linearly independent.

Proof:

Suppose that there are k positive components in $x \in P$ and they are positioned as the first k components of x, i.e., $x = \begin{bmatrix} x_p \\ 0 \end{bmatrix}$ where

$$x_p = \begin{bmatrix} x_1 \\ x_2 \\ \vdots \\ x_k \end{bmatrix} > 0.$$

Let B be the columns of A associated with the components of x_p. Then, $Ax = Bx_p = b$.

(Proof of forward direction =>)

Assume that $x \in P$ is an extreme point. Now suppose that B is singular (i.e., columns of B are linearly dependent), then there exists a non-zero vector w such that $Bw = 0$. For sufficiently small $\varepsilon > 0$, $x_p + \varepsilon w \geq 0$ and $x_p - \varepsilon w \geq 0$. Furthermore, $B(x_p + \varepsilon w) = Bx_p + \varepsilon Bw = b$ and $B(x_p - \varepsilon w) = Bx_p - \varepsilon Bw = b$ since $Bw = 0$. Therefore, the following two vectors

$$z^+ = \begin{bmatrix} (x_p + \varepsilon w) \\ 0 \end{bmatrix} \text{ and } z^- = \begin{bmatrix} (x_p - \varepsilon w) \\ 0 \end{bmatrix}$$

are in the set P since $Az^+ = b$ and $Az^- = b$. However, $\frac{1}{2}z^+ + \frac{1}{2}z^- = x$, which means x is a convex combination of z^+ and z^-, contradicting that it is an extreme point.

(Proof of reverse direction <=)

Suppose that the columns of B are linearly independent and that x is not an extreme point. Then x can be written as the convex combination of two distinct points v_1 and v_2 both in P (and different from x) i.e $x = \begin{bmatrix} x_p \\ 0 \end{bmatrix} = \lambda v_1 + (1 - \lambda)v_2$ for some positive λ such that $0 < \lambda < 1$. Now v_1 and v_2 are both non-negative since they are in P and λ is positive, so the last $n - k$ components of v_1 and v_2 must be zeros, i.e., v_1 and v_2 can be written as

$$v_1 = \begin{bmatrix} v_p^1 \\ 0 \end{bmatrix} \text{ and } v_2 = \begin{bmatrix} v_p^2 \\ 0 \end{bmatrix}$$

where v_p^1 and v_p^2 are the first k components of v_1 and v_2, respectively. Thus, $B(x - v_1) = Bx_p - Bv_p^1 = b - b = 0$, but $x_p - v_p^1 \neq 0$ since $x \neq v_1$. So the columns of B are linearly dependent, which is a contradiction. ■

Theorem 2.13 allows access to extreme points through using the associated linear algebra matrix constructs, e.g., the matrix B. Further to this development, we use the characterization from Theorem 2.13 to define the full algebraic representation of an extreme point in standard form (i.e., basic feasible solutions), will prove to be very useful in developing practical algorithms for solving LPs, as will be seen in Chapter 3.

We assume that the matrix A has m rows, n columns, and $m \leq n$. Further, we assume that A has full row rank (i.e., the m rows are linearly independent). Let x_B = the set of variables of x corresponding to the columns of an $m \times m$ submatrix B of A. Denote by N the submatrix of A of dimension $m \times (n - m)$ consisting of columns not in B, and let x_N = the set of variables of x corresponding to the columns of N.

Definition 2.14

A vector $x \in P = \{x \in R^n | Ax = b, x \geq 0\}$ is a basic feasible solution (BFS) is there is a partition of the matrix A into an non-singular $m \times m$ square submatrix B and an $m \times (n-m)$ submatrix N such that $x = \begin{bmatrix} x_B \\ x_N \end{bmatrix}$ with $x_B \geq 0$ and $x_N = 0$ and $Ax = Bx_B = b$.

B is called the basis matrix, N is called the non-basis (or non-basic) matrix, x_B the set of basic variables, and x_N is the set of non-basic variables.

Example 2.15

Consider from the first row of Table 2.1 the extended feasible solution

$$x = \begin{bmatrix} x_1 \\ x_2 \\ x_3 \\ x_4 \end{bmatrix} = \begin{bmatrix} 0 \\ 0 \\ 20 \\ 30 \end{bmatrix}$$

corresponding to the corner point $v_1 = \begin{bmatrix} x_1 \\ x_2 \end{bmatrix} = \begin{bmatrix} 0 \\ 0 \end{bmatrix}$, which can be rewritten by letting $x_B = \begin{bmatrix} x_3 \\ x_4 \end{bmatrix} = \begin{bmatrix} 20 \\ 30 \end{bmatrix}$ and $x_N = \begin{bmatrix} x_1 \\ x_2 \end{bmatrix} = \begin{bmatrix} 0 \\ 0 \end{bmatrix}$ so that $x = \begin{bmatrix} x_B \\ x_N \end{bmatrix}$.

Let the matrix B consist of the columns associated with the variables in x_B, then

$$B = \begin{bmatrix} 1 & 0 \\ 0 & 1 \end{bmatrix}$$

and let N consist of columns associated with the variables in x_N i.e.

$$N = [A_1 \; A_2] = \begin{bmatrix} 1 & 1 \\ 2 & 1 \end{bmatrix}.$$

Observe that B is non-singular and that $Bx_B = b$ (i.e., $x_B = B^{-1}b$), $x_B > 0$, and if $x_N = 0$. Thus, x is a basic feasible solution which corresponds to the extreme point v_1 of P. In fact, all of the extended feasible solutions in Table 2.1 are basic feasible solutions and thus correspond to extreme points.

Further, if one were to start with the matrix $B = \begin{bmatrix} 1 & 0 \\ 0 & 1 \end{bmatrix}$ and let $x_B = B^{-1}b$ and $x_N = 0$, then the vector $x = \begin{bmatrix} x_B \\ x_N \end{bmatrix}$ is a basic feasible solution.

In general, we have the following correspondence, which is a direct consequence of Theorem 2.13.

Corollary 2.16
A vector $x \in P = \{x \in R^n | Ax = b, x \geq 0\}$ *is an extreme point if and only if there is some matrix* B *so that* x *is a basic feasible solution with* B *as the basis matrix.*

2.3.1 Generating Basic Feasible Solutions

Theorem 2.13 says that any m columns of A that form a submatrix B that is invertible and whose corresponding variables x_B are such that $Bx_B = b$ and $x_B > 0$ will correspond to a basic feasible solution and therefore to an extreme point. The total number of possible ways to select m columns out of a total n is $C(n, m) = \frac{n!}{m!(n-m)!}$, which is a finite number. It is possible that for a particular selection of m columns, either B is not invertible or $x_B = B^{-1}b$ has negative components. Thus, we have the following.

Corollary 2.17
The feasible set $P = \{x \in R^n | Ax = b, x \geq 0\}$ *has at most* $\frac{n!}{m!(n-m)!}$ *extreme points.*

Given a linear program in standard form, it is possible to generate all basic feasible solutions and thereby access the extreme points (corner points) of the LP without the need to graph the feasible set.

The procedure amounts to choosing m columns out of the n columns of the A matrix to form the basis matrix B. A particular choice of m columns will generate an extreme point if (1) the matrix B is non-singular and (2) x_B is non-negative. Note that if B is non-singular, then one can set $x_N = 0$ so that $Ax = [B|N]\begin{bmatrix} x_B \\ x_N \end{bmatrix} = Bx_B + Nx_N = Bx_B = b$.

Example 2.18
Consider the feasible set defined by the constraints

$$\begin{aligned} x_1 + x_2 &\leq 1 \\ x_1 \qquad &\leq 1 \\ x_2 &\leq 1 \\ x_1 \geq 0, x_2 \geq 0. & \end{aligned} \tag{2.4}$$

In standard form, the constraints are

$$\begin{aligned} &x_1 + x_2 + x_3 = 1 \\ &x_1 + \qquad\qquad + x_4 = 1 \\ &\quad x_2 + \qquad\qquad + x_5 = 1 \\ &x_1, x_2, x_3, x_4, x_5 \geq 0 \end{aligned}$$

with

$$A = \begin{bmatrix} 1 & 1 & 1 & 0 & 0 \\ 1 & 0 & 0 & 1 & 0 \\ 0 & 1 & 0 & 0 & 1 \end{bmatrix} \text{ and } b = \begin{bmatrix} 1 \\ 1 \\ 1 \end{bmatrix}.$$

There are $\begin{pmatrix} 5 \\ 3 \end{pmatrix} = \frac{5!}{3!(2)!} = 10$ possible extreme points. For example, one can select columns associated with variables x_3, x_4, and x_5 to form the submatrix

$$B = \begin{bmatrix} 1 & 0 & 0 \\ 0 & 1 & 0 \\ 0 & 0 & 1 \end{bmatrix}.$$

Now B is invertible, and with selected variables $x_B = \begin{bmatrix} x_3 \\ x_4 \\ x_5 \end{bmatrix}$ we have

$$x_B = B^{-1}b = \begin{bmatrix} 1 & 0 & 0 \\ 0 & 1 & 0 \\ 0 & 0 & 1 \end{bmatrix} \begin{bmatrix} 1 \\ 1 \\ 1 \end{bmatrix} = \begin{bmatrix} 1 \\ 1 \\ 1 \end{bmatrix},$$

so with

$$x_N = \begin{bmatrix} x_1 \\ x_2 \end{bmatrix} = \begin{bmatrix} 0 \\ 0 \end{bmatrix}$$

then,

$$x = \begin{bmatrix} x_B \\ x_N \end{bmatrix} = \begin{bmatrix} 1 \\ 1 \\ 1 \\ 0 \\ 0 \end{bmatrix}$$

is basic feasible solution (note the re-ordering of the variables) and thus an extreme point. We now show all possibilities for generating an extreme point for selection of $m = 3$ columns of A in the following tables. The first column shows the partition of the variables of x into x_B and x_N where the variables x_B are selected so that the corresponding columns will form the matrix B which is entered in the second column, the third column computes the values of $x_B = B^{-1}b$ if possible (if B is invertible), and the fourth column indicates whether the computed values for components of x represent an extreme point. Table 2.2 lists those partitions that result in basic feasible solutions and Table 2.3 lists partitions that do not result in basic feasible solutions, either due to infeasibility or non-negativity of basic variables.

Table 2.2 All basic feasible solutions for Example 2.16

Partition $x = \begin{bmatrix} x_B \\ x_N \end{bmatrix}$	Basis matrix B	$x_B = B^{-1}b$	x bfs?
$x_B = \begin{bmatrix} x_3 \\ x_4 \\ x_5 \end{bmatrix}, x_N = \begin{bmatrix} x_1 \\ x_2 \end{bmatrix}$	$\begin{bmatrix} 1 & 0 & 0 \\ 0 & 1 & 0 \\ 0 & 0 & 1 \end{bmatrix}$	$\begin{bmatrix} 1 \\ 1 \\ 1 \end{bmatrix}$	yes
$x_B = \begin{bmatrix} x_1 \\ x_4 \\ x_5 \end{bmatrix}, x_N = \begin{bmatrix} x_3 \\ x_2 \end{bmatrix}$	$\begin{bmatrix} 1 & 0 & 0 \\ 1 & 1 & 0 \\ 0 & 0 & 1 \end{bmatrix}$	$\begin{bmatrix} 1 \\ 0 \\ 1 \end{bmatrix}$	yes
$x_B = \begin{bmatrix} x_3 \\ x_1 \\ x_5 \end{bmatrix}, x_N = \begin{bmatrix} x_4 \\ x_2 \end{bmatrix}$	$\begin{bmatrix} 1 & 1 & 0 \\ 0 & 1 & 0 \\ 0 & 0 & 1 \end{bmatrix}$	$\begin{bmatrix} 0 \\ 1 \\ 1 \end{bmatrix}$	yes
$x_B = \begin{bmatrix} x_2 \\ x_4 \\ x_5 \end{bmatrix}, x_N = \begin{bmatrix} x_1 \\ x_3 \end{bmatrix}$	$\begin{bmatrix} 1 & 0 & 0 \\ 0 & 1 & 0 \\ 1 & 0 & 1 \end{bmatrix}$	$\begin{bmatrix} 1 \\ 1 \\ 0 \end{bmatrix}$	yes
$x_B = \begin{bmatrix} x_3 \\ x_4 \\ x_2 \end{bmatrix}, x_N = \begin{bmatrix} x_1 \\ x_5 \end{bmatrix}$	$\begin{bmatrix} 1 & 0 & 1 \\ 0 & 1 & 0 \\ 0 & 0 & 1 \end{bmatrix}$	$\begin{bmatrix} 0 \\ 1 \\ 1 \end{bmatrix}$	yes
$x_B = \begin{bmatrix} x_1 \\ x_2 \\ x_4 \end{bmatrix}, x_N = \begin{bmatrix} x_3 \\ x_5 \end{bmatrix}$	$\begin{bmatrix} 1 & 1 & 0 \\ 1 & 0 & 1 \\ 0 & 1 & 0 \end{bmatrix}$	$\begin{bmatrix} 0 \\ 1 \\ 1 \end{bmatrix}$	yes
$x_B = \begin{bmatrix} x_1 \\ x_2 \\ x_5 \end{bmatrix}, x_N = \begin{bmatrix} x_3 \\ x_4 \end{bmatrix}$	$\begin{bmatrix} 1 & 1 & 0 \\ 1 & 0 & 0 \\ 0 & 1 & 1 \end{bmatrix}$	$\begin{bmatrix} 1 \\ 0 \\ 1 \end{bmatrix}$	yes

Table 2.3 Infeasible partitions of Example 2.16

Partition $x = \begin{bmatrix} x_B \\ x_N \end{bmatrix}$	Basis matrix B	$x_B = B^{-1}b$	x bfs?
$x_B = \begin{bmatrix} x_3 \\ x_4 \\ x_1 \end{bmatrix}, x_N = \begin{bmatrix} x_5 \\ x_2 \end{bmatrix}$	$\begin{bmatrix} 1 & 0 & 1 \\ 0 & 1 & 1 \\ 0 & 0 & 0 \end{bmatrix}$	B singular	no
$x_B = \begin{bmatrix} x_3 \\ x_2 \\ x_5 \end{bmatrix}, x_N = \begin{bmatrix} x_1 \\ x_4 \end{bmatrix}$	$\begin{bmatrix} 1 & 1 & 0 \\ 0 & 0 & 0 \\ 0 & 1 & 1 \end{bmatrix}$	B singular	no
$x_B = \begin{bmatrix} x_1 \\ x_2 \\ x_3 \end{bmatrix}, x_N = \begin{bmatrix} x_4 \\ x_5 \end{bmatrix}$	$\begin{bmatrix} 1 & 1 & 1 \\ 1 & 0 & 0 \\ 0 & 1 & 0 \end{bmatrix}$	$\begin{bmatrix} 1 \\ 1 \\ -1 \end{bmatrix}$	no

2.3.1.1 Generating Basic Feasible Solutions with MATLAB®

We give an example in MATLAB on how a basic feasible solution can be obtained for a linear program in standard form. Consider the linear program in Example 2.16. Recall that the constraint matrix in standard form and right-hand side vector is

$$A = \begin{bmatrix} 1 & 1 & 1 & 0 & 0 \\ 1 & 0 & 0 & 1 & 0 \\ 0 & 1 & 0 & 0 & 1 \end{bmatrix} \text{ and } b = \begin{bmatrix} 1 \\ 1 \\ 1 \end{bmatrix},$$

To generate the basic feasible solution

$$x_B = \begin{bmatrix} x_3 \\ x_4 \\ x_5 \end{bmatrix}, \; x_N = \begin{bmatrix} x_1 \\ x_2 \end{bmatrix},$$

the last three columns of A are chosen as the basis matrix B.

In MATLAB we write

```
>> A=[1,1,1,0,0;1,0,0,1,0;0,1,0,0,1]; % Enter A matrix
>> b=[1;1;1]; % Enter right hand side vector

>> B=A(:,3:5) % basis matrix is last three columns of A
B =
1 0 0
0 1 0
```

```
0 0 1
>> x_b=B\b  %Solve for basic variables
x_b =
1
1
1

>> x_n=zeros(2,1);  % set non-basic variables to 0
>>x=[x_b;x_n]  % Constructing the basic feasible solution
x =
1
1
1
0
0
```

Now, to get the basic feasible solution

$$x_B = \begin{bmatrix} x_1 \\ x_4 \\ x_5 \end{bmatrix}, x_N = \begin{bmatrix} x_3 \\ x_2 \end{bmatrix}$$

from the previous basic feasible solution, we can replace the first column in the current basis matrix B with the first column in A.

```
>>B(:,1)=A(:,1);
>>B
B =
1 0 0
1 1 0
0 0 1

>>x_b=B\b  % new basic variable values
x_b =
1
0
1
>>x=[x_b;x_n]  % Constructing the new basic feasible solution
x =
1
0
1
0
0
```

In general, one can select any 3 columns of A and check in MATLAB if the resulting choice results in a basic feasible solution and repeat until all possible choices of selecting 3 columns is exhausted.

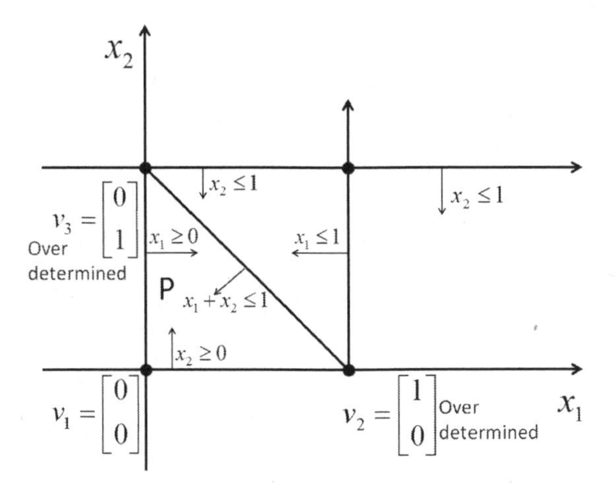

FIGURE 2.13
Feasible set (2.4).

2.3.2 Degeneracy

An extreme point $x \in P = \{x \in R^n | Ax = b, x \geq 0\}$ is a geometric entity and has a corresponding representation as a basic feasible solution. A natural question to ask is whether this correspondence is one to one. It is clear that a given basic feasible solution will correspond to a unique extreme point. But, for an extreme point x, is there a unique representation as a basic feasible solution? This is not the case as Example 2.17 above illustrates that several basic feasible solutions can correspond to the same extreme point. The feasible set in Example 2.17 is graphed in Figure 2.13.

There are three extreme points $v_1 = \begin{bmatrix} 0 \\ 0 \end{bmatrix}$, $v_2 = \begin{bmatrix} 1 \\ 0 \end{bmatrix}$, and $v_3 = \begin{bmatrix} 0 \\ 1 \end{bmatrix}$. The basic feasible solution in row 1 of Table 2.2 is the only BFS that corresponds to the extreme point v_1. However, the basic feasible solutions in rows 2, 3, and 5 in Table 2.2 correspond to v_2 since $x_1 = 1$ and $x_2 = 0$ in all of these solutions and the basic feasible solutions in rows 4, 5, and 6 in Table 2.2 correspond to v_3 since the $x_1 = 0$ and $x_2 = 1$ in all of these solutions. Thus, there is not in general a one-to-one correspondence between extreme points and basic feasible solutions.

There are three basic feasible solutions associated with v_2 and v_3. This arises since one of the variables in the basic set x_B of each of these basic feasible solutions is zero. This makes the three basic feasible solutions indistinguishable in terms of representing the corresponding extreme points. This motivates the following definitions.

Definition 2.19
A basic feasible solution $x \in P = \{x \in R^2 | Ax = b, x \geq 0\}$ is degenerate if

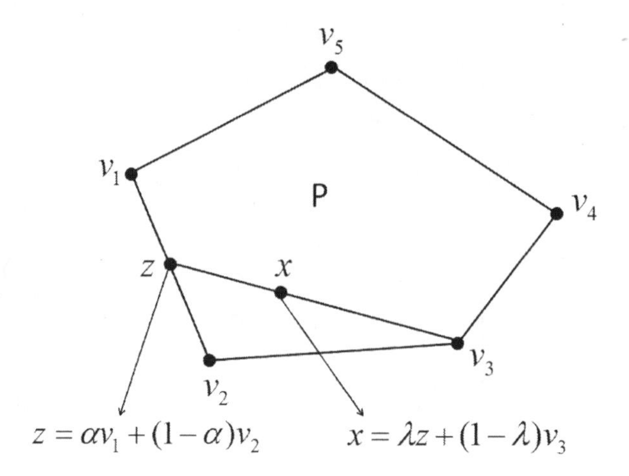

FIGURE 2.14
A polytope with 5 extreme points.

at least one of the variables in the basic set x_B is zero. $x \in P$ is said to be non-degenerate if all m of the basic variables are positive.

Degeneracy arises in a linear program in standard form at a extreme point because the constraints that give rise to the extreme point are "over-determined". For example, v_2 is determined by the intersection of three constraints (other than the non-negativity constraints), where only two constraints are needed; see Figure 2.13.

2.4 Resolution (Representation) Theorem

In this section, a characterization of the feasible set of $P = \{x \in R^2 | Ax = b, x \geq 0\}$ is first given when P is bounded and the general case when P can be unbounded is then considered. In particular, a representation of any $x \in P$ is sought in terms of the extreme points of P and recession directions.

Example 2.20

Consider the case when P is bounded, i.e., a polytope; see Figure 2.14. P has 5 extreme points v_1, v_2, v_3, v_4, and v_5.

The point x can be represented as the convex combination of the extreme point v_3 and the point z, which is not an extreme point, i.e., $x = \lambda z + (1-\lambda)v_3$ for some $\lambda \in [0,1]$. Furthermore, the point z can be represented as the convex combination of the extreme points v_1 and v_2, i.e., $z = \alpha v_1 + (1-\alpha)v_2$ for some $\alpha \in [0,1]$. Thus, $x = \lambda(\alpha v_1 + (1-\alpha)v_2) + (1-\lambda)v_3 = \lambda\alpha v_1 + \lambda(1-\alpha)v_2 + (1-\lambda)v_3$

where $\lambda\alpha, \lambda(1-\alpha)$, and $(1-\lambda)$ are all in $[0,1]$ and $\lambda\alpha+\lambda(1-\alpha)+(1-\lambda)=1$. Thus, $x \in P$ is a convex combination of extreme points in P. In general, any $x \in P$ in a polytope can be represented as a combination of extreme points in P.

Example 2.21
Consider the polytope

$$P = \{(x_1, x_2, x_3) \in R^3 | \ x_1 + x_2 + x_3 \leq 1, x_1 \geq 0, x_2 \geq 0, x_3 \geq 0\}.$$

Then, the extreme points are $v^1 = \begin{bmatrix} 1 \\ 0 \\ 0 \end{bmatrix}$, $v^2 = \begin{bmatrix} 1 \\ 0 \\ 0 \end{bmatrix}$, $v^3 = \begin{bmatrix} 0 \\ 0 \\ 1 \end{bmatrix}$, and $v^4 = \begin{bmatrix} 0 \\ 0 \\ 0 \end{bmatrix}$. Any point $x \in P$ can be written as a convex combination of these extreme points. For example, the point

$$x = \begin{bmatrix} \frac{1}{5} \\ \frac{1}{3} \\ \frac{2}{7} \end{bmatrix} = \tfrac{1}{5}v^1 + \tfrac{1}{3}v^2 + \tfrac{2}{7}v^3 + (1 - (\tfrac{1}{5} + \tfrac{1}{3} + \tfrac{2}{7}))v^4$$
$$= \tfrac{1}{5}v^1 + \tfrac{1}{3}v^2 + \tfrac{2}{7}v^3 + (\tfrac{29}{105})v^4.$$

In general, for any feasible vector $x = \begin{bmatrix} x_1 \\ x_2 \\ x_3 \end{bmatrix} \in P$ we can write

$$x = x_1v^1 + x_2v^2 + x_3v^3 + (1 - (x_1 + x_2 + x_3))v^4.$$

Example 2.22
We now consider the case when P is unbounded. As an example, let P be defined by the following set of inequalities

$$\begin{aligned} x_2 - x_1 &\leq 3 \\ x_1 \geq 0, x_2 &\geq 0. \end{aligned} \tag{2.5}$$

Figure 2.15 shows the graph of P, which is an unbounded set. The extreme points of P are $v_1 = \begin{bmatrix} 0 \\ 0 \end{bmatrix}$ and $v_2 = \begin{bmatrix} 3 \\ 0 \end{bmatrix}$. The point $x = \begin{bmatrix} 5 \\ 3 \end{bmatrix}$ cannot be represented as a convex combination of v_1 and v_2. Additional constructs are needed to develop a representation of x.

Definition 2.23
A ray is a set of the form $\{x \in R^n | x = x_0 + \lambda d$ for $\lambda \geq 0\}$ where x_0 is a given point and d is a non-zero vector called the direction vector.

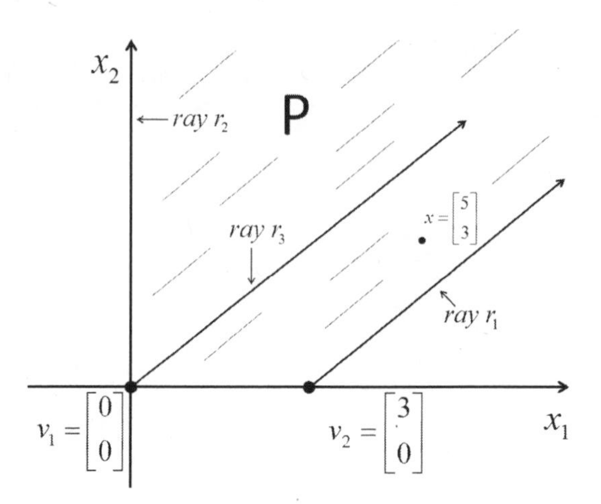

FIGURE 2.15
Some rays of feasible set (2.5).

For example, the ray $r_1 = \{x \in R^n | x = \left[\begin{smallmatrix}3\\0\end{smallmatrix}\right] + \lambda\left[\begin{smallmatrix}1\\1\end{smallmatrix}\right] \ for \ \lambda \geq 0\}$ extends indefinitely from the point $x_0 = \left[\begin{smallmatrix}3\\0\end{smallmatrix}\right]$ in the direction of $d_1 = \left[\begin{smallmatrix}1\\1\end{smallmatrix}\right]$ and the ray $r_2 = \{x \in R^n | x = \left[\begin{smallmatrix}0\\0\end{smallmatrix}\right] + \lambda\left[\begin{smallmatrix}0\\1\end{smallmatrix}\right] \ for \ \lambda \geq 0\}$ extends indefinetely from the point $x_0 = \left[\begin{smallmatrix}0\\0\end{smallmatrix}\right]$ in the direction of $d_2 = \left[\begin{smallmatrix}0\\1\end{smallmatrix}\right]$. Another ray is $r_3 = \{x \in R^n | x = \left[\begin{smallmatrix}0\\0\end{smallmatrix}\right] + \lambda\left[\begin{smallmatrix}1\\1\end{smallmatrix}\right] \ for \ \lambda \geq 0\}$. The three rays are in P; see Figure 2.15.

In turns out that for any starting point $x_0 \in P$, the rays $\{x \in R^n | x = x_0 + \lambda\left[\begin{smallmatrix}1\\1\end{smallmatrix}\right] \ for \ \lambda \geq 0\} \subset P$ and $\{x \in R^n | x = x_0 + \lambda\left[\begin{smallmatrix}0\\1\end{smallmatrix}\right] \ for \ \lambda \geq 0\} \subset P$. In such cases, the directions d_1 and d_2 are called recession directions, which are defined formally in the next definition.

Definition 2.24

Let P be a non-empty feasible set of a LP. A non-zero direction d is called a recession direction if for any $x_0 \in P$, the ray $\{x \in R^n | x = x_0 + \lambda d$ for $\lambda \geq 0\} \subset P$.

It is clear that a bounded feasible set P cannot have such directions. In fact, it can be proven that a feasible set P is bounded if and only if it does not contain a recession direction; see Exercise 2.13. It is also the case that these particular recession directions d_1 and d_2 cannot be written as a positive multiple of the other, i.e., $ad_1 \neq d_2$ for any $a > 0$. Such recession directions are called *extreme directions*.

Now with the concept of extreme directions, the point $x = \left[\begin{smallmatrix}5\\3\end{smallmatrix}\right]$ can be written as

$$x = \left[\begin{smallmatrix}5\\3\end{smallmatrix}\right] = v_2 + 2d_1 + 1d_2 = 0v_1 + 1v_2 + 2d_1 + 1d_2.$$

In other words, x can be written as a convex combination of extreme points of P and a non-negative linear combination of extreme directions. In general, the insights above lead to the following result, which we state without proof.

Theorem 2.25 (Resolution Theorem)
Let $P = \{x \in R^n | Ax = b, x \geq 0\}$ be a non-empty set P . Let $v_1, v_2, ..., v_k$ be the extreme points of P.

(Case 1) If P is bounded, then any $x \in P$ can be represented as the convex combination of extreme points, i.e., $x = \sum_{i=1}^{k} \lambda_i v_i$ for some $\lambda_1, ..., \lambda_k \geq 0$ and $\sum_{i=1}^{k} \lambda_i = 1$.

(Case 2) If P is unbounded, then there exists at least one extreme direction. Let $d_1, ..., d_l$ be the extreme directions of P. Then, any $x \in P$ can be represented as $x = \sum_{i=1}^{k} \lambda_i v_i + \sum_{i=1}^{l} \mu_i d_i$, where $\lambda_1, ..., \lambda_k \geq 0$ and $\sum_{i=1}^{k} \lambda_i = 1$ and $\mu_i \geq 0$ for $i = 1, ..., l$.

Note: The constraint set P in the Resolution Theorem was assumed to be in standard form, but holds for other representations e.g. $P = \{x \in R^n | Ax \leq b, x \geq 0\}$. The example preceding the Resolution Theorem was given in inequality form to facilitate graphing.

2.4.1 Fundamental Theorem of Linear Programming

In this section, we prove the Fundamental Theorem of Linear Programming which formalizes the geometric intuition that if a linear program has a finite optimal solution, then it can occur at a corner point. The proof of the result will rely on the Resolution Theorem and the following results.

Theorem 2.26
For a feasible set $P = \{x \in R^n | Ax = b, x \geq 0\}$, a non-zero vector d is a recession vector if and only if $Ad = 0$ and $d \geq 0$.
Proof: see Exercise 2.7.

Corollary 2.27
A non-negative linear combination of recession directions of a feasible set P is a recession direction of P.
Proof: Let $d_1, ..., d_l$ be the recession directions of P and let $d = \sum_{i=1}^{l} \mu_i d_i$ for $\mu_i \geq 0$ for $i = 1, ..., l$. Since d_i is a recession direction by Theorem 2.26 above, we have that $Ad_i = 0$ and so $Ad = A \sum_{i=1}^{l} \mu_i d_i = \mu_i \sum_{i=1}^{l} Ad_i = 0$. Also we have that $d_i \geq 0$ for each $i = 1, ..., l$ and so $d = \sum_{i=1}^{l} \mu_i d_i \geq 0$. Therefore, by Theorem 2.26 d is a recession direction. ■

Theorem 2.28 (Fundamental Theorem of Linear Programming)
Consider an LP in standard form and suppose that P is non-empty. Then, either the LP is unbounded over P or an optimal solution for the LP can be attained at an extreme point of P.
Proof: Let $v_1, v_2, ..., v_k$ be the extreme points of P and let $d_1, ..., d_l$ be

the extreme directions of P. Then, by the Resolution Theorem, every point $x \in P$ can be expressed as $x = \sum_{i=1}^{k} \lambda_i v_i + \sum_{i=1}^{l} \mu_i d_i$ where $\lambda_1, ..., \lambda_k \geq 0$ and $\sum_{i=1}^{k} \lambda_i = 1$ and $\mu_i \geq 0$ for $i = 1, ..., l$. Without loss of generality, let $d = \sum_{i=1}^{l} \mu_i d_i$, which is a recession direction by Corollary 2.27. There are two cases.

Case (1) d is such that $c^T d < 0$. In this case, for any $x_0 \in P$, the ray $r = \{x \in R^n | x_0 + \lambda d \text{ for } \lambda \in [0,1]\} \subset P$ will be such that $c^T x = c^T x_0 + \lambda c^T d$ and this can be made to diverge toward $-\infty$ as $\lambda \to \infty$ since $c^T d < 0$ and $\lambda \geq 0$.

Case (2) d is such that $c^T d \geq 0$. So $x = \sum_{i=1}^{k} \lambda_i v_i + d$ where $\lambda_1, ..., \lambda_k \geq 0$ and $\sum_{i=1}^{k} \lambda_i = 1$. Now let $v_{\min}$ be that extreme point that results in the minimum value of $c^T v_i$ over $i = 1, ..., k$. Then, for any $x \in P$, $c^T x = c^T(\sum_{i=1}^{k} \lambda_i v_i + d) = c^T(\sum_{i=1}^{k} \lambda_i v_i) + c^T d \geq c^T(\sum_{i=1}^{k} \lambda_i v_i) = \sum_{i=1}^{k} \lambda_i c^T v_i \geq \sum_{i=1}^{k} \lambda_i c^T v_{\min} = c^T v_{\min}(\sum_{i=1}^{k} \lambda_i) = c^T v_{\min}$. The first inequality holds since $c^T d \geq 0$. Thus, the minimum value for the LP is attained at $v_{\min}$ an extreme point. ■

The Fundamental Theorem of Linear Program will serve as the basis of algorithmic development for linear programs. The major implication is that the search for an optimal solution for a linear program can be restricted to extreme points, i.e., basic feasible solutions. Since the number of extreme points is finite, one obvious strategy is to simply enumerate all possible basic feasible solutions and then select the one that gives the minimum objective value. It is important to note that this enumeration can be done algebraically and without resorting to use of graphical methods. Exercise 2.15 asks the reader to develop MATLAB code that will solve a linear program by generating all possible basic feasible solutions.

For example, consider the LP from Chapter 1 (in standard form)

$$\begin{array}{ll} \textit{minimize} & -x_1 - x_2 \\ \textit{subject to} & x_1 + \qquad x_3 = 1 \\ & \qquad x_2 + \qquad x_4 = 1 \\ & x_1 \geq 0, x_2 \geq 0, x_3 \geq 0, x_4 \geq 0. \end{array}$$

There are four basic feasible solutions

$$x^{(0)} = \begin{bmatrix} x_B \\ x_N \end{bmatrix} = \begin{bmatrix} x_3 \\ x_4 \\ x_1 \\ x_2 \end{bmatrix} = \begin{bmatrix} 1 \\ 1 \\ 0 \\ 0 \end{bmatrix},$$

$$x^{(1)} = \begin{bmatrix} x_B \\ x_N \end{bmatrix} = \begin{bmatrix} x_3 \\ x_2 \\ x_1 \\ x_4 \end{bmatrix} = \begin{bmatrix} 1 \\ 1 \\ 0 \\ 0 \end{bmatrix},$$

$$x^{(2)} = \begin{bmatrix} x_B \\ x_N \end{bmatrix} = \begin{bmatrix} x_1 \\ x_4 \\ x_3 \\ x_2 \end{bmatrix} = \begin{bmatrix} 1 \\ 1 \\ 0 \\ 0 \end{bmatrix},$$

and

$$x^{(3)} = \begin{bmatrix} x_B \\ x_N \end{bmatrix} = \begin{bmatrix} x_1 \\ x_2 \\ x_3 \\ x_4 \end{bmatrix} = \begin{bmatrix} 1 \\ 1 \\ 0 \\ 0 \end{bmatrix}.$$

The objective function values are $-2, -1, -1$, and 0, respectively. We can then deduce that the extreme point associated with $x^{(1)}$ is the optimal solution since its objective function value -2 is the minimum objective function value among all of the basic feasible solutions.

Unfortunately, this approach is not practical since the quantity $C(n, m) = \frac{n!}{m!(n-m)!}$ can be very large for practical sized problems with hundreds, thousands, or even millions of variables and constraints.

This motivates other strategies for solving LPs that exploit the Fundamental Theorem of LP. In Chapter 3, the simplex method is developed where basic feasible solutions are explored systematically and does not, except in the most exceptional cases, require the examination of all basic feasible solutions. In Chapter 6, another important class of strategies called interior point methods are introduced where most of the search occurs in the interior of the feasible set, but where the search ultimately gravitates and converges to an optimal basic feasible solution.

2.5 Exercises

Exercise 2.1
Consider the constraint

$$-2x_1 + x_2 \leq 2.$$

(a) Express this constraint as a closed-halfspace of the form $H_{\leq} = \{x \in R^n | a^T x \leq \beta\}$, i.e., determine a and β.

(b) Sketch the closed halfspace in (a) showing any vector that is normal to the hyperplane that is contained in $H_{\leq}$.

(c) Show that the closed halfspace in (a) is a convex set.

Exercise 2.2
Consider a linear program in standard form

$$\begin{array}{ll} \textit{minimize} & c^T x \\ \textit{subject to} & Ax = b \\ & x \geq 0. \end{array}$$

(a) Prove that the feasible set $P = \{x \in R^n | \ Ax = b, x \geq 0\}$ of the linear program is a convex set directly using the definition of convex set.

(b) Prove that the set of optimal solutions for the linear program in standard form $P^* = \{x \in R^n | \ x \text{ is an optimal solution for LP}\}$ is a convex set.

Exercise 2.3

Solve the following linear programs graphically by using the sketch of the feasible set and illustrate the hyperplane characterization of optimality when a finite optimal solution(s) exists, else illustrate the unboundedness of the linear program using hyperplanes.

$$\text{(a)} \quad \begin{array}{lll} \textit{minimize} & -x_1 - 2x_2 & \\ \textit{subject to} & -2x_1 + x_2 & \leq 2 \\ & -x_1 + x_2 & \leq 3 \\ & x_1 & \leq 2 \\ & x_1 \geq 0, x_2 \geq 0 & \end{array}$$

$$\text{(b)} \quad \begin{array}{lll} \textit{minimize} & -x_1 - 2x_2 & \\ \textit{subject to} & x_1 - 2x_2 & \geq 2 \\ & x_1 + x_2 & \leq 4 \\ & x_1 \geq 0, x_2 \geq 0 & \end{array}$$

$$\text{(c)} \quad \begin{array}{lll} \textit{maximize} & x_1 + x_2 & \\ \textit{subject to} & x_1 - x_2 & \geq 1 \\ & x_1 - 2x_2 & \geq 2 \\ & x_1 \geq 0, x_2 \geq 0 & \end{array}$$

Exercise 2.4

(a) For the linear program (a) in Exercise 2.3, find all basic feasible solutions by converting the constraints into standard form.

(b) For each linear program in Exercise 2.3, find two linearly independent directions d_1 and d_2 of unboundedness if they exist .

Exercise 2.5

Consider the constraints

$$\begin{array}{ll} 2x_1 + x_2 & \leq 5 \\ x_1 + x_2 & \leq 4 \\ x_1 & \leq 2 \\ x_1 \geq 0, x_2 \geq 0. & \end{array}$$

(a) Sketch the feasible region.

(b) Convert the constraints to standard form and find all basic feasible solutions.

(c) Identify the extreme points in the original constraints.

Exercise 2.6

Consider the linear program

$$\begin{array}{lll} \textit{maximize} & x_1 + x_2 & \\ \textit{subject to} & x_1 - x_2 & \geq 1 \\ & x_1 - 2x_2 & \geq 2 \\ & x_1 \geq 0, x_2 \geq 0. & \end{array}$$

(a) Sketch the feasible region.

(b) Convert the constraints to standard form, find all basic feasible solutions, and find two extreme directions d_1 and d_2 (i.e., two linearly independent directions of unboundedness).

(c) Show that the extreme directions d_1 and d_2 from (b) satisfy $Ad = 0$ and $d \geq 0$.

Exercise 2.7

Suppose that a linear program is in standard form

$$\begin{array}{ll} \textit{minimize} & c^T x \\ \textit{subject to} & Ax = b \\ & x \geq 0. \end{array}$$

Show that if a vector $d \neq 0$ is such that $Ad = 0$ and $d \geq 0$, then d must be a direction of unboundedness.

Exercise 2.8

Consider the following system of constraints

$$\begin{array}{ll} x_1 + x_2 & \leq 6 \\ x_1 - x_2 & \leq 0 \\ x_1 & \leq 3 \\ x_1 \geq 0, x_2 \geq 0. & \end{array}$$

(a) Sketch the feasible region.

(b) Convert to standard form and find all basic feasible solutions.

(c) Is there a one-to-one correspondence between basic feasible solutions and extreme points? If not, which extreme points can be represented by multiple basic feasible solutions.

Exercise 2.9

(a) Solve the linear program in Exercise 2.3 (a) by generating all basic feasible solutions.

(b) Solve the linear program in Exercise 2.3 (b) by generating all basic feasible solutions. Also, illustrate Exercise 2.2 (b), i.e., show the set of optimal solutions is convex.

Exercise 2.10
Consider the following polyhedron

$$P = \{(x_1, x_2, x_3, x_4) \in R^4 |\ x_1 - x_2 - 2x_3 \leq 1, -3x_1 - x_3 + 2x_4 \leq 1,\\ x_1 \geq 0, x_2 \geq 0, x_3 \geq 0, x_4 \geq 0\}.$$

Find all extreme points and extreme directions of P and represent the point

$$x = \begin{bmatrix} 2 \\ 1 \\ 1 \\ 1 \end{bmatrix}$$

as a convex combination of the extreme points plus a non-negative combination of extreme directions.

Exercise 2.11
Consider a set $P = \{x \in R^n | Ax < b\}$. Prove that the problem

$$\begin{array}{ll} \textit{maximize} & c^T x \\ \textit{subject to} & x \in P \end{array}$$

does not have an optimal solution. Assume that $c \neq 0$.

Exercise 2.12
Explain what happens to a linear program when a constraint is deleted. In particular, what happens to the feasible set and objective function?

Exercise 2.13
Prove that a feasible set P is bounded if and only if it contains no extreme directions.

Exercise 2.14
Another concept of a corner point of a feasible set P of a linear program is through the notion of a vertex x of P. We say that $x \in P$ is a vertex of P if there is a vector q such that $q^T x < q^T y$ for all $y \in P$ and $x \neq z$. Without loss of generality, assume the $P = \{x \in R^n | A_1 x \geq b_1, A_2 x = b_2\}$ where A_1 is $m_1 \times n$ and A_2 is $m_2 \times n$ and $m_1 + m_2 \geq n$.

(a) Prove that if x is a vertex, then x is an extreme point.
(b) Prove that if x is a basic feasible solution, then x is a vertex. (Note:

that with P defined as above, a vector x is a basic feasible solution if there is a set of at least n constraints in P at equality and linearly independent.

Exercise 2.15

Write MATLAB code that takes a linear program in standard form and solves for the optimal solution by generating all possible basic feasible solutions. Assume that the linear program has a finite optimal solution.

Notes and References

The geometry of linear programs in low dimensions gives insights for characterizing the nature of feasible and optimal solutions for higher dimensions. The corresponding algebraic concepts are important as it will enable computationally practical algorithms for solving linear programs. The feasible set of a linear program was found to be a convex set. Convexity plays an important role in optimization and more details about the convex sets and its role in optimization in general can be found in Mangasarian (1969), Rockafellar (1970), and Boyd and Vandenberghe (2004). The Representation Theorems began with the work of Minkowski (1896). For further geometric insights on linear programs see Murty (1983), Bazarra, Jarvis, and Sherali (1977), and Bertsimas and Tsitsiklis (1997).

3

The Simplex Method

3.1 Introduction

In this chapter the simplex method, which is an important and well-known method to solve linear programming problems, is developed. The simplex method was conceived by Dantzig (1948), still remains a powerful class of methods, and is often the main strategy for solving linear programs in commercial software. We know by the Fundamental Theorem of Linear Programming in Chapter 2 that if an LP has a finite optimal solution, then it can be attained at an extreme point and therefore at some basic feasible solution. The basic strategy of the simplex method is to explore the extreme points of the feasible region of an LP to find the optimal extreme point. However, in practice, the simplex method will in most cases not need to explore all possible extreme points before finding an optimal one. The strategy of the simplex method is as follows: given an initial basic feasible solution, the simplex method determines whether the basic feasible solution is optimal. If it is optimal, then the method terminates, else, another basic feasible solution is generated whose objective function value is better or no worse than the previous one, optimality is checked and so on, until an optimal basic feasible solution is obtained. If the LP is unbounded, then the simplex method will be able to detect this and return with the recession direction along which the objective function is unbounded. The simplex method requires that a linear program is in standard form. A high-level description of the method is summarized below (the finer details of each step will be subsequently developed).

Simplex Method

Step 0: Generate an initial basic feasible solution $x^{(0)}$. Let $k = 0$ and go to Step 1.

Step 1: Check optimality of $x^{(k)}$. If $x^{(k)}$ is optimal, then STOP and return $x^{(k)}$ as the optimal solution, else go to Step 2.

Step 2: Check whether the LP is unbounded; if so STOP, else go to Step 3.

Step 3: Generate another basic feasible solution $x^{(k+1)}$ so $c^T x^{(k+1)} \leq c^T x^{(k)}$.

Let $k = k + 1$ and go to Step 1.

The simplex method is not intended to be an exhaustive brute-force search method. Recall that it is possible to have an algorithm for solving linear programs that is based on enumerating all basic feasible solutions. A critical difference between such an exhaustive brute-force method and the simplex method is that for the latter, given an extreme point, it will be possible to efficiently determine whether it is an optimal solution or not without having to compare with all of the extreme points. In an exhaustive brute-force approach, one needs to examine all extreme points in order to guarantee that a given basic feasible solution is optimal. The number of possible extreme points is $\binom{n}{m}$, which can be quite large for large instances of linear programs.

3.1.1 Example of Simplex Method

We give a high-level illustration of the simplex method. Consider the linear program

$$\begin{array}{ll} \textit{minimize} & -x_1 - x_2 \\ \textit{subject to} & x_1 \leq 1 \\ & x_2 \leq 1 \\ & x_1 \geq 0, x_2 \geq 0. \end{array} \tag{3.1}$$

The simplex method will require an LP to be in standard form. For the LP above, the standard form is

$$\begin{array}{lllll} \textit{minimize} & -x_1 - x_2 & & & \\ \textit{subject to} & x_1 + & & x_3 & = 1 \\ & & x_2 + & & x_4 = 1 \\ & x_1 \geq 0, x_2 \geq 0, x_3 \geq 0, x_4 \geq 0. & & & \end{array}$$

There are, in total, 4 basic feasible solutions listed below, each corresponding to a different extreme point.

$$v^{(0)} = \begin{bmatrix} x_B \\ x_N \end{bmatrix} = \begin{bmatrix} x_3 \\ x_4 \\ x_1 \\ x_2 \end{bmatrix} = \begin{bmatrix} 1 \\ 1 \\ 0 \\ 0 \end{bmatrix} \quad v^{(1)} = \begin{bmatrix} x_B \\ x_N \end{bmatrix} = \begin{bmatrix} x_3 \\ x_2 \\ x_1 \\ x_4 \end{bmatrix} = \begin{bmatrix} 1 \\ 1 \\ 0 \\ 0 \end{bmatrix}$$

$$v^{(2)} = \begin{bmatrix} x_B \\ x_N \end{bmatrix} = \begin{bmatrix} x_1 \\ x_4 \\ x_3 \\ x_2 \end{bmatrix} = \begin{bmatrix} 1 \\ 1 \\ 0 \\ 0 \end{bmatrix} \quad v^{(3)} = \begin{bmatrix} x_B \\ x_N \end{bmatrix} = \begin{bmatrix} x_1 \\ x_2 \\ x_3 \\ x_4 \end{bmatrix} = \begin{bmatrix} 1 \\ 1 \\ 0 \\ 0 \end{bmatrix}$$

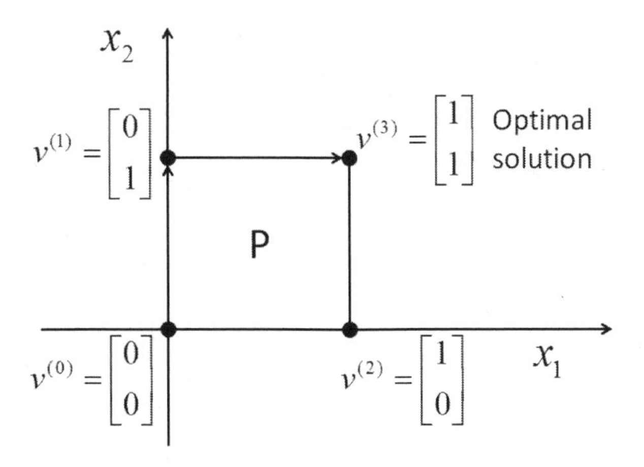

FIGURE 3.1
Possible trajectory of the simplex method for linear program (3.1).

A possible sequence of iterations that the simplex method may undergo is as follows (note the finer details of the steps are omitted). The basic feasible solution $x^{(0)} = v^{(0)}$ is generated as the initial basic feasible solution in Step 0. The first iteration begins, and in Step 1 $x^{(0)}$ is determined to be suboptimal, and then Step 2 is executed to generate the basic feasible solution $x^{(1)} = v^{(1)}$.

Iteration 2 starts, in which $x^{(1)}$ is determined to be suboptimal in Step 1, and then in Step 2, basic feasible solution $x^{(2)} = v^{(3)}$ is generated.

Iteration 3 starts, in which $x^{(2)}$ is found to be an optimal solution, and so the simplex method terminates and returns $v^{(3)}$ as an optimal solution. This sequence of points $v^{(0)}, v^{(1)}, v^{(3)}$ in the original space of the problem is shown in Figure 3.1.

It is also possible for the simplex method to generate the following sequence $v^{(0)}, v^{(2)}, v^{(3)}$. In general, the simplex method can have multiple options in terms of what basic feasible solution to move to next and must select exactly one of these at such an iteration.

3.1.2 Adjacent Basic Feasible Solutions

The movement from a basic feasible solution to another in the simplex method is such that only neighboring or "adjacent" basic feasible solutions are considered. In the illustration above, movements from $v^{(0)}$ to either $v^{(1)}$ or $v^{(2)}$ are permitted, but a movement from $v^{(0)}$ directly to $v^{(3)}$is prohibited. Intuitively, from Figure 3.1, $v^{(0)}$ and $v^{(1)}$ are adjacent extreme points (as are $v^{(0)}$ and $v^{(2)}$) since the line segment between them is on the edge (boundary) of the feasible set, and thus one can be reached from the other directly along this edge. The points $v^{(0)}$ and $v^{(3)}$ are not adjacent since the line segment between them is not on the edge of the feasible set and therefore one cannot move from one to the other along the edges of the feasible set without first going through $v^{(1)}$ or $v^{(2)}$.

An algebraic definition of adjacency is now given.

Definition 3.1

Two different basic feasible solutions x and y are said to be adjacent if they have exactly $m - 1$ basic variables in common.

Example 3.2

$$v^{(0)} = \begin{bmatrix} x_B \\ x_N \end{bmatrix} = \begin{bmatrix} x_3 \\ x_4 \\ x_1 \\ x_2 \end{bmatrix} = \begin{bmatrix} 1 \\ 1 \\ 0 \\ 0 \end{bmatrix} \text{ and } v^{(1)} = \begin{bmatrix} x_B \\ x_N \end{bmatrix} = \begin{bmatrix} x_3 \\ x_2 \\ x_1 \\ x_4 \end{bmatrix} = \begin{bmatrix} 1 \\ 1 \\ 0 \\ 0 \end{bmatrix}$$

are adjacent since they have $m - 1 = 2 - 1 = 1$ basic variables in common, i.e., the basic variable x_3. $v^{(0)}$ and $v^{(3)}$ are not adjacent since they have 0 basic variables in common.

3.2 Simplex Method Development

The description and illustration of the simplex method above gives only a high level description of the dynamics of the simplex method and many important details remain. In particular, an efficient test for determining whether a basic feasible solution is optimal needs to be developed, as well as a test for checking for unboundedness of a linear program. Furthermore, details on how to generate a new and improved adjacent basic feasible solution is required. In this section, we develop the remaining constructs needed to fully describe the simplex method with enough detail so that a MATLAB implementation is possible. MATLAB® code for the simplex method is given later in this chapter.

3.2.1 Checking Optimality of a Basic Feasible Solution

Suppose we have a linear program in standard form:

$$\begin{array}{ll} \textit{minimize} & c^T x \\ \textit{subject to} & Ax = b \\ & x \geq 0. \end{array}$$

Given a current basic feasible solution x^*, we would like to be able to efficiently determine whether it is optimal for the linear program. Let $P = \{x \in R^n | Ax = b, x \geq 0\}$ be the feasible set. Any such test is equivalent to checking if the condition $c^T x \geq c^T x^*$ holds for all $x \in P$. We wish to develop an efficient test that does not rely on explicitly using other feasible solutions, e.g., basic feasible solutions, to compare with x^*.

The key to the development of an efficient test is to use a representation of the linear programming problem that reflects the partition of variables into basic and non-basic, implied by the basic feasible solution $x^*$where the partition is $x^* = \begin{bmatrix} x_B^* \\ x_N^* \end{bmatrix}$ and x_B^* are the basic variables and x_N^* the non-basic variables. Then, the constraint matrix can be partitioned as $A = [B|N]$ where B is the basis matrix that contains the columns in A associated with variables in x_B^*, and N is the non-basis matrix consisting of columns associated with the variables in x_N^*. Finally, the cost vector c can be partitioned as $c = \begin{bmatrix} c_B \\ c_N \end{bmatrix}$ where c_B contains the cost coefficients associated with the variables in x_B^* and c_N contains the cost coefficients associated with the variables in x_N^*.

Now $x_B^* \geq 0$ and $x_N^* = 0$ since x^* is a basic feasible solution. Consider any $x \in P$, then the variables can be partitioned according to the basic and non-basic partition of x^*. That is, x can be re-arranged so that $x = \begin{bmatrix} x_B \\ x_N \end{bmatrix}$ where x_B are the components (variables) in x that are in x_B^*, and x_N are components of x that are in x_N^*. This does not mean that x is a basic feasible solution, just that the variables are re-arranged to match the variable partition of x^*.

Now we represent the linear program in terms of the partition implied by x^*. First, for any $x \in P$ it is assumed that the components are re-arranged so that $x = \begin{bmatrix} x_B \\ x_N \end{bmatrix}$. The objective function $c^T x$ can be re-written as

$$c^T x = \begin{bmatrix} c_B \\ c_N \end{bmatrix}^T \begin{bmatrix} x_B \\ x_N \end{bmatrix} = c_B^T x_B + c_N^T x_N.$$

The constraints $Ax = b$ can be written as

$$A = \begin{bmatrix} B & N \end{bmatrix} \begin{bmatrix} x_B \\ x_N \end{bmatrix} = Bx_B + Nx_N = b.$$

Finally, the requirement that $x \geq 0$ is equivalent to $x_B \geq 0$ and $x_N \geq 0$. Thus, the linear program becomes

$$\begin{array}{ll} minimize & z = c_B^T x_B + c_N^T x_N \\ subject\ to & Bx_B + Nx_N = b \\ & x_B \geq 0, x_N \geq 0. \end{array} \tag{3.2}$$

Now the idea is to first solve for x_B using the first constraint and then substitute into the objective function equation z. After the substitution, some terms will be re-arranged from which a necessary and sufficient condition for optimality will be obtained.

From the first constraint we have $Bx_B = b - Nx_N$ and so $x_B = B^{-1}(b - Nx_N)$ and then substituting into

$$z = c_B^T x_B + c_N^T x_N$$

we get

$$\begin{aligned} z &= c_B^T B^{-1}(b - Nx_N) + c_N^T x_N \\ &= c_B^T B^{-1}b - c_B^T B^{-1}Nx_N + c_N^T x_N \\ &= c_B^T B^{-1}b + (c_N^T - c_B^T B^{-1}N)x_N \\ &= c_B^T x_B^* + (c_N^T - c_B^T B^{-1}N)x_N \end{aligned}$$

since $x_B^* = B^{-1}b$. Now z is the objective function value associated with $x \in P$ and so $z = c^T x$. Also, $c^T x^* = c_B^T x_B^* + c_N^T x_N^* = c_B^T x_B^*$ since $x_N^* = 0$, so

$$c^T x = c_B^T x_B^* + (c_N^T - c_B^T B^{-1}N)x_N$$

or

$$c^T x - c^T x^* = (c_N^T - c_B^T B^{-1}N)x_N.$$

It is clear that if the right-hand side of the last equation is non-negative, then $c^T x - c^T x^* \geq 0$ for any $x \in P$ or equivalently $c^T x \geq c^T x^*$ for all $x \in P$. That is, x^* would be an optimal solution. It is clear that $x_N \geq 0$ since x is a feasible solution, and so if all of the m components of $(c_N^T - c_B^T B^{-1}N)$ are non-negative, then x^* is an optimal solution. This leads us to define the vector

$$r_N = (c_N^T - c_B^T B^{-1}N)$$

called the vector of *reduced costs* associated with x^*. Note that each component of r_N is associated with a non-basic variable. This optimality condition is summarized as follows.

Theorem 3.1

If the reduced cost vector r_N corresponding to a basic feasible solution $x^* = \begin{bmatrix} x_B^* \\ x_N^* \end{bmatrix}$ is non-negative, then x^* is an optimal solution.

Optimality Check for the Simplex Method

Let x^* be a basic feasible solution and $\bar{N}$ the set of indices of non-basic variables of x^*. Then, the qth component of r_N is $r_q = c_q - c_B^T B^{-1} N_q$ where N_q is the column in N associated with the non-basic variable x_q. So, by Theorem 3.1, to check optimality of x^*, compute r_q for all $q \in \bar{N}$. There are two cases.

Case (1) If $r_q \geq 0$ for all $q \in \bar{N}$, then x^* is optimal.

Case (2) There is at least one non-basic variable $r_q < 0$, then x^* is not optimal.

Example 3.2
Consider the LP

$$\begin{array}{ll} \textit{minimize} & -x_1 - x_2 \\ \textit{subject to} & x_1 + \quad x_3 \quad = 1 \\ & \quad x_2 + \quad x_4 = 1 \\ & x_1 \geq 0, x_2 \geq 0, x_3 \geq 0, x_4 \geq 0 \end{array}$$

and the basic feasible solution

$$v^{(1)} = \begin{bmatrix} x_B \\ x_N \end{bmatrix} = \begin{bmatrix} x_3 \\ x_2 \\ x_1 \\ x_4 \end{bmatrix} = \begin{bmatrix} 1 \\ 1 \\ 0 \\ 0 \end{bmatrix}.$$

The basic variables are x_3 and x_2 with corresponding basis matrix as $B = \begin{bmatrix} 1 & 0 \\ 0 & 1 \end{bmatrix}$, cost vector $c_B = \begin{bmatrix} c_3 \\ c_2 \end{bmatrix} = \begin{bmatrix} 0 \\ -1 \end{bmatrix}$. The non-basic variables are x_1 and x_4 with corresponding non-basic matrix $N = \begin{bmatrix} 1 & 0 \\ 0 & 1 \end{bmatrix}$ where $N_1 = \begin{bmatrix} 1 \\ 0 \end{bmatrix}$ is the column in N associated with the non-basic variable x_1 and $N_4 = \begin{bmatrix} 0 \\ 1 \end{bmatrix}$ is the column in N associated with non-basic variable x_4, non-basic index set $\bar{N} = \{1, 4\}$, and cost vector $c_N = \begin{bmatrix} c_1 \\ c_4 \end{bmatrix} = \begin{bmatrix} -1 \\ 0 \end{bmatrix}$.

To check optimality, we compute r_1 and r_4 since 1 and $4 \in \bar{N}$. Then,

$$r_1 = c_1 - c_B^T B^{-1} N_1 = -1 - (0, -1) \begin{bmatrix} 1 & 0 \\ 0 & 1 \end{bmatrix}^{-1} \begin{bmatrix} 1 \\ 0 \end{bmatrix} = -1 < 0$$

and

$$r_4 = c_4 - c_B^T B^{-1} N_4 = 0 - (0, -1) \begin{bmatrix} 1 & 0 \\ 0 & 1 \end{bmatrix}^{-1} \begin{bmatrix} 0 \\ 1 \end{bmatrix} = 1 > 0.$$

Therefore, $v^{(1)}$ is not an optimal basic feasible solution since there is at least one reduced cost, i.e., r_1 that is negative.

Consider the basic feasible solution $v^{(3)} = \begin{bmatrix} x_B \\ x_N \end{bmatrix} = \begin{bmatrix} x_1 \\ x_2 \\ x_3 \\ x_4 \end{bmatrix} = \begin{bmatrix} 1 \\ 1 \\ 0 \\ 0 \end{bmatrix}$.

Computing the reduced costs

$$r_3 = c_3 - c_B^T B^{-1} N_3 = 0 - (-1, -1) \begin{bmatrix} 1 & 0 \\ 0 & 1 \end{bmatrix}^{-1} \begin{bmatrix} 1 \\ 0 \end{bmatrix} = 1 \geq 0$$

and

$$r_4 = c_4 - c_B^T B^{-1} N_4 = 0 - (-1, -1) \begin{bmatrix} 1 & 0 \\ 0 & 1 \end{bmatrix}^{-1} \begin{bmatrix} 0 \\ 1 \end{bmatrix} = 1 \geq 0,$$

we see that there are no negative reduced costs indicating that $v^{(3)}$ is an optimal solution.

3.2.2 Moving to an Improved Adjacent Basic Feasible Solution

If for a basic feasible solution x^*, at least one of the reduced costs is negative, then by Theorem 3.1, x^* cannot be an optimal solution. In fact, it will be possible (as will be shown later) in this case to move to an improved adjacent basic feasible solution $\widetilde{x}$ such that $c^T x^* > c^T \widetilde{x}$.

Feasible Search Directions

The dynamics of movement from a current non-optimal basic feasible solution $x^{current}$ to a new adjacent basic feasible solution x^{new} in the simplex method can be described by the following equation

$$x^{new} = x^{current} + \alpha d,$$

where d is a direction and $\alpha \geq 0$ is a step length; see Figure 3.2.

It remains to choose the form of d and α so that x^{new} will be an adjacent basic feasible solution that has a lower objective function value than the objective function value of $x_{current}$.

In particular, adjacency requires that the choice of d and α will select one non-basic variable, say x_q in $x^{current}$, to become a basic variable in x^{new}, which means that one current basic variable in $x^{current}$, say x_l, must be set to non-basic, i.e., $x_l = 0$ in x^{new}. The starting point is the equation

$$Bx_B + Nx_N = b \text{ or equivalently } x_B = B^{-1}b - B^{-1}Nx_N.$$

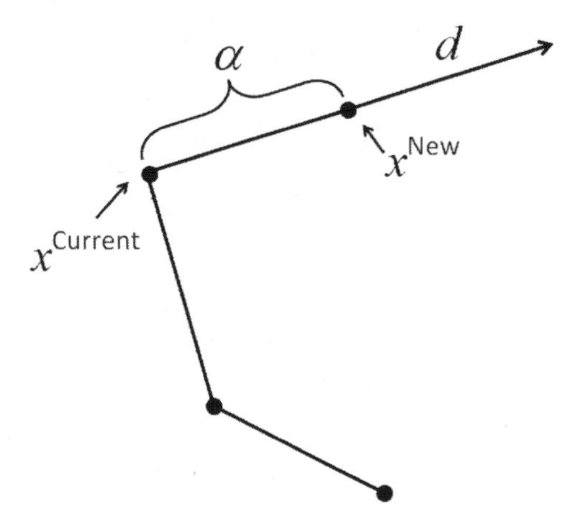

FIGURE 3.2
Dynamics of movement in the simplex method along a direction d and step length α.

Only one non-basic variable x_q will be increased to some positive value, and the other non-basic variables will remain at 0 so the equation simplifies to

$$x_B = B^{-1}b - B^{-1}N_q x_q.$$

Since $B^{-1}b$ are the basic variables of the current basic feasible solution (i.e., $x_B^{current} = B^{-1}b$), we can interpret x_B to be the new basic variable set for x^{new} after the suitable increase in x_q is made (i.e., $x_B^{new} = x_B = B^{-1}b - B^{-1}N_q x_q = x_B^{current} - B^{-1}N_q x_q$). Expanding to the full n dimensions of the vectors in

$$x^{new} = x^{current} + \alpha d$$

gives

$$\begin{bmatrix} x_B^{new} \\ x_N^{new} \end{bmatrix} = \begin{bmatrix} x_B^{current} \\ x_N^{current} \end{bmatrix} + \alpha d$$

$$= \begin{bmatrix} B^{-1}b \\ 0 \end{bmatrix} + \alpha d.$$

This suggests that the direction

$$d = \begin{bmatrix} -B^{-1}N_q \\ e_q \end{bmatrix}$$

is suitable where e_q is a vector of dimension $n - m$ of all zeros except a one at the position in the current basic feasible solution $x^{current}$ corresponding to x_q. Furthermore, $\alpha = x_q$. There are m components in the subvector $-B^{-1}N_q$ and so d has dimension n. From here on, we denote the direction d as d^q, which indicates its dependence on a selection of a non-basic variable x_q from $x^{current}$.

Improving Directions

It is not clear if moving in the direction d starting from $x^{current}$ for some step length α will result in an improvement in the objective function value. We now find the conditions under which improvement is guaranteed.

Let $x^{new} = x^{current} + \alpha d^q$, then we wish to find when

$$c^T x^{new} < c^T x^{current}$$

or

$$c^T (x^{current} + \alpha d^q) < c^T x^{current},$$

which reduces to the following condition

$$c^T d < 0 \text{ since } \alpha \geq 0.$$

Next, by the partition into basic and non-basic variables implied by the basic feasible solution $x^{current}$ we have

$$c^T d = \begin{bmatrix} c_B \\ c_N \end{bmatrix}^T \begin{bmatrix} -B^{-1}N_q \\ e_q \end{bmatrix}$$

$$= -c_B^T B^{-1} N_q + c_N^T e_q = c_q - c_B^T B^{-1} N_q$$

$$= r_q,$$

which is the reduced cost associated with non-basic variable x_q. Therefore, the objective function value will improve (decrease) along any direction d^q associated with a non-basic variable x_q that has negative reduced cost. In this case, we say that d^q is a descent direction. We summarize the improvement condition as follows.

Theorem 3.3

Suppose x^* is basic feasible solution with basis matrix B and non-basis matrix N. For any non-basic variable x_q with $r_q < 0$ the direction

$$d^q = \begin{bmatrix} -B^{-1}N_q \\ e_q \end{bmatrix}$$

will lead to a decrease in the objective function.

Feasibility of a Descent Direction

A fundamental issue not addressed yet is whether the resultant vector $x^{new} = x^{current} + \alpha d^q$ is feasible for the linear program in standard form. In particular, x^{new} is feasible if the following conditions hold

$$(1)\ Ax^{new} = b \text{ and } (2)\ x^{new} \geq 0.$$

To show that (1) holds, consider that $Ax^{new} = A(x^{current} + \alpha d^q) = Ax^{current} + \alpha A d^q$. Thus, if $Ad^q = 0$, then (1) will hold. Now, by the partition implied by $x^{current}$,

$$Ad^q = \begin{bmatrix} B & N \end{bmatrix} \begin{bmatrix} -B^{-1}N_q \\ e_q \end{bmatrix}$$

$$= -BB^{-1}N_q + Ne_q$$

$$= -N_q + N_q = 0.$$

To show (2) without loss of generality, assume that d^q is a descent direction. Then, we need

$$x^{new} = x^{current} + \alpha d^q \geq 0.$$

Case 1: $d^q \geq 0$, then $x^{new} = x^{current} + \alpha d^q \geq 0$ for any positive value of α since $x^{current} \geq 0$ so α can be increased indefinitely, but $c^T d < 0$ since $r_q < 0$ and so $c^T(x^{new}) = c^T(x^{current} + \alpha d^q) = c^T x^{current} + \alpha c^T d^q$ will go to negative infinity as α goes to infinity. In this case, the linear program is unbounded.

Case 2: There is at least one component of d^q that is negative, then from the requirement that $x^{current} + \alpha d^q \geq 0$ we get $\alpha d^q \geq -x^{current}$ and thus the largest that α can be is the minimum of the ratios in the set $\{-\frac{x_j^{current}}{d_j^q} | d_j^q < 0\}$ where $x_j^{current}$ is the jth basic variable in $x^{current}$ and d_j^q is the component of d^q corresponding to the basic variable $x_j^{current}$. In other words,

$$\alpha = \min_{j \in \bar{B}} \{-\frac{x_j^{current}}{d_j^q} | d_j^q < 0\}$$

where $\bar{B}$ is index set of basic variables in $x^{current}$. The determination of α is called the minimum ratio test.

We summarize the discussion and results above as follows.

Theorem 3.4 (Direction-Step Length)

Consider a linear program in standard form and assume that all basic feasible solutions are non-degenerate, and suppose that $x^{current}$ is a basic feasible solution that is not optimal and x_q is a non-basic variable such that its reduced cost $r_q < 0$. Then, $x^{new} = x^{current} + \alpha d^q$ is feasible for the linear program where $d^q = \begin{bmatrix} -B^{-1}N_q \\ e_q \end{bmatrix}$ and $\alpha = \min_{j \in \bar{B}} \{-\frac{x_j^{current}}{d_j^q} | d_j^q < 0\}$.

If $d^q \geq 0$, then the linear program is unbounded, else x^{new} is a basic feasible solution that is adjacent to $x^{current}$ and has a lower objective function value, i.e., $c^T x^{new} < c^T x^{current}$.

Example 3.5

Consider the linear program (3.1) and let

$$x^{current} = v^{(1)} = \begin{bmatrix} x_B \\ x_N \end{bmatrix} = \begin{bmatrix} x_3 \\ x_2 \\ x_1 \\ x_4 \end{bmatrix} = \begin{bmatrix} 1 \\ 1 \\ 0 \\ 0 \end{bmatrix}.$$

From Example 3.2 we know that x^{new} is not optimal and that $r_1 < 0$. Then, we construct

$$d^1 = \begin{bmatrix} -B^{-1}N_1 \\ e_1 \end{bmatrix} = \begin{bmatrix} -\begin{bmatrix} 1 & 0 \\ 0 & 1 \end{bmatrix}^{-1} \begin{bmatrix} 1 \\ 0 \end{bmatrix} \\ 1 \\ 0 \end{bmatrix}$$

$$= \begin{bmatrix} -1 \\ 0 \\ 1 \\ 0 \end{bmatrix} = \begin{bmatrix} d_3^1 \\ d_2^1 \\ d_1^1 \\ d_4^1 \end{bmatrix}.$$

Since there is at least one component of d^1 that is negative, we cannot conclude at this point that the LP is unbounded. So, we compute α by using the minimum ration test, i.e.,

$$\alpha = \min_{j \in \bar{B} = \{3,2\}} \{-\frac{x_j^{current}}{d_j^q} | d_j^q < 0\} = \{-\frac{x_3^{current}}{d_3^1}\} = \{-\frac{1}{-1}\} = 1$$

(the set contains only one ratio since d_3^1 is the only component associated with a basic variable that is negative in d^1).

Then, $x^{new} = x^{current} + \alpha d^q = \begin{bmatrix} 1 \\ 1 \\ 0 \\ 0 \end{bmatrix} + (1)\begin{bmatrix} -1 \\ 0 \\ 1 \\ 0 \end{bmatrix} = \begin{bmatrix} 0 \\ 1 \\ 1 \\ 0 \end{bmatrix} = \begin{bmatrix} x_3^{new} \\ x_2^{new} \\ x_1^{new} \\ x_4^{new} \end{bmatrix}$.

The basic variables are $x_B^{new} = \begin{bmatrix} x_1^{new} \\ x_2^{new} \end{bmatrix} = \begin{bmatrix} 1 \\ 1 \end{bmatrix}$ and non-basic variables $x_N^{new} = \begin{bmatrix} x_3^{new} \\ x_4^{new} \end{bmatrix} = \begin{bmatrix} 0 \\ 0 \end{bmatrix}$. Rearranging $x^{new} = \begin{bmatrix} x_B^{new} \\ x_N^{new} \end{bmatrix} = \begin{bmatrix} x_1^{new} \\ x_2^{new} \\ x_3^{new} \\ x_4^{new} \end{bmatrix}$.
Observe that $c^T x^{new} = -2 < -1 = c^T x^{current}$. Also, we know from Example 3.2 that x^{new} is an optimal solution.

3.2.3 Simplex Method (Detailed Steps)

We can now develop in more detail the steps of the simplex method with Theorem 3.1, Theorem 3.3, and the Direction and Step Length Theorem 3.4.

Simplex Method

Step 0: (Initialization)

Generate an initial basic feasible solution $x^{(0)} = \begin{bmatrix} x_B \\ x_N \end{bmatrix}$. Let B be the basis matrix and N the non-basis matrix with corresponding partition $c = (c_B, c_N)^T$. Let $\bar{B}$ and $\bar{N}$ be the index sets of the basic and non-basic variables. Let $k = 0$ and go to Step 1.

Step 1: (Optimality Check)

Compute the reduced costs $r_q = c_q - c_B^T B^{-1} N_q$ for all $q \in \bar{N}$. If $r_q \geq 0$ for all $q \in \bar{N}$, then $x^{(k)}$ is an optimal solution for the linear program STOP, else select one x_q non-basic such that $r_q < 0$ and go to Step 2.

Step 2: (Descent Direction Generation)

Construct $d^q = \begin{bmatrix} -B^{-1}N_q \\ e_q \end{bmatrix}$.

If $d^q \geq 0$, then the linear program is unbounded STOP, else go to Step 3.

Step 3: (Step Length Generation)

Compute the step length $\alpha = \min_{j \in \bar{B}}\{-\frac{x_j^{current}}{d_j^q} | d_j^q < 0\}$ (the minimum ratio test). Let j^* be the index of the basic variable that attains the minimum ratio α. Go to Step 4.

Step 4: (Improved Adjacent Basic Feasible Solution Computation) Now let $x^{(k+1)} = x^{(k)} + \alpha d^q$. Go to Step 5.

Step 5: (Basis Update)

Let B_{j^*} be the column in B associated with the leaving basic variable x_{j^*} Update the basis matrix B by removing B_{j^*} and adding the column N_q, thus

$$\bar{B} = \bar{B} - \{j^*\} \cup \{q\}.$$

Update the non-basis matrix N by removing N_q and adding B_{j^*}, thus

$$\bar{N} = \bar{N} - \{q\} \cup \{j^*\}.$$

Let $k = k + 1$ and go to Step 1.

Example 3.6

Consider again the linear program (3.1)

$$\begin{array}{ll} \textit{minimize} & -x_1 - x_2 \\ \textit{subject to} & x_1 + \quad x_3 \quad = 1 \\ & \quad x_2 + \quad x_4 = 1 \\ & x_1 \geq 0, x_2 \geq 0, x_3 \geq 0, x_4 \geq 0. \end{array}$$

We start (Step 0) the simplex method with the basic feasible solution

$$x^{(0)} = \begin{bmatrix} x_B \\ x_N \end{bmatrix} = \begin{bmatrix} x_3 \\ x_4 \\ x_1 \\ x_2 \end{bmatrix} = \begin{bmatrix} 1 \\ 1 \\ 0 \\ 0 \end{bmatrix} \text{ with } B = \begin{bmatrix} 1 & 0 \\ 0 & 1 \end{bmatrix} \text{ and } N =$$

$\begin{bmatrix} 1 & 0 \\ 0 & 1 \end{bmatrix}$ and

$$c_B = \begin{bmatrix} c_3 \\ c_4 \end{bmatrix} = \begin{bmatrix} 0 \\ 0 \end{bmatrix}, c_N = \begin{bmatrix} c_1 \\ c_2 \end{bmatrix} = \begin{bmatrix} -1 \\ -1 \end{bmatrix}, \bar{B} = \{3, 4\}, \text{ and } \bar{N} =$$

$\{1, 2\}$.

First Iteration

Step 1: Check the optimality of $x^{(0)}$ by computing the reduced costs of the non-basic variables r_1 and r_2. Now

$$r_1 = c_1 - c_B^T B^{-1} N_1 = -1 - (0, 0)^T \begin{bmatrix} 1 & 0 \\ 0 & 1 \end{bmatrix}^{-1} \begin{bmatrix} 1 \\ 0 \end{bmatrix} = -1 < 0$$

and

$$r_2 = c_2 - c_B^T B^{-1} N_2 = -1 - (0,0)^T \begin{bmatrix} 1 & 0 \\ 0 & 1 \end{bmatrix}^{-1} \begin{bmatrix} 0 \\ 1 \end{bmatrix} = -1 < 0 .$$

Thus, $x^{(0)}$ is not optimal. Choose x_1 as the non-basic variable to enter the basis. Go to Step 2.

Step 2: Construct $d^1 = \begin{bmatrix} -B^{-1}N_1 \\ e_1 \end{bmatrix} = \begin{bmatrix} -\begin{bmatrix} 1 & 0 \\ 0 & 1 \end{bmatrix}^{-1} \begin{bmatrix} 1 \\ 0 \end{bmatrix} \\ 1 \\ 0 \end{bmatrix} = \begin{bmatrix} -1 \\ 0 \\ 1 \\ 0 \end{bmatrix}.$

Since $d_1 \ngeq 0$ the linear program cannot be determined to be unbounded at this point. Go to Step 3.

Step 3: Compute the step length $\alpha = \min\limits_{j \in \bar{B}=\{3,4\}} \{-\frac{x_j^{current}}{d_j^1} | d_j^1 < 0\}$ $= \{-\frac{x_3^{current}}{d_3^1}\} = \{-\frac{1}{-1}\} = 1.$

Go to Step 4.

Step 4: So $x^{(1)} = x^{(0)} + \alpha d^1 = \begin{bmatrix} 1 \\ 1 \\ 0 \\ 0 \end{bmatrix} + (1) \begin{bmatrix} -1 \\ 0 \\ 1 \\ 0 \end{bmatrix} = \begin{bmatrix} 0 \\ 1 \\ 1 \\ 0 \end{bmatrix}.$

Observe that the variable x_3 leaves the basis (i.e., becomes non-basic). Go to Step 5.

Step 5: For $x^{(1)}$, the basic variables are $x_B = \begin{bmatrix} x_1 \\ x_4 \end{bmatrix} = \begin{bmatrix} 1 \\ 1 \end{bmatrix}$ and the non-basic variables are $x_N = \begin{bmatrix} x_3 \\ x_2 \end{bmatrix} = \begin{bmatrix} 0 \\ 0 \end{bmatrix}$.

The updated basis matrix $B = \begin{bmatrix} 1 & 0 \\ 0 & 1 \end{bmatrix}$, the updated non-basis matrix

$N = \begin{bmatrix} 1 & 0 \\ 0 & 1 \end{bmatrix}$, $c_B = \begin{bmatrix} c_1 \\ c_4 \end{bmatrix} = \begin{bmatrix} -1 \\ 0 \end{bmatrix}$, $c_N = \begin{bmatrix} c_3 \\ c_2 \end{bmatrix} = \begin{bmatrix} 0 \\ -1 \end{bmatrix}$,

$\bar{B} = \{1,4\}$ and $\bar{N} = \{3,2\}$. Go to Step 1.

Second Iteration

Step 1: Check the optimality of $x^{(1)}$ by computing the reduced costs of the non-basic variables r_3 and r_2. Now

$$r_3 = c_3 - c_B^T B^{-1} N_3 = 0 - (-1,0)^T \begin{bmatrix} 1 & 0 \\ 0 & 1 \end{bmatrix}^{-1} \begin{bmatrix} 1 \\ 0 \end{bmatrix} = 1 \geq 0$$

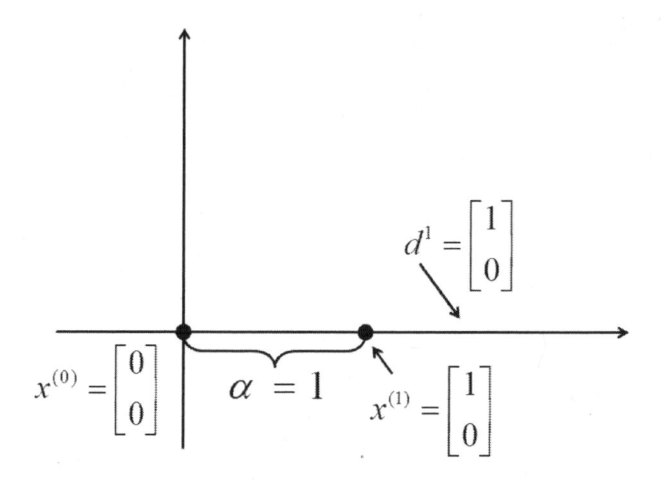

FIGURE 3.3
Direction and step length for first iteration of simplex method for Example 3.6.

and

$$r_2 = c_2 - c_B^T B^{-1} N_2 = -1 - (-1,0)^T \begin{bmatrix} 1 & 0 \\ 0 & 1 \end{bmatrix}^{-1} \begin{bmatrix} 0 \\ 1 \end{bmatrix} = -1 < 0\,.$$

Thus, $x^{(1)}$ is not optimal. Select x_2 as the non-basic variable to enter the basis and go to Step 2.

Step 2: Construct $d^2 = \begin{bmatrix} -B^{-1}N_2 \\ e_2 \end{bmatrix} = \begin{bmatrix} -\begin{bmatrix} 1 & 0 \\ 0 & 1 \end{bmatrix}^{-1} \begin{bmatrix} 0 \\ 1 \end{bmatrix} \\ 0 \\ 1 \end{bmatrix} = \begin{bmatrix} 0 \\ -1 \\ 0 \\ 1 \end{bmatrix}$.

Since $d^2 \ngeq 0$ the linear program cannot be determined to be unbounded at this point. Go to Step 3.

Step 3: Compute the step length $\alpha = \min_{j \in \bar{B}=\{1,4\}} \{-\frac{x_j^{current}}{d_j^2} | d_j^2 < 0\}$ $= \{-\frac{x_4^{current}}{d_4^2}\} = \{-\frac{1}{-1}\} = 1$.

Go to Step 4.

Step 4: Then, $x^{(2)} = x^{(1)} + \alpha d^2 = \begin{bmatrix} 1 \\ 1 \\ 0 \\ 0 \end{bmatrix} + (1) \begin{bmatrix} 0 \\ -1 \\ 0 \\ 1 \end{bmatrix} = \begin{bmatrix} 1 \\ 0 \\ 0 \\ 1 \end{bmatrix}$. The

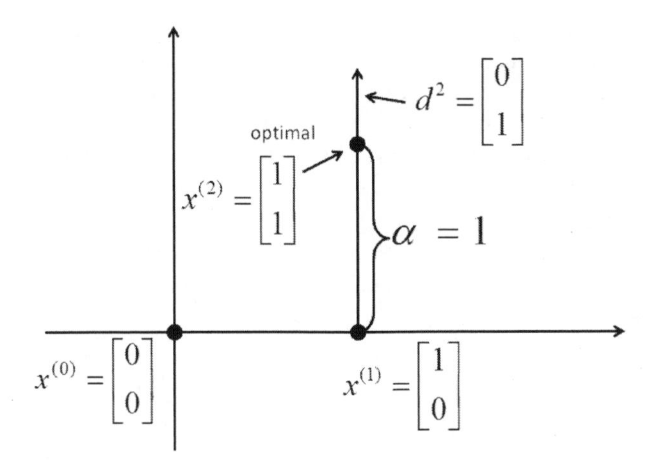

FIGURE 3.4
Direction and step length for second iteration of simplex method for Example 3.6.

variable x_4 leaves the basis. Go to Step 5.

Step 5: For $x^{(2)}$, the basic variables are $x_B = \begin{bmatrix} x_1 \\ x_2 \end{bmatrix} = \begin{bmatrix} 1 \\ 1 \end{bmatrix}$ and the non-basic variables are $x_N = \begin{bmatrix} x_3 \\ x_4 \end{bmatrix} = \begin{bmatrix} 0 \\ 0 \end{bmatrix}$.

The updated basis matrix $B = \begin{bmatrix} 1 & 0 \\ 0 & 1 \end{bmatrix}$ and $N = \begin{bmatrix} 0 & 1 \\ 1 & 0 \end{bmatrix}$ and $c_B = \begin{bmatrix} c_1 \\ c_2 \end{bmatrix} = \begin{bmatrix} -1 \\ -1 \end{bmatrix}$ and $c_N = \begin{bmatrix} c_3 \\ c_4 \end{bmatrix} = \begin{bmatrix} 0 \\ 0 \end{bmatrix}$ with $\bar{B} = \{1, 2\}$ and $\bar{N} = \{3, 4\}$.

Go to Step 1.

Third Iteration
Step 1: Check the optimality of $x^{(2)}$. From Example 3.5, we know that $x^{(2)}$ has non-negative reduced costs r_N and so $x^{(2)} = \begin{bmatrix} x_B \\ x_N \end{bmatrix} = \begin{bmatrix} x_1 \\ x_2 \\ x_3 \\ x_4 \end{bmatrix} = \begin{bmatrix} 1 \\ 1 \\ 0 \\ 0 \end{bmatrix}$ is an optimal solution and the simplex method terminates.

Figures 3.3 and 3.4 show the progress of the simplex method for the first two iterations where the directions in the space of the original problem are shown along with the step lengths.

Observations: In Step 0 we provided an initial basic feasible solution and obtaining this was not hard. In general, it might be a challenge and a more systematic method needs to be employed. This issue will be developed below. Also, in the first iteration there were two reduced costs, r_1 and r_2, that were negative, and only one of the associated non-basic variables, i.e., x_1 was selected to enter the basis. It would have been equally valid to select x_2 to enter the basis instead. In this case, the simplex method would take a different trajectory toward the optimal solution. In general, when there are several negative reduced costs, there is freedom to select any one of the associated non-basic variables to enter the basis. The activity of selecting a non-basic variable to enter the basis is called pivoting. Finally, in Step 3 it could be possible that there is more than one index $j \in \bar{B}$ that attains the minimum ratio; in this case ties can be broken arbitrarily to determine the basic variable that will leave the basis.

Example 3.7

Consider the linear program

$$\begin{array}{lll} \textit{maximize} & -2x_1 + x_2 & \\ \textit{subject to} & -x_1 + \quad x_2 & \leq 4 \\ & 2x_1 \; + \; x_2 & \leq 6 \\ & x_1 \geq 0, x_2 \geq 0. & \end{array}$$

Solve using the simplex method.

Solution:

First convert the objective to a minimization problem to maximize $-2x_1 + x_2 = -$minimize $2x_1 - x_2$.

We start (Step 0) the simplex method with the basic feasible solution

$x^{(0)} = \begin{bmatrix} x_B \\ x_N \end{bmatrix} = \begin{bmatrix} x_3 \\ x_4 \\ x_1 \\ x_2 \end{bmatrix} = \begin{bmatrix} 4 \\ 6 \\ 0 \\ 0 \end{bmatrix}$ with $B = \begin{bmatrix} 1 & 0 \\ 0 & 1 \end{bmatrix}$ and $N = \begin{bmatrix} -1 & 1 \\ 2 & 1 \end{bmatrix}$ and

$c_B = \begin{bmatrix} c_3 \\ c_4 \end{bmatrix} = \begin{bmatrix} 0 \\ 0 \end{bmatrix}$, $c_N = \begin{bmatrix} c_1 \\ c_2 \end{bmatrix} = \begin{bmatrix} 2 \\ -1 \end{bmatrix}$, $\bar{B} = \{3, 4\}$, and $\bar{N} = \{1, 2\}$.

First Iteration

Step 1: Check the optimality of $x^{(0)}$ by computing the reduced costs of the non-basic variables r_1 and r_2. Now

$$r_1 = c_1 - c_B^T B^{-1} N_1 = 2 - (0,0)^T \begin{bmatrix} 1 & 0 \\ 0 & 1 \end{bmatrix}^{-1} \begin{bmatrix} -1 \\ 2 \end{bmatrix} = 2 \geq 0$$

and

$$r_2 = c_2 - c_B^T B^{-1} N_2 = -1 - (0,0)^T \begin{bmatrix} 1 & 0 \\ 0 & 1 \end{bmatrix}^{-1} \begin{bmatrix} 1 \\ 1 \end{bmatrix} = -1 < 0\,.$$

Thus, $x^{(0)}$ is not optimal and x_2 must be selected as the non-basic variable to enter the basis. Go to Step 2.

Step 2: Construct $d^2 = \begin{bmatrix} -B^{-1}N_2 \\ e_2 \end{bmatrix} = \begin{bmatrix} -\begin{bmatrix} 1 & 0 \\ 0 & 1 \end{bmatrix}^{-1} \begin{bmatrix} 1 \\ 1 \end{bmatrix} \\ 0 \\ 1 \end{bmatrix} = \begin{bmatrix} -1 \\ -1 \\ 0 \\ 1 \end{bmatrix}$.

Since $d_2 \ngeq 0$, the linear program cannot be determined to be unbounded at this point. Go to Step 3.

Step 3: Compute the step length $\alpha = \min_{j \in \bar{B}=\{3,4\}} \{-\frac{x_j^{current}}{d_j^2} | d_j^2 < 0\}$

$= \min\{-\frac{x_3^{current}}{d_3^2}, -\frac{x_4^{current}}{d_4^2}\} = \{-\frac{4}{-1}, -\frac{6}{-1}\} = 4$. Go to Step 4.

Step 4: So $x^{(1)} = x^{(0)} + \alpha d^1 = \begin{bmatrix} 4 \\ 6 \\ 0 \\ 0 \end{bmatrix} + (4) \begin{bmatrix} -1 \\ -1 \\ 0 \\ 1 \end{bmatrix} = \begin{bmatrix} 0 \\ 2 \\ 0 \\ 4 \end{bmatrix}$.

Observe that the variable x_3 leaves the basis (i.e., becomes non-basic). Go to Step 5.

Step 5: For $x^{(1)}$, the basic variables are $x_B = \begin{bmatrix} x_2 \\ x_4 \end{bmatrix} = \begin{bmatrix} 4 \\ 2 \end{bmatrix}$ and the non-basic variables are $x_N = \begin{bmatrix} x_1 \\ x_3 \end{bmatrix} = \begin{bmatrix} 0 \\ 0 \end{bmatrix}$.

The updated basis matrix $B = \begin{bmatrix} 1 & 0 \\ 1 & 1 \end{bmatrix}$, the updated non-basis matrix

$N = \begin{bmatrix} -1 & 1 \\ 2 & 0 \end{bmatrix}$, $c_B = \begin{bmatrix} c_2 \\ c_4 \end{bmatrix} = \begin{bmatrix} -1 \\ 0 \end{bmatrix}$, $c_N = \begin{bmatrix} c_1 \\ c_3 \end{bmatrix} = \begin{bmatrix} 0 \\ -1 \end{bmatrix}$,

$\bar{B} = \{2, 4\}$, and $\bar{N} = \{1, 3\}$. Go to Step 1.

Second Iteration

Step 1: Check the optimality of $x^{(1)}$ by computing the reduced costs of the non-basic variables r_1 and r_3. Now

$$r_1 = c_1 - c_B^T B^{-1} N_1 = 2 - (-1, 0)^T \begin{bmatrix} 1 & 0 \\ 1 & 1 \end{bmatrix}^{-1} \begin{bmatrix} -1 \\ 2 \end{bmatrix} = 1 \geq 0$$

and

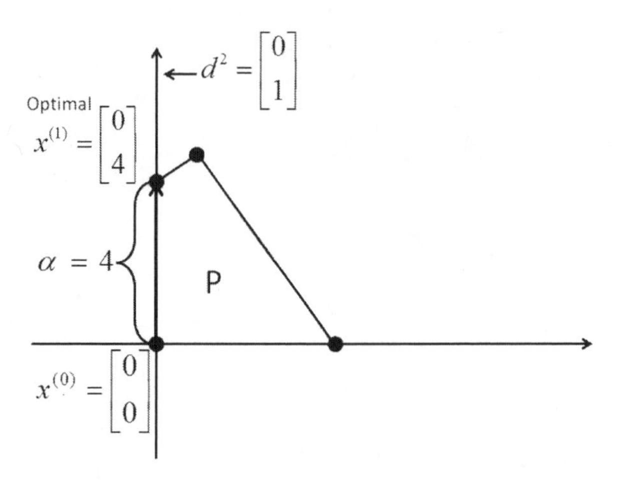

FIGURE 3.5
Movement of the simplex method in the space of original variables in Example 3.7.

$$r_3 = c_3 - c_B^T B^{-1} N_3 = 0 - (-1, 0)^T \begin{bmatrix} 1 & 0 \\ 0 & 1 \end{bmatrix}^{-1} \begin{bmatrix} 1 \\ 0 \end{bmatrix} = 1 \geq 0.$$

Thus, $x^{(1)} = \begin{bmatrix} x_B \\ x_N \end{bmatrix} = \begin{bmatrix} x_2 \\ x_4 \\ x_1 \\ x_3 \end{bmatrix} = \begin{bmatrix} 4 \\ 2 \\ 0 \\ 0 \end{bmatrix}$ is an optimal solution.

Figure 3.5 gives the movement to the optimal solution that the simplex method takes in the space of the original variables.

Example 3.8 (Unboundedness)
Consider the linear program

$$\begin{array}{ll} \textit{minimize} & -x_1 - x_2 \\ \textit{subject to} & -2x_1 + x_2 \leq 1 \\ & \quad x_1 - x_2 \leq 1 \\ & x_1 \geq 0, x_2 \geq 0. \end{array}$$

Solve using the simplex method.
Solution:
We add slack variables to bring the linear program in standard form

$$\begin{array}{lllll} \textit{minimize} & -x_1 - x_2 & & & \\ \textit{subject to} & -2x_1 + x_2 + & x_3 & & = 1 \\ & \quad x_1 - x_2 + & & x_4 & = 1 \\ & x_1 \geq 0, x_2 \geq 0, x_3 \geq 0, x_4 \geq 0. & & & \end{array}$$

We start (Step 0) the simplex method with the basic feasible solution

$$x^{(0)} = \begin{bmatrix} x_B \\ x_N \end{bmatrix} = \begin{bmatrix} x_3 \\ x_4 \\ x_1 \\ x_2 \end{bmatrix} = \begin{bmatrix} 1 \\ 1 \\ 0 \\ 0 \end{bmatrix} \text{ with } B = \begin{bmatrix} 1 & 0 \\ 0 & 1 \end{bmatrix} \text{ and } N =$$

$\begin{bmatrix} -2 & 1 \\ 1 & -1 \end{bmatrix}$ and

$$c_B = \begin{bmatrix} c_3 \\ c_4 \end{bmatrix} = \begin{bmatrix} 0 \\ 0 \end{bmatrix}, \; c_N = \begin{bmatrix} c_1 \\ c_2 \end{bmatrix} = \begin{bmatrix} -1 \\ -1 \end{bmatrix}, \; \bar{B} = \{3,4\}, \text{ and } \bar{N} =$$

$\{1,2\}$.

First Iteration

Step 1: Check the optimality of $x^{(0)}$ by computing the reduced costs of the non-basic variables r_1 and r_2. Now

$$r_1 = c_1 - c_B^T B^{-1} N_1 = -1 - (0,0)^T \begin{bmatrix} 1 & 0 \\ 0 & 1 \end{bmatrix}^{-1} \begin{bmatrix} -2 \\ 1 \end{bmatrix} = -1 < 0$$

and

$$r_2 = c_2 - c_B^T B^{-1} N_2 = -1 - (0,0)^T \begin{bmatrix} 1 & 0 \\ 0 & 1 \end{bmatrix}^{-1} \begin{bmatrix} 1 \\ -1 \end{bmatrix} = -1 < 0 \, .$$

Thus, $x^{(0)}$ is not optimal. Choose x_1 as the non-basic variable to enter the basis. Go to Step 2.

Step 2: Construct

$$d^1 = \begin{bmatrix} -B^{-1}N_1 \\ e_1 \end{bmatrix} = \begin{bmatrix} -\begin{bmatrix} 1 & 0 \\ 0 & 1 \end{bmatrix}^{-1} \begin{bmatrix} -2 \\ 1 \end{bmatrix} \\ 1 \\ 0 \end{bmatrix} = \begin{bmatrix} 2 \\ -1 \\ 1 \\ 0 \end{bmatrix}.$$

Since $d_1 \ngeq 0$, the linear program cannot be determined to be unbounded at this point. Go to Step 3.

Step 3: Compute the step length $\alpha = \min\limits_{j \in \bar{B} = \{3,4\}} \{-\frac{x_j^{current}}{d_j^1} | d_j^1 < 0\}$

$= \{-\frac{x_4^{current}}{d_4^1}\} = \{-\frac{1}{-1}\} = 1.$

Go to Step 4.

Step 4: So $x^{(1)} = x^{(0)} + \alpha d^1 = \begin{bmatrix} 1 \\ 1 \\ 0 \\ 0 \end{bmatrix} + (1) \begin{bmatrix} 2 \\ -1 \\ 1 \\ 0 \end{bmatrix} = \begin{bmatrix} 3 \\ 0 \\ 1 \\ 0 \end{bmatrix}.$

Observe that the variable x_4 leaves the basis. Go to Step 5.

Step 5: For $x^{(1)}$ the basic variables are $x_B = \begin{bmatrix} x_3 \\ x_1 \end{bmatrix} = \begin{bmatrix} 3 \\ 1 \end{bmatrix}$ and the non-basic variables are $x_N = \begin{bmatrix} x_4 \\ x_2 \end{bmatrix} = \begin{bmatrix} 0 \\ 0 \end{bmatrix}$.

The updated basis matrix $B = \begin{bmatrix} 1 & -2 \\ 0 & 1 \end{bmatrix}$, the updated non-basis matrix

$$N = \begin{bmatrix} 0 & 1 \\ 1 & -1 \end{bmatrix}, c_B = \begin{bmatrix} c_3 \\ c_1 \end{bmatrix} = \begin{bmatrix} 0 \\ -1 \end{bmatrix}, c_N = \begin{bmatrix} c_4 \\ c_2 \end{bmatrix} = \begin{bmatrix} 0 \\ -1 \end{bmatrix},$$

$\bar{B} = \{3, 1\}$ and $\bar{N} = \{4, 2\}$. Go to Step 1.

Second Iteration

Step 1: Check the optimality of $x^{(1)}$ by computing the reduced costs of the non-basic variables r_4 and r_2. Now

$$r_4 = c_4 - c_B^T B^{-1} N_4 = 0 - (0, -1)^T \begin{bmatrix} 1 & -2 \\ 0 & 1 \end{bmatrix}^{-1} \begin{bmatrix} 0 \\ 1 \end{bmatrix} = 1 \geq 0$$

and

$$r_2 = c_2 - c_B^T B^{-1} N_2 = -1 - (0, -1)^T \begin{bmatrix} 1 & -2 \\ 0 & 1 \end{bmatrix}^{-1} \begin{bmatrix} 1 \\ -1 \end{bmatrix} = -2 < 0\,.$$

Thus, $x^{(1)}$ is not optimal. Select x_2 as the non-basic variable to enter the basis and go to Step 2.

Step 2: Construct

$$d^2 = \begin{bmatrix} -B^{-1}N_2 \\ e_2 \end{bmatrix} = \begin{bmatrix} -\begin{bmatrix} 1 & -2 \\ 0 & 1 \end{bmatrix}^{-1} \begin{bmatrix} 1 \\ -1 \end{bmatrix} \\ 0 \\ 1 \end{bmatrix} = \begin{bmatrix} 1 \\ 1 \\ 0 \\ 1 \end{bmatrix}.$$

Since $d^2 \geq 0$, the linear program is unbounded at this point and the simplex method is terminated.

In particular, $\alpha = x_2$ can be increased indefinitely in

$$x^{(2)} = x^{(1)} + \alpha d^2 = \begin{bmatrix} 3 \\ 1 \\ 0 \\ 0 \end{bmatrix} + (x_2) \begin{bmatrix} 1 \\ 1 \\ 0 \\ 1 \end{bmatrix} = \begin{bmatrix} 3 + x_2 \\ 1 + x_2 \\ 0 \\ x_2 \end{bmatrix}$$

since $x^{(2)}$ is feasible for any $x_2 \geq 0$. Also, observe that the objective function is $-x_1 - x_2 = -(1 + x_2\ ) - x_2$, so it diverges toward negative infinity as x_2 increases indefinitely. Thus, the linear program is unbounded along the ray

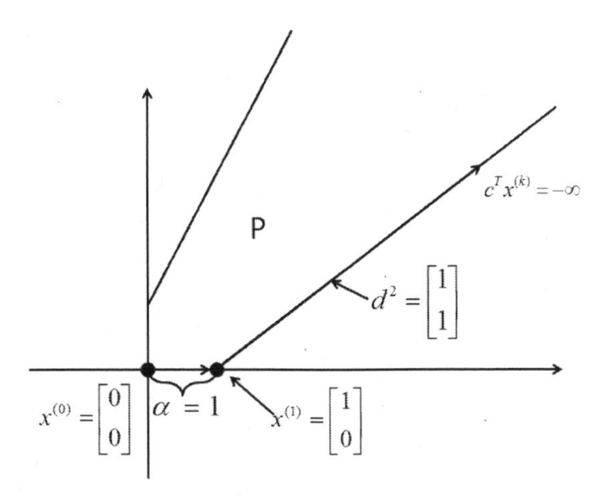

FIGURE 3.6
Movement of simplex method towards unboundedness in Example 3.8.

with vertex $x^{(1)}$ and the direction of the ray d^2. This is depicted in the Figure 3.6 in the space of original variables.

Example 3.9 (Multiple Optimal Solutions)
Consider the linear program

$$\begin{array}{lll} \textit{maximize} & x_1 + x_2 & \\ \textit{subject to} & x_1 + \ x_2 & \leq 4 \\ & 2x + \ x_2 & \leq 6 \\ & x_1 \geq 0, x_2 \geq 0. & \end{array}$$

Solution:
We negate the objective function and add slack variables to bring the linear program into standard form

$$\begin{array}{lll} \textit{minimize} & -x_1 - x_2 & \\ \textit{subject to} & x_1 + \ x_2 + \quad x_3 & = 4 \\ & 2x_1 + x_2 + \qquad\qquad x_4 & = 6 \\ & x_1 \geq 0, x_2 \geq 0, x_3 \geq 0, x_4 \geq 0. & \end{array}$$

We start (Step 0) the simplex method with the basic feasible solution

$$x^{(0)} = \begin{bmatrix} x_B \\ x_N \end{bmatrix} = \begin{bmatrix} x_3 \\ x_4 \\ x_1 \\ x_2 \end{bmatrix} = \begin{bmatrix} 4 \\ 6 \\ 0 \\ 0 \end{bmatrix}$$ with $B = \begin{bmatrix} 1 & 0 \\ 0 & 1 \end{bmatrix}$ and $N = \begin{bmatrix} 1 & 1 \\ 2 & 1 \end{bmatrix}$ and

$c_B = \begin{bmatrix} c_3 \\ c_4 \end{bmatrix} = \begin{bmatrix} 0 \\ 0 \end{bmatrix}$, $c_N = \begin{bmatrix} c_1 \\ c_2 \end{bmatrix} = \begin{bmatrix} -1 \\ -1 \end{bmatrix}$, $\bar{B} = \{3,4\}$, and $\bar{N} = \{1,2\}$.

First Iteration

Step 1: Check the optimality of $x^{(0)}$ by computing the reduced costs of the non-basic variables r_1 and r_2. Now

$$r_1 = c_1 - c_B^T B^{-1} N_1 = -1 - (0,0)^T \begin{bmatrix} 1 & 0 \\ 0 & 1 \end{bmatrix}^{-1} \begin{bmatrix} 1 \\ 2 \end{bmatrix} = -1 < 0$$

and

$$r_2 = c_2 - c_B^T B^{-1} N_2 = -1 - (0,0)^T \begin{bmatrix} 1 & 0 \\ 0 & 1 \end{bmatrix}^{-1} \begin{bmatrix} 1 \\ 1 \end{bmatrix} = -1 < 0 .$$

Thus, $x^{(0)}$ is not optimal. Choose x_1 as the non-basic variable to enter the basis. Go to Step 2.

Step 2: Construct $d^1 = \begin{bmatrix} -B^{-1}N_1 \\ e_1 \end{bmatrix} = \begin{bmatrix} -\begin{bmatrix} 1 & 0 \\ 0 & 1 \end{bmatrix}^{-1} \begin{bmatrix} 1 \\ 2 \end{bmatrix} \\ 1 \\ 0 \end{bmatrix} = \begin{bmatrix} -1 \\ -2 \\ 1 \\ 0 \end{bmatrix}$.

Since $d_1 \ngeq 0$, the linear program cannot be determined to be unbounded at this point. Go to Step 3.

Step 3: Compute the step length $\alpha = \min\limits_{j \in \bar{B} = \{3,4\}} \{-\frac{x_j^{current}}{d_j^1} | d_j^1 < 0\}$

$= \min\{-\frac{x_3^{current}}{d_3^1}, -\frac{x_4^{current}}{d_4^1}\} = \min\{-\frac{4}{-1}, -\frac{6}{-2}\} = 3$. Go to Step 4.

Step 4: So $x^{(1)} = x^{(0)} + \alpha d^1 = \begin{bmatrix} 4 \\ 6 \\ 0 \\ 0 \end{bmatrix} + (3) \begin{bmatrix} -1 \\ -2 \\ 1 \\ 0 \end{bmatrix} = \begin{bmatrix} 1 \\ 0 \\ 3 \\ 0 \end{bmatrix}$.

Observe that the variable x_4 leaves the basis (i.e. becomes non-basic). Go to Step 5.

Step 5: For $x^{(1)}$ the basic variables are $x_B = \begin{bmatrix} x_3 \\ x_1 \end{bmatrix} = \begin{bmatrix} 1 \\ 3 \end{bmatrix}$ and the non-basic variables are $x_N = \begin{bmatrix} x_4 \\ x_2 \end{bmatrix} = \begin{bmatrix} 0 \\ 0 \end{bmatrix}$.

The updated basis matrix $B = \begin{bmatrix} 1 & 1 \\ 0 & 2 \end{bmatrix}$, the updated non-basis matrix $N = \begin{bmatrix} 0 & 1 \\ 1 & 1 \end{bmatrix}$, $c_B = \begin{bmatrix} c_3 \\ c_1 \end{bmatrix} = \begin{bmatrix} 0 \\ -1 \end{bmatrix}$, $c_N = \begin{bmatrix} c_4 \\ c_2 \end{bmatrix} = \begin{bmatrix} 0 \\ -1 \end{bmatrix}$, $\bar{B} = \{3, 1\}$, and $\bar{N} = \{4, 2\}$. Go to Step 1.

Second Iteration

Step 1: Check the optimality of $x^{(1)}$ by computing the reduced costs of the non-basic variables r_4 and r_2. Now

$$r_4 = c_4 - c_B^T B^{-1} N_4 = 0 - (0, -1)^T \begin{bmatrix} 1 & 1 \\ 0 & 2 \end{bmatrix}^{-1} \begin{bmatrix} 0 \\ 1 \end{bmatrix} = 1/2 \geq 0$$

and

$$r_2 = c_2 - c_B^T B^{-1} N_2 = -1 - (0, -1)^T \begin{bmatrix} 1 & 1 \\ 0 & 2 \end{bmatrix}^{-1} \begin{bmatrix} 1 \\ 1 \end{bmatrix} = -1/2 < 0 \; .$$

Thus, $x^{(1)}$ is not optimal. Select x_2 as the non-basic variable to enter the basis and go to Step 2.

Step 2: Construct $d^2 = \begin{bmatrix} -B^{-1}N_2 \\ e_2 \end{bmatrix} = \begin{bmatrix} -\begin{bmatrix} 1 & 1 \\ 0 & 1 \end{bmatrix}^{-1} \begin{bmatrix} 1 \\ 1 \end{bmatrix} \\ 0 \\ 1 \end{bmatrix} = \begin{bmatrix} -1/2 \\ -1/2 \\ 0 \\ 1 \end{bmatrix}$.

Since $d^2 \ngeq 0$, the linear program cannot be determined to be unbounded at this point. Go to Step 3.

Step 3: Compute the step length $\alpha = \min\limits_{j \in \bar{B} = \{3,1\}} \{-\frac{x_j^{current}}{d_j^2} | d_j^2 < 0\}$

$$= \min\{-\frac{x_3^{current}}{d_3^2}, -\frac{x_1^{current}}{d_1^2}\} = \{-\frac{1}{-1/2}, -\frac{3}{-1/2}\} = 2.$$

Go to Step 4.

Step 4: Then, $x^{(2)} = x^{(1)} + \alpha d^2 = \begin{bmatrix} 1 \\ 3 \\ 0 \\ 0 \end{bmatrix} + (2) \begin{bmatrix} -1/2 \\ -1/2 \\ 0 \\ 1 \end{bmatrix} = \begin{bmatrix} 0 \\ 2 \\ 0 \\ 2 \end{bmatrix}$.

The variable x_3 leaves the basis. Go to Step 5.

Step 5: For $x^{(2)}$ the basic variables are $x_B = \begin{bmatrix} x_2 \\ x_1 \end{bmatrix} = \begin{bmatrix} 2 \\ 2 \end{bmatrix}$ and the non-basic variables are $x_N = \begin{bmatrix} x_4 \\ x_3 \end{bmatrix} = \begin{bmatrix} 0 \\ 0 \end{bmatrix}$.

The updated basis matrix $B = \begin{bmatrix} 1 & 1 \\ 1 & 2 \end{bmatrix}$, the updated non-basis matrix is

$N = \begin{bmatrix} 0 & 1 \\ 1 & 0 \end{bmatrix}$, $c_B = \begin{bmatrix} c_2 \\ c_1 \end{bmatrix} = \begin{bmatrix} -1 \\ -1 \end{bmatrix}$, and $c_N = \begin{bmatrix} c_4 \\ c_3 \end{bmatrix} = \begin{bmatrix} 0 \\ 0 \end{bmatrix}$.

Go to Step 1.

Third Iteration

Step 1: Check the optimality of $x^{(2)}$. The reduced costs $r_4 = c_4 - c_B^T B^{-1} N_4 = 0 \geq 0$ and $r_3 = c_3 - c_B^T B^{-1} N_3 = 1 \geq 0$

so $x^{(2)} = \begin{bmatrix} x_B \\ x_N \end{bmatrix} = \begin{bmatrix} x_2 \\ x_1 \\ x_4 \\ x_3 \end{bmatrix} = \begin{bmatrix} 2 \\ 2 \\ 0 \\ 0 \end{bmatrix}$ is an optimal solution with objective function value $-(-2-2) = 4$ and the simplex method stops.

It turns out that $x^{(2)}$ is not the only optimal solution. For instance, if x_4 is increased while x_3 remains at 0, we get

$$x_B = B^{-1}b - B^{-1}N_4 x_4 =$$

$$\begin{bmatrix} x_2 \\ x_1 \end{bmatrix} = \begin{bmatrix} 2 \\ 2 \end{bmatrix} - \begin{bmatrix} -1 \\ 1 \end{bmatrix} x_4 = \begin{bmatrix} 2 + x_4 \\ 2 - x_4 \end{bmatrix}.$$

Thus, for any $x_4 \leq 2$, any point of the form

$$\begin{bmatrix} x_2 \\ x_1 \\ x_4 \\ x_3 \end{bmatrix} = \begin{bmatrix} 2 + x_4 \\ 2 - x_4 \\ x_4 \\ 0 \end{bmatrix}$$

is also an optimal solution with objective function value 4. Note that when $x_4 = 2$, then the point is another optimal basic feasible solution. This situation of an infinite number of alternative optimal solutions arises due to the fact that the objective function contours are parallel to the hyperplane defined by the first constraint $x_1 + x_2 = 4$. Figure 3.7 depicts the movement of the simplex method in the space of original variables.

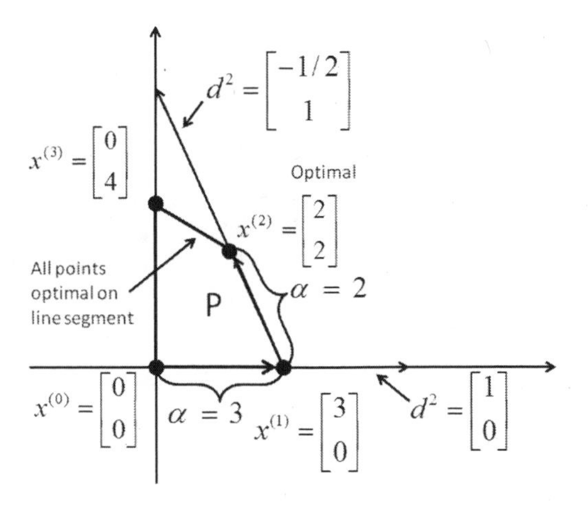

FIGURE 3.7
Movement of simplex method in Example 3.9 with multiple optimal solutions.

3.2.4 Finite Termination under Non-Degeneracy

If it is assumed that all basic feasible solutions are non-degenerate, then there is a one-to-one correspondence between the basic feasible solutions and the extreme points of the linear program. If there is not a one-to-one correspondence, i.e., there is degeneracy, then it is possible that the simplex method may move to a different basic feasible solution that represents the current extreme point and may then subsequently revisit a basic feasible solution generated previously. In this case the iterations can get trapped in a never-ending sequence of re-visiting an extreme point via the different basic feasible solutions that all represent the same extreme point. This phenomenon is called cycling and is illustrated later in this chapter. It is clear that cycling cannot happen when there is a one-to-one correspondence between extreme points and basic feasible solutions.

Then in the absence of degeneracy, at each iteration of the simplex method, one of the following occurs: (1) an optimal basic feasible solution is found, (2) the linear program will be determined to be unbounded, or (3) an improved adjacent basic feasible solution will be generated. Since there are a finite number of basic feasible solutions, the simplex method will terminate in a finite number of steps with an optimal basic feasible solution or with the conclusion that the problem is unbounded.

3.3 Generating an Initial Basic Feasible Solution (Two-Phase and Big M Methods)

The simplex method requires an initial basic feasible solution to start. Step 0 is where an initial basic feasible is to be generated. The examples of the simplex method above involved linear programs in which an initial basic feasible solution was easily obtained.

For example, in the linear program

$$\begin{array}{ll} \textit{minimize} & -x_1 - x_2 \\ \textit{subject to} & x_1 \leq 1 \\ & x_2 \leq 1 \\ & x_1 \geq 0, x_2 \geq 0, \end{array}$$

after adding slack variables x_3 and x_4 we get

$$\begin{array}{ll} \textit{minimize} & -x_1 - x_2 \\ \textit{subject to} & x_1 + \quad x_3 = 1 \\ & \quad x_2 + \quad x_4 = 1 \\ & x_1 \geq 0, x_2 \geq 0, x_3 \geq 0, x_4 \geq 0. \end{array}$$

An initial basic feasible solution can be obtained by setting the basic variables to be the slack variables at the non-negative right-hand side values, i.e., $x_B = \begin{bmatrix} x_3 \\ x_4 \end{bmatrix} = \begin{bmatrix} 1 \\ 1 \end{bmatrix}$ and the original variables to be non-basic, i.e., $x_N = \begin{bmatrix} x_1 \\ x_2 \end{bmatrix} = \begin{bmatrix} 0 \\ 0 \end{bmatrix}$. One can easily verify that this is a basic feasible solution. Observe that the basis B corresponding to the slack variables form an $m \times m$ identity matrix and so is invertible and $x_B = B^{-1}b \geq 0$.

In general, for a linear program if (after the possible addition of slack and surplus variables to put in standard form) the constraint set $Ax = b$ and the right-hand side vector b is non-negative (if a component is originally negative then the associated equality constraint can be multiplied by -1 on both sides of the constraint) contains an $m \times m$ identity submatrix, then the variables associated with this matrix can be set to be the initial basic variables and the original non-slack variables can be set to the initial non-basic variables.

However, many linear programs may be such that after conversion to standard form an $m \times m$ submatrix is not present or readily apparent.

Example 3.10
Consider the constraint set

$$\begin{array}{ll} x_1 + 2x_2 + x_3 & \leq 4 \\ -x_1 \qquad -x_3 & \geq 3 \\ x_1 \geq 0, x_2 \geq 0, x_3 \geq 0. & \end{array}$$

We can convert the constraints to standard form by adding a slack variable to the first constraint and a surplus variable to the second constraint to get

$$\begin{array}{ll} x_1 + 2x_2 + x_3 + x_4 & = 4 \\ -x_1 + \qquad -x_3 - \qquad x_5 & = 3 \\ x_1 \geq 0, x_2 \geq 0, x_3 \geq 0, x_4 \geq 0, x_5 \geq 0. & \end{array}$$

The constraint matrix does not contain an identity submatrix, and thus it is not immediately clear what an initial starting basic feasible solution is.

Example 3.11
Consider the constraints

$$\begin{array}{l} x_1 + \quad x_2 \geq 2 \\ -x_1 + 2x_2 \geq 3 \\ 2x_1 + x_2 \leq 4 \\ x_1 \geq 0, x_2 \geq 0. \end{array}$$

After adding slack and surplus variables to get in standard form we get

$$\begin{array}{ll} x_1 + \quad x_2 \qquad -x_3 & = 2 \\ -x_1 + 2x_2 \qquad\qquad -x_4 & = 3 \\ 2x_1 + x_2 \qquad\qquad\qquad +x_5 & = 4 \\ x_1 \geq 0, x_2 \geq 0, x_3 \geq 0, x_4 \geq 0, x_5 \geq 0. & \end{array}$$

This is another case where there is no identity submatrix.

Artificial Variables
In the case where it is impossible or difficult to determine an identity submatrix in the constraint set $Ax = b$, one can add artificial variables to create an identity submatrix. For example, in the constraint set from Example 3.10, one can add an artificial variable x_6 to the first constraint and x_7 to the second constraint to get

$$\begin{array}{l} x_1 + \quad 2x_2 + \quad x_3 + \quad x_4 \qquad\qquad +x_6 = 4 \\ -x_1 \qquad\qquad -x_3 \qquad\qquad -x_5 \qquad\quad +x_7 = 3 \\ x_1 \geq 0, x_2 \geq 0, x_3 \geq 0, x_4 \geq 0, x_5 \geq 0, x_6 \geq 0, x_7 \geq 0. \end{array}$$

The constraint set above now admits an identity submatrix associated with variables x_6 and x_7. This constraint set can be represented in matrix form as $Ax + x_a = b$ and with $x \geq 0$ and $x_a \geq 0$ where

$$A = \begin{bmatrix} 1 & 2 & 1 & 1 & 0 \\ -1 & 0 & -1 & 0 & -1 \end{bmatrix}, b = \begin{bmatrix} 4 \\ 3 \end{bmatrix}$$

$$x = (x_1, x_2, x_3, x_4, x_5)^T$$

$$x_a = (x_6, x_7)^T.$$

And for the constraints in Example 3.11 we can add an artificial variable x_6 to the first constraint and an artificial variable x_7 to the second constraint, and an artificial variable x_8 to the third constraint to get the following

$$\begin{array}{llllllll} x_1 + & x_2 - x_3 & & & +x_6 & & & = 2 \\ -x_1 + 2x_2 & & -x_4 & & & +x_7 & & = 3 \\ 2x_1 + x_2 & & & +x_5 & & & +x_8 & = 4 \\ \multicolumn{8}{l}{x_1 \geq 0, x_2 \geq 0, x_3 \geq 0, x_4 \geq 0, x_5 \geq 0, x_6 \geq 0, x_7 \geq 0, x_8 \geq 0.} \end{array}$$

An identity submatrix can be readily formed by considering the columns of this constraint matrix corresponding to the variables x_6, x_7, and x_8.

The term artificial variable refers to the fact that such a variable is added on top of the original set of variables and any slack or surplus variables used to get the linear program in standard form. The main motivation for introducing artificial variables is to create an identity submatrix. However, a linear program that uses the constraints with the artificial variables may not be equivalent to the original linear program. In particular, the identity submatrix obtained from using artificial variables will not be a valid basis for the original linear program.

Suppose a linear program is converted to standard form (with the possible use of slack and surplus variables)

$$\begin{array}{lll} \textit{minimize} & c^T x & \text{SFLP} \\ \textit{subject to} & Ax = b & \\ & x \geq 0 & \end{array}$$

and consider the addition of artificial variables

$$\begin{array}{lll} \textit{minimize} & c^T x & \text{ALP} \\ \textit{subject to} & Ax + x_a = b & \\ & x \geq 0. & \end{array}$$

Then, the feasible set for SFLP is not in general equivalent to the feasible set of ALP. Nevertheless, there are some important relationships between the feasible sets. For example,

(1) If SFLP has a feasible solution, then ALP has a solution with $x_a = 0$.

(2) If ALP has a feasible solution with $x_a = 0$, then SFLP has a feasible solution.

In other words, there is a one-to-one correspondence between feasible solutions of SFLP and feasible solutions of ALP with $x_a = 0$.

This motivates the following idea: develop an auxiliary problem that attempts to remove the artificial variables by attempting to set $x_a = 0$ while generating a basic feasible solution for the original linear program. There are two major approaches in implementing this idea.

3.3.1 The Two-Phase Method

In this approach, an auxiliary problem (called the Phase I problem) is first solved to attempt to generate an initial basic feasible solution for the original linear program. Then, the Phase II problem is the original linear programming problem using as an initial basic feasible solution the one generated (if possible) from the Phase I problem. Let x_a^i be the *ith* component of x_a and $\mathbf{1}$ is a vector of n dimensions where all components have value 1.

Phase I is the following auxiliary problem

$$\begin{array}{lll} \textit{minimize} & \mathbf{1}^T x_a = \sum_{i=1}^{m} x_a^i & \text{Phase I} \\ \textit{subject to} & Ax + x_a = b & \\ & x \geq 0, x_a \geq 0. & \end{array}$$

The Phase I problem can be initialized with the basic feasible solution $x = 0$ and $x_a = b$.

Let $\tilde{x} = \begin{bmatrix} x^* \\ x_a^* \end{bmatrix}$ be the optimal solution to Phase I.

Case (1): If $x_a^* \neq 0$, then the original linear program is infeasible.

Proof: If the original linear program is feasible, then there is an x such that $Ax = b$ and $x \geq 0$ and thus $\bar{x} = \begin{bmatrix} x \\ 0 \end{bmatrix}$ would be feasible for the Phase I problem. But the objective function value of $\bar{x}$ is 0, which is lower than $1^T x_a^*$ since $x_a^* \neq 0$, contradicting the optimality of x^*. ■

Case (2):

Otherwise, $x_a^* = 0$.

Subcase (1): All of the artificial variables are non-basic in $\tilde{x}$. Then, discard the artificial variables and partition x^* into basic x_B and non-basic x_N variables corresponding to the B and N matrices obtained in the final iteration in solving Phase I, and use as a starting solution for the Phase II problem

$$\begin{array}{lll} \textit{minimize} & c_B^T x_B + c_N^T x_N & \text{Phase II} \\ \textit{subject to} & Bx_B + Nx_N = b & \\ & x_B \geq 0, x_N \geq 0. & \end{array}$$

Subcase (2): Some of the artificial variables are basic variables with zero value. Let x_a^{*i} denote an artificial variable that is basic and suppose that it occurs in the *jth* position among the basic variables. Then, the idea is to exchange such an artificial basic variable with a current non-basic and non-artificial variable x_q. There are two further cases:

(1) If there is an x_q such that $e_j^T B^{-1} N_q \neq 0$, then x_q can replace x_a^{*i} in the Phase I optimal basis. If it is possible to perform this exchange for all such artificial variables that are in the optimal Phase I basis, then the basis that results after all exchanges are done will constitute a valid starting basis for Phase II without any artificial variables.

(2) If $e_j^T B^{-1} N_q = 0$ for all non-basic x_q, then the *jth* row of the constraint matrix A is redundant and so can be removed and Phase I restarted.

The proofs are left as an exercise for the reader.

Example 3.12
Consider the following linear program:

$$\begin{array}{ll} \textit{minimize} & x_1 + 2x_2 \\ \textit{subject to} & x_1 + x_2 \geq 2 \\ & -x_1 + 2x_2 \geq 3 \\ & 2x_1 + x_2 \leq 4 \\ & x_1 \geq 0, x_2 \geq 0. \end{array}$$

The Phase I problem is

$$\begin{array}{lllllllll} \textit{minimize} & x_6 + x_7 + x_8 \\ \textit{subject to} & x_1 + x_2 - x_3 & & & +x_6 & & & = 2 \\ & -x_1 + 2x_2 & -x_4 & & & +x_7 & & = 3 \\ & 2x_1 + x_2 & & +x_5 & & & +x_8 & = 4 \\ & \multicolumn{8}{l}{x_1 \geq 0, x_2 \geq 0, x_3 \geq 0, x_4 \geq 0, x_5 \geq 0, x_6 \geq 0, x_7 \geq 0, x_8 \geq 0.} \end{array}$$

The initial basic variable set for the Phase I problem is $x_B = \begin{bmatrix} x_6 \\ x_7 \\ x_8 \end{bmatrix} = b = \begin{bmatrix} 2 \\ 3 \\ 4 \end{bmatrix}$ with corresponding basis matrix $B = \begin{bmatrix} 1 & 0 & 0 \\ 0 & 1 & 0 \\ 0 & 0 & 1 \end{bmatrix}$ and the initial non-basic variable set is $x_N = \begin{bmatrix} x_1 \\ x_2 \\ x_3 \\ x_4 \\ x_5 \end{bmatrix} = \begin{bmatrix} 0 \\ 0 \\ 0 \\ 0 \\ 0 \end{bmatrix}$ with corresponding non-basis matrix

$$N = \begin{bmatrix} 1 & 1 & -1 & 0 & 0 \\ -1 & 2 & 0 & -1 & 0 \\ 2 & 1 & 0 & 0 & 5 \end{bmatrix}.$$

Solving the Phase I problem using the simplex method with this initial basic feasible solution gives the optimal solution

$$x_1 = 1, x_2 = 2, x_3 = 1, x_4 = 0, x_5 = 0, x_6 = 0, x_7 = 0, x_8 = 0$$

with basic variables

$$x_B = [x_1, x_2, x_3]^T$$

and non-basic variables

$$x_N = [x_4, x_5, x_6, x_7, x_8].$$

Since the artificial variables x_6, x_7, x_8 are zero and are not in the basis we discard the columns associated with these variables and proceed to Phase II using as an initial basis B the columns associated with x_1, x_2, x_3 and non-basis N the columns associated with x_4 and x_5. Solving the Phase II problem with this initial basic feasible solution gives the optimal solution

$$x_1 = 1/3, x_2 = 5/3, x_3 = 1, x_4 = 0, x_5 = 5/3.$$

3.3.2 Big M Method

The second approach for generating an initial basic feasible solution involves only one optimization problem that simultaneously involves artificial variables as a mechanism to generate an initial basic feasible solution while penalizing the artificial variables in the objective function by a suitably large penalty to attempt to drive out these variables in an optimal solution.

The Big M problem is

$$\begin{array}{ll} \textit{minimize} & c^T x + M\mathbf{1}^T x_a \\ \textit{subject to} & Ax + x_a = b \\ & x \geq 0, x_a \geq 0 \end{array}$$

where $M > 0$ is a large parameter. Thus, the second term in the objective function $M\mathbf{1}^T x_a$ penalizes any solution where $x_a \neq 0$. The Big M problem can be solved by the simplex method with the initial basic feasible solution $x_a = b$ as basic variables and $x = 0$ as the non-basic variables.

Example 3.13

The Big M problem corresponding to the LP in Example 3.12 is

$$\begin{array}{llllllllll} \textit{minimize} & x_1 + 2x_2 + & & & & Mx_6 + Mx_7 + Mx_8 & & & \\ \textit{subject to} & x_1 + \quad x_2 - x_3 & & & & +x_6 & & & = 2 \\ & -x_1 + 2x_2 & -x_4 & & & & +x_7 & & = 3 \\ & 2x_1 + x_2 & & +x_5 & & & & +x_8 & = 4 \\ & \multicolumn{8}{l}{x_1 \geq 0, x_2 \geq 0, x_3 \geq 0, x_4 \geq 0, x_5 \geq 0, x_6 \geq 0, x_7 \geq 0, x_8 \geq 0.} \end{array}$$

An initial basic feasible solution is $x_6 = 2, x_7 = 3$, and $x_8 = 4$, and non-basic variables are $x_i = 0$ for $i = 1, ..., 5$. Exercise 3.12 asks the reader use this initial basic feasible solution to solve this problem using the simplex method.

Using the simplex method to solve the Big M model with a suitably large penalty parameter $M > 0$ will result in one of two cases (infeasibility is not a possibility).

Case (1) The Big M model results in a finite optimal solution $\tilde{x} = \begin{bmatrix} x^* \\ x_a^* \end{bmatrix}$.

Subcase (1):

If $x_a^* = 0$ *, then* x^* *is an optimal solution for the original linear program.*

Proof: Since $\tilde{x}$ is an optimal solution for the Big M problem, then

$$c^T x^* + M\mathbf{1}^T x_a^* = c^T x^* \leq c^T x + M\mathbf{1}^T x_a$$

for any feasible solution $x' = \begin{bmatrix} x \\ x_a \end{bmatrix}$ for the Big M problem. In particular the inequality holds for $x' = \begin{bmatrix} x \\ 0 \end{bmatrix}$ and so $c^T x^* \leq c^T x + M\mathbf{1}^T 0 = c^T x$ for any x such that $Ax = b$ and $b \geq 0$. ■

Subcase (2):
If $x_a^* \neq 0$, then the original linear program is infeasible.
Proof: Exercise 3.21.

Case (2) The Big M problem is unbounded below.

Subcase (1):
If $x_a = 0$, then the original linear program is also unbounded below.

Proof: The objective of the Big M problem $c^T x + M\mathbf{1}^T x_a$ is diverging toward $-\infty$, but the second term in the Big M objective is $M\mathbf{1}^T 0 = 0$ since $x_a = 0$ and so the diverging Big M problem simplifies to minimize $c^T x$ subject to $Ax = b$ ■

Subcase (2):
If at least one artificial variable is non-zero, then the original problem is infeasible.

Proof: Proof is left to the reader.

3.4 Degeneracy and Cycling

We have seen in Chapter 2 that it is possible for basic feasible solutions to have some basic variables with zero values, i.e., these basic feasible solutions are degenerate. Let x_d be a degenerate basic feasible solution, then in the simplex method a descent direction d^q and a step length α is generated so that $x_{new} = x_d + \alpha d^q$ is feasible, i.e.,

$$\begin{aligned} Ax_{new} &= b \\ x_{new} &= x_d + \alpha d^q \geq 0. \end{aligned}$$

It is possible that since some of the basic variables in x_d are zero, then the minimum ratio test may set α to 0. In this case, x_{new} will represent a basic feasible solution that is distinct from and adjacent to x_d. However, the extreme points corresponding to both x_d and x_{new} are the same. Also, the objective function values $c^T x_d = c^T x_{new} = c^T(x_d + \alpha d^q) = c^T x_d$, so there is no improvement in the objective function value. Note that if all basic feasible solutions are non-degenerate, then the simplex method generates a strictly monotonically improving sequence of basic feasible solutions.

Cycling

Thus, in the presence of degeneracy the simplex method can proceed with the iterations with $\alpha = 0$ and in practice degeneracy is quite prevalent. The concern is that there might be a return to a basic feasible solution that was visited in the previous iterations. When this occurs, the simplex method may not terminate since it may be trapped in a cycle of iterations starting from a degenerate basic feasible solution and returning back to it and repeating the cycle over and over again. Cycling is very rare in practice, but is possible as the following example from Beale (1955) illustrates.

Example 3.14

Consider the linear program

$$\begin{array}{ll} \textit{minimize} & -\frac{3}{4}x_4 + 20x_5 - \frac{1}{2}x_6 + 6x_7 \\ \textit{subject to} & x_1 + \ \frac{1}{4}x_4 - 8x_5 \ -x_6 + 9x_7 = 0 \\ & x_2 + \frac{1}{2}x_4 \ -12x_5 - \frac{1}{2}x_6 + 3x_7 = 0 \\ & x_3 + x_6 = 1 \\ & x_1 \geq 0, x_2 \geq 0, x_3 \geq 0, x_4 \geq 0, x_5 \geq 0, x_6 \geq 0, x_7 \geq 0. \end{array}$$

Suppose the simplex method is used to solve the linear program above where

(1) the rule for entering a non-basic variable into the basis is based on the variable with the most negative reduced cost r_q;

(2) if there is a tie in the minimum ratio test, then select as the leaving variable the one with the smallest subscript.

Using as an initial basis B_0 the columns associated with x_1, x_2, and x_3, the following summary of the iterations is obtained:

Table 3.1 Example of cycling

Iteration	Entering	Leaving	Basic variables	Obj value
0			$x_1 = 0,\ x_2 = 0,\ x_3 = 1$	0
1	x_4	x_1	$x_4 = 0,\ x_2 = 0,\ x_3 = 1$	0
2	x_5	x_2	$x_4 = 0,\ x_5 = 0,\ x_3 = 1$	0
3	x_6	x_4	$x_6 = 0,\ x_5 = 0,\ x_3 = 1$	0
4	x_7	x_5	$x_6 = 0,\ x_7 = 0,\ x_3 = 1$	0
5	x_1	x_6	$x_1 = 0,\ x_7 = 0,\ x_3 = 1$	0
6	x_2	x_7	$x_1 = 0,\ x_2 = 0,\ x_3 = 1$	0

At the end of iteration 6, the simplex method revisits the initial basis B_0. Thus, the simplex method will cycle indefinitely and never terminate in this case if the entering and leaving variable rules remain as in (1) and (2). The simplex method is stuck at an extreme point whose objective function value is 0. In general, when degenerate basic feasible solutions are successively generated, the objective function value does not improve over these iterations and all of the basic feasible solutions generated in a cycle are degenerate.

3.4.1 Anti-Cycling Rules (Bland's Rule and the Lexicographic Method)

Cycling can be prevented by carefully choosing the entering and leaving variables during the simplex method.

3.4.1.1 Bland's Rule

The first method we present is due to Bland (1977) and assumes that the variables are ordered in some sequence, e.g., $x_1, x_2, x_3, ..., x_n$.

Bland's Rule

(1) For non-basic variables with negative reduced costs, select the variable with the smallest index to enter the basis.

(2) If there is a tie in the minimum ratio test, select the variable with the smallest index to leave the basis.

If Bland's rule is implemented in the simplex method, then it maintains the following property.

Lemma 3.15

If a non-basic variable x_q enters the basis, then it will not leave the basis until another variable x_k that was non-basic when x_q entered and has a larger index, i.e., $k > q$ enters the basis.

Proof: Exercise 3.17.

Theorem 3.16

If the simplex method uses Bland's rule, then the simplex method will not cycle.

Proof: Suppose that the simplex method cycles. Then, any non-basic variable that enters during the cycle into the basis must also eventually leave the basis. Since there are a finite number of variables, there is a largest index l among those variables that enter and leave the basis. But by Lemma 3.15 above, when variable x_l leaves, there must be an entering variable x_k which was non-basic when x_l entered with index $k > l$, which is a contradiction. ■

Example 3.17

Consider the linear program from Beale (1955) in Example 3.14. If Bland's rule is used where the starting basis consists of the columns associated with x_1, x_2, and x_3, then the simplex method terminates after 6 iterations with an optimal basic feasible solution where the basic variables are $x_6 = 1, x_1 = 3/4$, and $x_4 = 1$ with optimal objective function value $= -5/4$.

Table 3.2 Bland's rule

Iteration	Entering	Leaving	Basic variables	Obj value
0			$x_1 = 0$, $x_2 = 0$, $x_3 = 1$	0
1	x_4	x_1	$x_4 = 0$, $x_2 = 0$, $x_3 = 1$	0
2	x_5	x_2	$x_4 = 0$, $x_5 = 0$, $x_3 = 1$	0
3	x_6	x_4	$x_6 = 0$, $x_5 = 0$, $x_3 = 1$	0
4	x_1	x_5	$x_6 = 0$, $x_1 = 0$, $x_3 = 1$	0
5	x_2	x_3	$x_6 = 1$, $x_1 = 1$, $x_2 = 1/2$	-1/2
6	x_4	x_2	$x_6 = 1$, $x_1 = 3/4$, $x_4 = 1$	-5/4

The first three iterations are the same as in Example 3.14. Starting from iteration 4, the iterations are different when using Bland's rule. At the end of iteration 3, the basis B corresponds to the columns associated with x_6, x_5, and x_3, i.e.,

$$B = \begin{bmatrix} -1 & -8 & 0 \\ -1/2 & -12 & 0 \\ 1 & 0 & 1 \end{bmatrix},$$

and the non-basis matrix corresponds to columns associated with non-basic variables x_1, x_2, x_4, and x_7, i.e.,

$$N = \begin{bmatrix} 1 & 0 & 1/2 & 9 \\ 0 & 1 & 1/2 & 3 \\ 0 & 0 & 0 & 0 \end{bmatrix}.$$

Thus, reduced costs are

$$r_1 = c_1 - c_B^T B^{-1} N_1 = 0 - (-1/2, 20, 0) \begin{bmatrix} -1 & -8 & 0 \\ -1/2 & -12 & 0 \\ 1 & 0 & 1 \end{bmatrix}^{-1} \begin{bmatrix} 1 \\ 0 \\ 0 \end{bmatrix} = -2,$$

$$r_2 = c_2 - c_B^T B^{-1} N_2 = 0 - (-1/2, 20, 0) \begin{bmatrix} -1 & -8 & 0 \\ -1/2 & -12 & 0 \\ 1 & 0 & 1 \end{bmatrix}^{-1} \begin{bmatrix} 0 \\ 1 \\ 0 \end{bmatrix} = 3,$$

$$r_4 = c_4 - c_B^T B^{-1} N_4 =$$
$$-3/4 - (-1/2, 20, 0) \begin{bmatrix} -1 & -8 & 0 \\ -1/2 & -12 & 0 \\ 1 & 0 & 1 \end{bmatrix}^{-1} \begin{bmatrix} 1/2 \\ 1/2 \\ 0 \end{bmatrix} = -1/4,$$

and

$$r_7 = c_7 - c_B^T B^{-1} N_4 = 6 - (-1/2, 20, 0) \begin{bmatrix} -1 & -8 & 0 \\ -1/2 & -12 & 0 \\ 1 & 0 & 1 \end{bmatrix}^{-1} \begin{bmatrix} 9 \\ 3 \\ 0 \end{bmatrix} = -3.$$

Bland's rule has variable x_1 entering since it has the smallest index among the variables x_1, x_4, and x_7 with negative reduced cost. Note that in Example 3.14, variable x_7 was selected instead since it has the reduced cost that is most negative.

With the selection of x_1 as the entering variable, the leaving variable is determined by the minimum ratio test $\alpha = \min_{j \in \bar{B} = \{6,5,3\}} \{-\frac{x_j^{current}}{d_j^1} | d_j^1 < 0\} = \min\{-\frac{x_5^{current}}{d_5^1} = 0, -\frac{x_3^{current}}{d_3^1} = 2/3\} = 0$ since

$$d_j^1 = \begin{bmatrix} -B^{-1}N_1 \\ e_1 \end{bmatrix} = \begin{bmatrix} -\begin{bmatrix} -1 & -8 & 0 \\ -1/2 & -12 & 0 \\ 1 & 0 & 1 \end{bmatrix}^{-1} \begin{bmatrix} 1 \\ 0 \\ 0 \end{bmatrix} \\ 1 \\ 0 \\ 0 \\ 0 \end{bmatrix} = \begin{bmatrix} 1.5000 \\ -0.0625 \\ -1.5000 \\ 1 \\ 0 \\ 0 \\ 0 \end{bmatrix}.$$

Thus, variable x_5 leaves the basis. In iteration 5 variable x_2 enters and variables x_3 leaves, and finally in the last iteration, variable x_4 enters and variable x_2 leaves.

3.4.1.2 Lexicographic (Perturbation) Method

Another method for dealing with cycling is called the lexicographic method by Dantzig, Orden, and Wolfe (1955) and is based on the perturbation strategy of Orden (1956) and Charnes (1952). The motivation of the method is to attempt to eliminate degeneracy since cycling can only occur in the presence of degeneracy. To combat degeneracy the idea is to perturb the right-hand side values of a linear program by positive constants so that no basic variable can attain the value of zero during the iterations of the simplex method.

For example, consider the linear program and assume the rows are linearly independent

$$\begin{array}{ll} \textit{minimize} & c^T x \\ \textit{subject to} & a_1^T x = b_1 \\ & a_2^T x = b_2 \\ & a_3^T x = b_3 \\ & x \geq 0, \end{array}$$

then a positive constant ε_i is added to the right-hand side of *ith* constraint to get the following perturbed version of the linear program

$$\begin{array}{ll} \textit{minimize} & c^T x \\ \textit{subject to} & a_1^T x = b_1 + \varepsilon_1 \\ & a_2^T x = b_2 + \quad \varepsilon_2 \\ & a_3^T x = b_3 + \quad\quad \varepsilon_3 \\ & x \geq 0. \end{array}$$

The constants must be such that each successive constant is "sufficiently" smaller than the previous constant where this requirement is denoted as

$$0 \ll \varepsilon_m \ll \cdots \ll \varepsilon_2 \ll \varepsilon_1.$$

One such possibility is to let $\varepsilon_1 = \varepsilon$,$\varepsilon_2 = \varepsilon_1^2, ..., \varepsilon_m = \varepsilon_{m-1}^2$ where ε is a small positive number. Adding such constants to the right-hand side will ensure that there is a unique variable to leave the basis during an iteration. In fact, it is only necessary to perturb the right-hand side of the constraints when there is a tie in the minimum ratio test.

When it comes to selecting an entering variable, if there is more than one variable with negative reduced costs, then selecting any one will suffice.

We illustrate how the perturbation method eliminates degeneracy in the following linear program

$$\begin{array}{lllll} \textit{minimize} & -x_1 - x_2 & & & \\ \textit{subject to} & 2x_1 + x_2 + x_3 & & & = 30 \\ & x_1 + x_2 & +x_4 & & = 20 \\ & x_1 & & +x_5 & = 15 \\ & \multicolumn{4}{l}{x_1 \geq 0, x_2 \geq 0, x_3 \geq 0, x_4 \geq 0, x_5 \geq 0.} \end{array}$$

Consider the initial iteration without perturbing the right-hand sides. Let the columns associated with the slack variables x_3, x_4, x_5 constitute an initial basis and then, $x_B = B^{-1}b = \begin{bmatrix} x_3 \\ x_4 \\ x_5 \end{bmatrix} = \begin{bmatrix} 30 \\ 20 \\ 15 \end{bmatrix}$. Then, the reduced costs are $r_1 = -1$ and $r_2 = -1$. Suppose we select x_1 to enter the basis, then

$$d^1 = \begin{bmatrix} -B^{-1}N_1 \\ e_1 \end{bmatrix} = \begin{bmatrix} -\begin{bmatrix} 1 & 0 & 0 \\ 0 & 1 & 0 \\ 0 & 0 & 1 \end{bmatrix} \begin{bmatrix} 2 \\ 1 \\ 1 \end{bmatrix} \\ 1 \\ 0 \end{bmatrix} = \begin{bmatrix} -2 \\ -1 \\ -1 \\ 1 \\ 0 \end{bmatrix}.$$

Thus, by the minimum ratio test $\alpha = \min\{\frac{30}{2}, \frac{20}{1}, \frac{15}{1}\} = 15$ and either x_3 or x_5 can leave. Suppose x_5 leaves the basis.

Then, the next basic feasible solution is

$$x_1 = \begin{bmatrix} 30 \\ 20 \\ 15 \\ 0 \\ 0 \end{bmatrix} + (15) \begin{bmatrix} -2 \\ -1 \\ -1 \\ 1 \\ 0 \end{bmatrix} = \begin{bmatrix} 0 \\ 5 \\ 0 \\ 15 \\ 0 \end{bmatrix},$$

which is degenerate where the basic variables are now $x_1 = 15, x_3 = 0$, and $x_4 = 5$.

Now consider where right-hand side values of linear program are perturbed so that we get

$$\begin{array}{lll} \textit{minimize} & -x_1 - x_2 & \\ \textit{subject to} & 2x_1 + x_2 + x_3 & = 30 + \varepsilon_1 \\ & x_1 + x_2 \quad\quad +x_4 & = 20 + \varepsilon_2 \\ & x_1 \quad\quad\quad\quad\quad +x_5 & = 15 + \varepsilon_3 \\ & x_1 \geq 0, x_2 \geq 0, x_3 \geq 0, x_4 \geq 0, x_5 \geq 0. & \end{array}$$

Then, using the same initial set of basic variables we get

$$x_B = B^{-1}b = \begin{bmatrix} x_3 \\ x_4 \\ x_5 \end{bmatrix} = \begin{bmatrix} 30 + \varepsilon_1 \\ 20 + \varepsilon_2 \\ 15 + \varepsilon_3 \end{bmatrix}.$$

The reduced costs are the same as before, i.e., $r_1 = r_2 = -1$, so select x_1 again to enter the basis. Then, the direction vector d^1 is also the same as before. But now, in the minimum ratio test, we get

$$\alpha = \min\{\tfrac{30+\varepsilon_1}{2}, \tfrac{20+\varepsilon_2}{1}, \tfrac{15+\varepsilon_3}{1}\} = 15 + \varepsilon_3 \text{ (why?)}$$

Thus, the leaving variable is uniquely determined to be x_5. Then,

$$x_1 = \begin{bmatrix} 30 + \varepsilon_1 \\ 20 + \varepsilon_2 \\ 15 + \varepsilon_3 \\ 0 \\ 0 \end{bmatrix} + (15 + \varepsilon_3) \begin{bmatrix} -2 \\ -1 \\ -1 \\ 1 \\ 0 \end{bmatrix} = \begin{bmatrix} \varepsilon_1 - 2\varepsilon_3 \\ 5 + \varepsilon_2 - \varepsilon_3 \\ 0 \\ 15 + \varepsilon_3 \\ 0 \end{bmatrix}.$$

The basic variables are then

$$x_B = B^{-1}b = \begin{bmatrix} x_3 \\ x_4 \\ x_1 \end{bmatrix} = \begin{bmatrix} \varepsilon_1 - 2\varepsilon_3 \\ 5 + \varepsilon_2 - \varepsilon_3 \\ 15 + \varepsilon_3 \end{bmatrix}$$

where

$$B = \begin{bmatrix} 1 & 0 & 2 \\ 0 & 1 & 1 \\ 0 & 0 & 1 \end{bmatrix}, \text{ so the reduced costs are}$$

$$r_2 = -1 - (0,0,-1)\begin{bmatrix} 1 & 0 & 2 \\ 0 & 1 & 1 \\ 0 & 0 & 1 \end{bmatrix}^{-1}\begin{bmatrix} 1 \\ 1 \\ 0 \end{bmatrix} = -1$$

and $r_5 = 1$ (in general, if a variable left the basis in the previous iteration, then it will not enter the basis in the next iteration), so x_2 is selected to enter the basis. Then,

$$d^2 = \begin{bmatrix} -B^{-1}N_2 \\ e_2 \end{bmatrix} = \begin{bmatrix} -\begin{bmatrix} 1 & 0 & 2 \\ 0 & 1 & 1 \\ 0 & 0 & 1 \end{bmatrix}^{-1}\begin{bmatrix} 1 \\ 1 \\ 0 \end{bmatrix} \\ 0 \\ 1 \end{bmatrix} = \begin{bmatrix} -1 \\ -1 \\ 0 \\ 0 \\ 1 \end{bmatrix}.$$

The minimum ratio test gives

$$\alpha = \min\{\tfrac{\varepsilon_1 - 2\varepsilon_3}{1}, \tfrac{5+\varepsilon_2-\varepsilon_3}{1}\} = \varepsilon_1 - 2\varepsilon_3 \text{ (why?)},$$

so x_3 leaves the basis. Then,

$$x_2 = \begin{bmatrix} \varepsilon_1 - 2\varepsilon_3 \\ 5 + \varepsilon_2 - \varepsilon_3 \\ 15 + \varepsilon_3 \\ 0 \\ 0 \end{bmatrix} + (\varepsilon_1 - 2\varepsilon_3)\begin{bmatrix} -1 \\ -1 \\ 0 \\ 0 \\ 1 \end{bmatrix} = \begin{bmatrix} 0 \\ 5 - \varepsilon_1 + \varepsilon_2 + \varepsilon_3 \\ 15 + \varepsilon_3 \\ 0 \\ \varepsilon_1 - 2\varepsilon_3 \end{bmatrix} \text{ with}$$

$$x_B = B^{-1}b = \begin{bmatrix} x_2 \\ x_4 \\ x_1 \end{bmatrix} = \begin{bmatrix} \varepsilon_1 - 2\varepsilon_3 \\ 5 - \varepsilon_1 + \varepsilon_2 + \varepsilon_3 \\ 15 + \varepsilon_3 \end{bmatrix}.$$

Now $r_5 = -(-1,0,-1)\begin{bmatrix} 1 & 0 & 2 \\ 1 & 1 & 1 \\ 0 & 0 & 1 \end{bmatrix}^{-1}\begin{bmatrix} 0 \\ 0 \\ 1 \end{bmatrix} = -1$, so x_5 enters the basis, then

$$d^5 = \begin{bmatrix} -B^{-1}N_5 \\ e_5 \end{bmatrix} = \begin{bmatrix} -\begin{bmatrix} 1 & 0 & 2 \\ 1 & 1 & 1 \\ 0 & 0 & 1 \end{bmatrix}^{-1} \begin{bmatrix} 0 \\ 0 \\ 1 \end{bmatrix} \\ 1 \\ 0 \end{bmatrix} = \begin{bmatrix} 2 \\ -1 \\ -1 \\ 1 \\ 0 \end{bmatrix},$$

so $\alpha = \min\{\frac{5-\varepsilon_1+\varepsilon_2+\varepsilon_3}{1}, \frac{15+\varepsilon_3}{1}\} = 5 - \varepsilon_1 + \varepsilon_2 + \varepsilon_3$. Thus, x_4 leaves the basis.

$$x_3 = \begin{bmatrix} \varepsilon_1 - 2\varepsilon_3 \\ 5 - \varepsilon_1 + \varepsilon_2 + \varepsilon_3 \\ 15 + \varepsilon_3 \\ 0 \\ 0 \end{bmatrix} + (5-\varepsilon_1+\varepsilon_2+\varepsilon_3) \begin{bmatrix} 2 \\ -1 \\ -1 \\ 1 \\ 0 \end{bmatrix} = \begin{bmatrix} 10 - \varepsilon_1 + 2\varepsilon_2 \\ 0 \\ 10 + \varepsilon_1 - \varepsilon_2 \\ 5 - \varepsilon_1 + \varepsilon_2 + \varepsilon_3 \\ 0 \end{bmatrix}$$

where

$$x_B = \begin{bmatrix} x_2 \\ x_5 \\ x_1 \end{bmatrix} = \begin{bmatrix} 10 - \varepsilon_1 + 2\varepsilon_2 \\ 5 - \varepsilon_1 + \varepsilon_2 + \varepsilon_3 \\ 10 + \varepsilon_1 - \varepsilon_2 \end{bmatrix}.$$

Now, $r_3 = 0$, and so we have an optimal solution for the perturbed problem. In the lexicographic method the terms with $\varepsilon_1, \varepsilon_2$, and ε_3 are dropped and then we get

$$x_B = \begin{bmatrix} x_2 \\ x_5 \\ x_1 \end{bmatrix} = \begin{bmatrix} 10 \\ 5 \\ 10 \end{bmatrix}$$

which is the optimal solution for the original (unperturbed) problem. It is observed that the iterations of the simplex method on the perturbed problem with the lexicographic method always generated non-degenerate basic feasible solutions and always produced a unique leaving variable, and in the final iteration, all terms with the perturbation constants are dropped to recover an optimal solution for the original problem. More generally, we have the following result.

Theorem 3.18

Assume that the constraint matrix of a linear program has full row rank m. If the leaving variable is determined by the lexicographic method, then the simplex method will always terminate.

Proof: It will suffice to show that degenerate basic feasible solutions can never be generated in the lexicographic method. The right-hand side vector of the perturbed linear program is initially of the form

$$\tilde{b} = b + \varepsilon$$

$$\text{where } b = \begin{bmatrix} b_1 \\ b_2 \\ \vdots \\ b_m \end{bmatrix} \text{ and } \varepsilon = \begin{bmatrix} \varepsilon_1 \\ \varepsilon_2 \\ \vdots \\ \varepsilon_m \end{bmatrix}.$$

Without loss of generality, assume that the initial basis B_0 is the $m \times m$ identity matrix and thus the initial basic variables take the form

$$x_{B_0} = B_0^{-1}(b + \varepsilon) = (b + \varepsilon),$$

where the ith basic variable is denoted $x_{B_0}^i$. Then, x_{B_0} can be written as

$$\begin{aligned} x_{B_0}^1 &= b_1 + \varepsilon_1 \\ x_{B_0}^2 &= b_2 + \varepsilon_2 \\ &\vdots \\ x_{B_0}^m &= b_m + \varepsilon_m, \end{aligned}$$

Any subsequent iteration produces basic variables of the form

$$\begin{aligned} x_B^1 &= \bar{b}_1 + q_{11}\varepsilon_1 + q_{12}\varepsilon_2 + \cdots + q_{1m}\varepsilon_m \\ x_B^2 &= \bar{b}_2 + q_{21}\varepsilon_1 + q_{22}\varepsilon_2 + \cdots + q_{2m}\varepsilon_m \\ &\vdots \\ x_B^m &= \bar{b}_m + q_{m1}\varepsilon_1 + q_{m2}\varepsilon_2 + \cdots + q_{mm}\varepsilon_m, \end{aligned}$$

where each basic variable can be seen to be the result of elementary row operations that are needed to generate the inverse of the current basis. Thus, each basic variable will possibly have one or more terms involving the original perturbation quantities $\varepsilon_1, \varepsilon_2, ..., \varepsilon_m$ as well as the quantities $\bar{b}_i$ which are the modified right- hand side quantities that arise as a result of the elementary row operations.

Consider the original linear system $B_0 x_{B_0} = b + \varepsilon$, from which x_{B_0} is obtained; then every subsequent iteration will solve the corresponding linear system to generate the current basic feasible solution and the system will have rank m since the original linear system has rank m. Therefore, for each basic variable x_B^i, at least one of $\bar{b}_i, q_{i1}$, q_{i2},or q_{im} will be non-zero. Furthermore, the basic variables will be non-negative since $0 \ll \varepsilon_m \ll \cdots \ll \varepsilon_2 \ll \varepsilon_1$, which means that a degenerate basic solution is impossible. ■

3.5 Revised Simplex Method

The simplex method requires, at each iteration, the availability of the inverse B^{-1} of the basis matrix B at several steps. In particular, B^{-1} is required in computing the reduced costs $r_N = c_N - c_B^T B^{-1} N$ and the search direction $d^q = \begin{bmatrix} -B^{-1}N_q \\ e_q \end{bmatrix}$. Explicitly forming the inverse of a non-singular matrix is well known to be undesirable due to numerical stability issues that arise from round-off and truncation errors; see Golub and Van Loan (1989). A better idea is to generate an inverse B^{-1} by solving an equivalent linear system of equations.

For example, to compute the reduced costs, the linear system

$$B^T \pi = c_B$$

is solved first for π, and then the vector $r_N = c_N - \pi^T N$ can be easily computed. To compute the search direction, the linear system

$$Bd = -N_q$$

is first solved for d and then the vector $d^q = \begin{bmatrix} d \\ e_q \end{bmatrix}$ is easily computed.

3.5.1 Detailed Steps of the Revised Simplex Method

One can modify the simplex method to accommodate these changes. The resulting method is called the revised simplex method and is summarized below.

Revised Simplex Method

Step 0: (Initialization)

Generate an initial basic feasible solution $x^{(0)} = \begin{bmatrix} x_B \\ x_N \end{bmatrix}$.

Let B be the basis matrix and N the non-basis matrix with corresponding partition of the cost vector $c = (c_B, c_N)^T$.

Let $\bar{B}$ and $\bar{N}$ be the index sets of the basic and non-basic variables.

Let $k = 0$ and go to Step 1.

Step 1: (Optimality Check)

Solve for π in the linear system $B^T \pi = c_B$.

Then, compute the reduced costs $r_q = c_q - \pi^T N_q$ for all $q \in \bar{N}$.

If $r_q \geq 0$ for all $q \in \bar{N}$, then $x^{(k)}$ is an optimal solution for the linear program STOP,

else select one x_q non-basic such that $r_q < 0$ and go to Step 2.

Step 2: (Descent Direction Generation)

Solve for d in the linear system $Bd = -N_q$.

Then, compute $d^q = \begin{bmatrix} -B^{-1}N_q \\ e_q \end{bmatrix}$.

If $d^q \geq 0$, then the linear program is unbounded STOP, else go to Step 3.

Step 3: (Step Length Generation)

Compute the step length $\alpha = \min_{j \in \bar{B}}\{-\frac{x_j^{current}}{d_j^q} | d_j^q < 0\}$ (the minimum ratio test). Let j^* be the index of the basic variable that attains the minimum ratio α. Go to Step 4.

Step 4: (Improved Adjacent Basic Feasible Solution Computation)

Now let $x^{(k+1)} = x^{(k)} + \alpha d^q$. Go to Step 5.

Step 5: (Basis Update)

Let B_{j^*} be the column in B associated with the leaving basic variable x_{j^*} Update the basis matrix B by removing B_{j^*} and adding the column N_q, thus

$$\bar{B} = \bar{B} - \{j^*\} \cup \{q\}.$$

Update the non-basis matrix N by the removing N_q and adding B_{j^*}, thus

$$\bar{N} = \bar{N} - \{q\} \cup \{j^*\}.$$

Let $k = k + 1$ and go to Step 1.

Example 3.19

Consider the revised simplex method for the linear program in Example 3.7

$$\begin{array}{ll} \textit{maximize} & -2x_1 + x_2 \\ \textit{subject to} & -x_1 + \quad x_2 \quad \leq 4 \\ & 2x_1 \quad + x_2 + \quad \leq 6 \\ & x_1 \geq 0, x_2 \geq 0. \end{array}$$

As before, we first convert the objective to a minimization problem, so the objective is now maximize $-2x_1 + x_2 = -$minimize $2x_1 - x_2$.

We start (Step 0) the simplex method with the basic feasible solution

$$x^{(0)} = \begin{bmatrix} x_B \\ x_N \end{bmatrix} = \begin{bmatrix} x_3 \\ x_4 \\ x_1 \\ x_2 \end{bmatrix} = \begin{bmatrix} 4 \\ 6 \\ 0 \\ 0 \end{bmatrix}$$ with $B = \begin{bmatrix} 1 & 0 \\ 0 & 1 \end{bmatrix}$ and $N = \begin{bmatrix} -1 & 1 \\ 2 & 1 \end{bmatrix}$ and

$c_B = \begin{bmatrix} c_3 \\ c_4 \end{bmatrix} = \begin{bmatrix} 0 \\ 0 \end{bmatrix}$, $c_N = \begin{bmatrix} c_1 \\ c_2 \end{bmatrix} = \begin{bmatrix} 2 \\ -1 \end{bmatrix}$, $\bar{B} = \{3, 4\}$, and $\bar{N} = \{1, 2\}$.

First Iteration

Step 1: Check the optimality of $x^{(0)}$ by computing the reduced costs of the non-basic variables r_1 and r_2.

First solve the linear system $B^T\pi = c_B$, i.e.,

$$\begin{bmatrix} 1 & 0 \\ 0 & 1 \end{bmatrix}^T \begin{bmatrix} \pi_1 \\ \pi_2 \end{bmatrix} = \begin{bmatrix} 0 \\ 0 \end{bmatrix} \text{ and so } \pi = \begin{bmatrix} \pi_1 \\ \pi_2 \end{bmatrix} = \begin{bmatrix} 0 \\ 0 \end{bmatrix}.$$

Now

$$r_1 = c_1 - \pi^T N_1 = 2 - (0,0)\begin{bmatrix} -1 \\ 2 \end{bmatrix} = 2 \geq 0$$

and

$$r_2 = c_2 - \pi^T N_2 = -1 - (0,0)\begin{bmatrix} 1 \\ 1 \end{bmatrix} = -1 < 0\,.$$

Thus, $x^{(0)}$ is not optimal and x_2 must be selected as the non-basic variable to enter the basis. Go to Step 2.

Step 2: Solve the system $Bd = -N_q$, i.e.,

$$\begin{bmatrix} 1 & 0 \\ 0 & 1 \end{bmatrix}\begin{bmatrix} d_1 \\ d_2 \end{bmatrix} = -\begin{bmatrix} 1 \\ 1 \end{bmatrix} \text{ so } d = \begin{bmatrix} -1 \\ -1 \end{bmatrix}.$$

Construct $d^2 = \begin{bmatrix} d \\ e_2 \end{bmatrix} = \begin{bmatrix} -\begin{bmatrix} 1 & 0 \\ 0 & 1 \end{bmatrix}^{-1}\begin{bmatrix} 1 \\ 1 \end{bmatrix} \\ 0 \\ 1 \end{bmatrix} = \begin{bmatrix} -1 \\ -1 \\ 0 \\ 1 \end{bmatrix}.$

Since $d_2 \ngeq 0$, the linear program cannot be determined to be unbounded at this point. Go to Step 3.

Step 3: Compute the step length $\alpha = \min\limits_{j \in \bar{B}=\{3,4\}} \{-\frac{x_j^{current}}{d_j^2} | d_j^2 < 0\}$

$= \min\{-\frac{x_3^{current}}{d_3^2}, -\frac{x_4^{current}}{d_4^2}\} = \{-\frac{4}{-1}, -\frac{6}{-1}\} = 4$. Go to Step 4.

Step 4: So $x^{(1)} = x^{(0)} + \alpha d^1 = \begin{bmatrix} 4 \\ 6 \\ 0 \\ 0 \end{bmatrix} + (4)\begin{bmatrix} -1 \\ -1 \\ 0 \\ 1 \end{bmatrix} = \begin{bmatrix} 0 \\ 2 \\ 0 \\ 4 \end{bmatrix}.$

Observe that the variable x_3 leaves the basis (i.e., becomes non-basic). Go to Step 5.

Step 5: For $x^{(1)}$, the basic variables are $x_B = \begin{bmatrix} x_2 \\ x_4 \end{bmatrix} = \begin{bmatrix} 4 \\ 2 \end{bmatrix}$ and the non-basic variables are $x_N = \begin{bmatrix} x_1 \\ x_3 \end{bmatrix} = \begin{bmatrix} 0 \\ 0 \end{bmatrix}$.

The updated basis matrix $B = \begin{bmatrix} 1 & 0 \\ 1 & 1 \end{bmatrix}$, the updated non-basis matrix

$$N = \begin{bmatrix} -1 & 1 \\ 2 & 0 \end{bmatrix}, c_B = \begin{bmatrix} c_2 \\ c_4 \end{bmatrix} = \begin{bmatrix} -1 \\ 0 \end{bmatrix}, c_N = \begin{bmatrix} c_1 \\ c_3 \end{bmatrix} = \begin{bmatrix} 0 \\ -1 \end{bmatrix},$$

$\bar{B} = \{2, 4\}$, and $\bar{N} = \{1, 3\}$. Go to Step 1.

Second Iteration

Step 1: Check the optimality of $x^{(1)}$ by computing the reduced costs of the non-basic variables r_1 and r_3.

Solve $B^T \pi = c_B$, i.e., $\begin{bmatrix} 1 & 1 \\ 0 & 1 \end{bmatrix} \begin{bmatrix} \pi_1 \\ \pi_2 \end{bmatrix} = \begin{bmatrix} -1 \\ 0 \end{bmatrix}$ so $\pi = \begin{bmatrix} -1 \\ 0 \end{bmatrix}$

Now

$$r_1 = c_1 - \pi^T N_1 = 2 - (-1, 0) \begin{bmatrix} -1 \\ 2 \end{bmatrix} = 1 \geq 0$$

and

$$r_3 = c_3 - c_B^T B^{-1} N_3 = 0 - (-1, 0)^T \begin{bmatrix} 1 \\ 0 \end{bmatrix} = 1 \geq 0.$$

Thus, $x^{(1)} = \begin{bmatrix} x_B \\ x_N \end{bmatrix} = \begin{bmatrix} x_2 \\ x_4 \\ x_1 \\ x_3 \end{bmatrix} = \begin{bmatrix} 4 \\ 2 \\ 0 \\ 0 \end{bmatrix}$ is an optimal solution.

3.5.2 Advantages of the Revised Simplex Method

In addition to the benefits of not explicitly computing the inverse of the basis matrix B, the revised simplex method often requires substantially less computer memory when the number of variables n is much larger than the number of constraints m and when the matrix B is large and sparse. Further gains in performance can be had by an appropriate choice to solve the linear systems in involving B. In particular, state-of-the-art approaches employ the use of a triangular factorization, such as LU decomposition to solve the square system and can greatly enhance both the numerical stability of the linear system and reduce space requirements, especially when B is large and sparse.

An LU decomposition of a basis matrix B provides a square lower triangular matrix L and a square upper triangular matrix U, so that B can be represented as the product of L and U, i.e., $B = LU$. A lower triangular matrix contains zeros in the entries above the diagonal and an upper triangular matrix contains zeros in entries below the diagonal. Such a factorization can be exploited to solve for π in the linear system $B^T\pi = c_B$. In particular, we have

$$B^T\pi = (LU)^T\pi = U^TL^T\pi = c_B$$

The strategy is to let $L^T\pi = y$ and then solve first for y in $U^Ty = c_B$ and then solve for π in the system $L^T\pi = y$. Each of these two systems can be solved easily and rapidly by substitution and do not require more numerically intense elementary row operations as in Gaussian elimination.

Since the basis changes in each iteration of the revised simplex method by one column, there are advanced numerical linear algebra approaches that efficiently update the factorization so that new upper and lower triangular factors do not have to be generated from scratch in each iteration. In addition, numerical scaling and partial pivoting methods are used as well to improve numerical stability. The full implementation details are beyond the scope of this book and the reader is urged to see Bartels and Golub (1969) and Murtagh (1981).

3.6 Complexity of the Simplex Method

An important consideration of any algorithm is its complexity, i.e., the amount of time or resources that it requires to solve a problem. In this section, we consider the complexity of the simplex method. We seek an upper bound on the time or number of arithmetic operations that it would take the simplex method to solve any instance of a particular size in the worst case. We introduce some notation to help with the classification of the complexity of algorithms.

Definition 3.20
A function $g(n) = O(f(n))$ (or Big-O of $f(n)$) if there is constant $C > 0$ such that for sufficiently large n

$$g(n) \leq Cf(n).$$

Example 3.21
The polynomial $g(n) = 2n^2 + 3n + 4 = O(n^2)$ since $g(n) \leq 4f(n) = 4n^2$ for $n \geq 3$ where $f(n) = n^2$.

There are two basic classes of algorithms to solve a problem (in this case a linear program). Those that are efficient are said to have polynomial time worst-case complexity in that the number of operations required in the worst case to solve any instance of a linear program is a polynomial function of the size of the problem. In this case, we say the complexity of the algorithm is $O(f(n))$ where $f(n)$ is a polynomial (n denotes the size of the problem instance). The other type of algorithm is said to be exponential if the worst-case complexity grows exponentially in the size of the problem, e.g., $f(n) = K^n$ for some constant $K > 1$.

Klee-Minty Problems

Klee and Minty (1972) showed that there is a class of linear programs with $2m$ variables and m constraints that has a feasible basis for every possible selection of m variables out of the $2m$ variables. A variation of the problems is as follows:

$$\begin{array}{lll} \textit{minimize} & \sum\limits_{j=1}^{m} 10^{m-j} x_j & \\ \textit{subject to} & 2\sum\limits_{j=1}^{i-1} 10^{i-j} x_j + x_i \leq 100^{i-j} & i = 1, ..., m \\ & x_1 \geq 0, ..., x_m \geq 0. & \end{array}$$

By converting to standard form, the problems will have $2m$ variables and m constraints. Now, if the simplex method is applied to an instance from this class of problems where the non-basic variable with the most negative reduced cost is always selected as the entering variable at each iteration, then the simplex method will visit all bases. Therefore, as m increases the number of bases grows exponentially, as we have seen from Chapter 2 that there will be $\binom{2m}{m}$ feasible bases. This demonstrates that in the worst case the simplex method has complexity that grows exponentially in the size of the problem. To get a feeling for this growth rate, let $m = 50$ and the number of variables $n = 2m = 100$ (which is by any measure a very small linear program), then $\binom{100}{50} = 10^{29}$. If we suppose that a computer can do 1 billion iterations of the simplex method in one second, then it will take the computer more than 3 trillion years to finish!

However, fortunately in practice the simplex method performs very well and has been observed not to require more iterations than a small multiple of the number of constraints in most cases; see Hoffman (1953).

3.7 Simplex Method MATLAB Code

This section contains MATLAB code for the simplex method. In particular, the function

function [xsol objval exitflag]=SimplexMethod(c, Aeq, beq, B_set)

is created where it takes as arguments the parameters of a linear program that is in standard form where c is the objective coefficient vector, Aeq the matrix of coefficients of the equality constraints, beq the vector of right-hand sides of equality constraints, and B_set is a set that contains theindices of the current basic variables. The function assumes that a set of initial basic variables is readily available and is indicated through the set B_set.

Example 3.22

Consider the linear program (3.1)

$$\begin{array}{ll} \textit{minimize} & -x_1 - x_2 \\ \textit{subject to} & x_1 \leq 1 \\ & x_2 \leq 1 \\ & x_1 \geq 0, x_2 \geq 0. \end{array}$$

Slack variables x_3 and x_4 must be added to get the linear program to standard form. Then the following MATLAB statements create the parameters for the function SimplexMethod.

```
>> c=[-1; -1; 0; 0;];
>> Aeq=[1 0 1 0;
        0 1 0 1];
>> beq=[1;1];
>> B_set=[3; 4]; % the subscript of initial basic variables
```

then the function can be called by writing (with output following)

```
>>[xsol fval exitflag]=SimplexMethod(c, Aeq, beq, B_set)

probelm solved

xsol =
1
1
0
0

fval =

-2

exitflag =

0
```

3.7.1 MATLAB Code

Below is the MATLAB code of the function SimplexMethod. It is written to correspond to the steps of the simple method from Section 3.2.3.

```
function [xsol objval exitflag]=SimplexMethod(c, Aeq, beq, B_set)
% Simplex_Method solves a linear program in standard form:
%                         min  c'*x
%                         s.t.  Aeq*x = beq
%                                  x >= 0
% by using the simplex method of George B. Dantzig
%
% Inputs:
%   c = n*1 vector of objective coefficients
%   Aeq = m*n matrix with m < n
%   beq = m*1 vector of right hand side (RHS) coefficients
%   B_set = m*1 vector that contains indices (subscripts) of basic variables
%
% Parameter and Variable Partitions
%
% c, Aeq, and beq are partitioned according to partition of x into
%   x' =[x_B' | x_N'] where
%    x_B is an m*1 vector of basic variables
%    x_N is an (n-m)*1 vector of non-basic variables
%   c' = [c_B' | c_N']
%    c_B is the objective coefficients of x_B, an m*1 vector
%    c_N is the objective coefficients of x_N, an (n-m)*1 vector
%   Aeq = [B | N]
%    B is the m*m basis matrix
%    N is m*(n-m) non-basis matrix
%   set = [B_set' | N_set']
%    set is a set of indices (subscripts) of x
%    N_set is an (n-m)*1 vector of indices (subscripts) of non-basic variables
%
% Output:
%   xsol = n*1 vector, contains final solution of LP
%   objval is a scalar, final objective value of LP
%   iter is a struct that includes for every iteration the following details:
%   B_Set, N_set, c_B, x_B, r_N, step, and d where step is a step length
%   and d is a search direction.
%
%   exitflag describes the exit condition of the problem as follows:
%            0  - optimal solution
%            1  - unbounded problem

xsol=[]; objval=[]; exitflag=[];
%% Step 0: Initialization
%%%%%%%%%%%%%%%%%%%%%%%%%%%%%
% Generate an initial basic feasible solution and partition c, x, and Aeq
% so that c=[c_B | c_N] x=[x_B | x_N] Aeq=[B | N]
```

```
set=[1:length(c)]';
set(find(ismember(set, B_set)==1))=[];
N_set=set;
B=Aeq(:,B_set);               %basis matrix B
c_B=c(B_set);                 %obj coefficients of current basic variables
x_B=B\beq;                    %compute basic variables
N=Aeq(:,N_set);               %non-basis matrix N
c_N=c(N_set);                 %obj coefficients of current non-basic variables
x_N=zeros(length(N_set),1); %x_N, non-basic variables equal 0
x=[x_B; x_N];                 %partition x according to basis
obj=[c_B; c_N]'*x;            %initial objective function value
k=0;
while k>=0
    %% Step 1: Optimality Check
    %%%%%%%%%%%%%%%%%%%%%%%%%%%%%%
    % Compute the reduced costs r_q=c_q-c_B'*B^(-1)*N_q for q in N_set
    % if r_q >= 0, STOP, current solution optimal, else go to STEP 2
    pie=B'\c_B;              %solve the system B^T*pie=c_B for simplex multipliers
    r_N=c_N'-pie'*N;         % compute reduced cost for non-basic variables
    ratioflag=find(r_N<0);
    if isempty(ratioflag) %if r_q >= 0, then STOP. Optimal
        disp('probelm solved')
        exitflag=0;
        objval=obj;
        %subscripts of x are in ascending order
        set_temp=[B_set; N_set];
        for a=1:length(c)
            xsol(a,1)=x(find(set_temp==a));
        end
        break
    else % if r_q < 0, GO TO Step 2
        %% Step 2: Descent Direction Generation
        %%%%%%%%%%%%%%%%%%%%%%%%%%%%%%%%%%%%%%%%%%%%%%
        % Construct d_q=[-B^(-1)*N; e_q].
        % If d_q >= 0, then LP is unbounded, STOP, else go to STEP 3.
        enter=ratioflag(1);          %choosing entering variable
        e=zeros(length(N_set),1);
        e(enter)=1;                  %construct vector e_q
        d=-B\N(:,enter);             %solve the system Bd=-N_q
        direction=[d; e];            %improved direction d
        d_flag=find(direction < 0);
        if isempty(d_flag) %if direction > 0, then STOP.(unbounded)
            disp('unbounded problem')
            exitflag=1;
            break
```

```
        else %if d_q < 0, GO TO Step 3
            %% Step 3: Step Length Generation
            %%%%%%%%%%%%%%%%%%%%%%%%%%%%%%%%%%%%%%%%
            % Compute step length by the minimum ratio test. Go to STEP 4.
            step_set=-x(d_flag)./direction(d_flag);
            step=min(step_set);
            %% Step 4: Improved Solution Generation
            %%%%%%%%%%%%%%%%%%%%%%%%%%%%%%%%%%%%%%%%%%%%%%%
            % Let x(k+1) = x(k) + alpha*d_q. Go to Step 5.
            x_d=x+step*direction;
            leave_set=find(x_d(1:length(B_set))==0);
            leave=leave_set(1);                %determining leaving variable
            %% Step 5: Basis Update
            %%%%%%%%%%%%%%%%%%%%%%%%%%%%
            % Generate the new basis B for next iteration,
            % Update c=[c_B|c_N],x=[x_B|x_N],& Aeq=[B|N]. Go to STEP 1.
            B_set_temp=B_set;
            N_set_temp=N_set;
            x_B=x_d(1:length(B_set));
            x_B_temp=x_d(1:length(B_set));
            x_N_temp=x_d(length(B_set)+1:end);
            %%%%%%%%%%%%%%%%%%%%%%%%%%%%%%%%%%%%%%%%%%%%%%%%%%%%%
            %exchange the entering and leaving variables in B_set
            B_set(find(B_set_temp==B_set_temp(leave)))=N_set_temp(enter);
            N_set(find(N_set_temp==N_set_temp(enter)))=B_set_temp(leave);
            x_B(find(x_B_temp==x_B_temp(leave)))=x_N_temp(enter);
            B=Aeq(:,B_set);                 %update basis B
            c_B=c(B_set);                   %update c_B
            N=Aeq(:,N_set);                 %update non-basis N
            c_N=c(N_set);                   %update c_N
            x=[x_B; x_N];                   %update x = [x_B | x_N]
            obj=[c_B; c_N]'*x;              %new objective value
            k=k+1; %GO TO Step 1
        end
    end
end
```

3.8 Exercises

Exercise 3.1

Consider the following linear program:

$$\begin{array}{ll} \textit{maximize} & 2x_1 + 2x_2 + 3x_3 + x_4 + 4x_5 \\ \textit{subject to} & 3x_1 + 7x_2 + 2x_3 + 3x_4 + 2x_5 \leq 40 \\ & x_1 \geq 0, x_2 \geq 0, x_3 \geq 0, x_4 \geq 0, x_5 \geq 0. \end{array}$$

(a) Solve by using the simplex method.
(b) Find a simple rule that obtains the optimal solution.

Exercise 3.2
Consider the following linear program:

$$\begin{array}{ll} \textit{minimize} & -4x_1 - 3x_2 \\ \textit{subject to} & x_1 + 2x_2 \leq 8 \\ & -2x_1 + x_2 \leq 5 \\ & 5x_1 + 3x_2 \leq 16 \\ & x_1 \geq 0, x_2 \geq 0. \end{array}$$

(a) Solve the LP using the simplex method.
(b) Solve the LP using the revised simplex method.
(c) Solve the LP using the linprog function from MATLAB.

Exercise 3.3
Consider the following linear program:

$$\begin{array}{ll} \textit{minimize} & -4x_1 - 3x_2 - 2x_3 \\ \textit{subject to} & 2x_1 - 3x_2 + 2x_3 \leq 6 \\ & -x_1 + x_2 + x_3 \leq 5 \\ & x_1 \geq 0, x_2 \geq 0, x_3 \geq 0. \end{array}$$

(a) Solve using the simplex method.
(b) Solve using the revised simplex method.
(c) Solve using the SimplexMethod MATLAB function from Section 3.7.

Exercise 3.4
Consider the following linear program:

$$\begin{array}{ll} \textit{minimize} & -2x_1 - 5x_2 \\ \textit{subject to} & -x_1 + 3x_2 \leq 2 \\ & -3x_1 + 2x_2 \leq 1 \\ & x_1 \geq 0, x_2 \geq 0. \end{array}$$

(a) Graph the feasible set.
(b) Solve the LP using the simplex method.
(c) Is the LP bounded? If not, find the ray along which the LP is unbounded.

Exercise 3.5
Consider the following linear program:

$$\begin{array}{ll} \textit{minimize} & -x_2 \\ \textit{subject to} & x_1 - 2x_2 \leq 2 \\ & x_1 - x_2 \leq 3 \\ & \quad\quad x_2 \leq 3 \\ & x_1 \geq 0, x_2 \geq 0. \end{array}$$

(a) Solve the LP using the simplex method.

(b) Does the LP bounded have a unique optimal solution? If not, derive an expression for the set of optimal solutions.

(c) Solve the LP using the SimplexMethod MATLAB function from section 3.7.

Exercise 3.6

Consider the following linear program

$$\begin{array}{ll} \textit{maximize} & 2x_1 - 3x_2 + x_3 \\ \textit{subject to} & x_1 + 3x_2 + x_3 \leq 12 \\ & -x_1 + x_2 + 2x_3 \leq 6 \\ & -x_1 + 3x_2 \leq 9 \\ & x_1 \geq 0, x_2 \geq 0, x_3 \geq 0. \end{array}$$

Solve using the simplex method.

Exercise 3.7

Consider the following systems of linear inequalities

(a)

$$\begin{array}{l} x_1 + 3x_2 \leq 5 \\ -x_1 + x_2 \leq 1 \\ x_1 \geq 0, x_2 \geq 0. \end{array}$$

Find a feasible solution.

(b)

$$\begin{array}{l} x_1 + 2x_2 + x_3 \leq 10 \\ -2x_1 + 3x_2 + 2x_3 \geq 3 \\ x_1 \text{ unrestricted}, x_2 \geq 0, x_3 \geq 0. \end{array}$$

Find a basic feasible solution by solving a Phase I problem.

Exercise 3.8

Consider the following linear program:

$$\begin{array}{ll} \textit{minimize} & -x_1 - 2x_2 \\ \textit{subject to} & 2x_1 - x_2 \geq 5 \\ & -2x_1 + x_2 = 2 \\ & x_1 \geq 0, x_2 \geq 0. \end{array}$$

(a) Solve the LP using the Two-Phase method
(b) Solve the LP using the Big-M method.
(c) Solve the LP using the linprog function from MATLAB.

Exercise 3.9
Consider the following linear program:

$$\begin{array}{ll} minimize & x_1 - 2x_2 \\ subject\ to & x_1 + x_2 \geq 3 \\ & -x_1 + x_2 \geq 2 \\ & x_2 \leq 4 \\ & x_1 \geq 0, x_2 \geq 0. \end{array}$$

Solve the LP using the Two-Phase method.

Exercise 3.10
Consider the following linear program:

$$\begin{array}{ll} minimize & 2x_1 - 2x_2 + x_3 \\ subject\ to & x_1 + 3x_2 - x_3 \geq 5 \\ & -3x_1 - 2x_2 + x_3 \leq 4 \\ & x_1 \geq 0, x_2 \geq 0, x_3 \geq 0. \end{array}$$

(a) Solve the LP using the Two-Phase method.
(b) Solve the LP using the Big-M method.
(c) Solve the LP using the linprog function from MATLAB.

Exercise 3.11
Consider the following linear program:

$$\begin{array}{ll} minimize & 2x_1 + 3x_2 - x_3 \\ subject\ to & 2x_1 + x_2 + x_3 \geq 2 \\ & -x_1 + x_2 \geq 1 \\ & -x_1 + 5x_2 + x_3 \leq 3 \\ & x_1 \geq 0, x_2 \geq 0, x_3 \geq 0. \end{array}$$

Solve the LP using (a) the Two-Phase method and (b) the Big-M method.

Exercise 3.12
Finish the Big-M iterations in Example 3.13. Assume that M is a very large constant.

Exercise 3.13
Finish the Bland's rule in Example 3.17.

Exercise 3.14
Consider the following linear program:

$$\begin{array}{ll} \textit{minimize} & -x_1 - 2x_2 - x_3 \\ \textit{subject to} & .5x_1 + 2x_2 + 3x_3 \leq 2 \\ & -x_1 + x_2 + 4x_3 \leq 1 \\ & x_1 + 3x_2 + x_3 \leq 6 \\ & x_1 \geq 0, x_2 \geq 0, x_3 \geq 0. \end{array}$$

Solve the LP using the lexicographic method.

Exercise 3.15
Consider the following linear program:

$$\begin{array}{ll} \textit{minimize} & x_3 \\ \textit{subject to} & -x_1 - x_2 + \varepsilon x_3 = \varepsilon \\ & x_1 \geq 0, x_2 \geq 0, x_3 \geq 0. \end{array}$$

(a) Formulate the Big-M problem for this linear program.

(b) Consider the vectors $x^1 = (0, 0, 1, 0)$ and $x^2 = (0, 0, 0, \varepsilon)$. Are these both feasible for the Big-M problem? What about for the original problem?

(c) Explain by using parts (a) and (b) that it is not always possible to choose an arbitrarily large M constant for the Big-M problem for an arbitrary linear program.

Exercise 3.16
Prove that if the reduced costs of a basic feasible solution are all strictly positive, then the optimal solution is unique.

Exercise 3.17
Prove the monotone property of Bland's rule, i.e., Lemma 3.15.

Exercise 3.18
Prove that the only way the simplex method fails to terminate is by cycling.

Exercise 3.19
Consider the following linear program:

$$\begin{array}{ll} \textit{minimize} & c^T x \\ \textit{subject to} & Ax \leq b \\ & x \geq 0. \end{array}$$

An interior point of the feasible set is a point x^* such that $Ax^* < b$ and $x^* > 0$. Prove that an interior point x^* cannot be an optimal solution for this linear program.

Exercise 3.20
Consider the following linear program:

$$\begin{array}{ll} \textit{minimize} & c^T x \\ \textit{subject to} & Ax \geq b \\ & x \geq 0. \end{array}$$

(a) Convert the LP to standard form.

(b) Show that if $x = \begin{bmatrix} x_B \\ x_N \end{bmatrix}$ is an optimal basic feasible solution, then the dual values are always non-negative.

Exercise 3.21

Suppose the Big M model results in a finite optimal solution $\tilde{x} = \begin{bmatrix} x^* \\ x_a^* \end{bmatrix}$ and $x_a^* \neq 0$. Prove that the original linear program is infeasible.

Exercise 3.22

Modify the MATLAB code of the function SimplexMethod to implement the revised simplex method.

Notes and References

The simplex method was developed by Dantzig and his book (1963) details many of the early developments in the simplex method and extensions. The revised simplex method is also due to Dantzig (1953) and Orchard-Hays (1954). The computational complexity of the simplex method in the worst-case sense was determined with the advent of the Klee-Minty (1972) examples. Average case complexity analysis, where a probability distribution on the space of problem instances of the simplex method is used, was examined by Borgwardt (1982) and Smale (1983). Cycling and its relationship to degeneracy was discovered by Hoffman (1953). Bland's rule is due to Bland (1977). The lexicographic method is due to Dantzig, Orden, and Wolfe (1955) and is based on the perturbation technique of Charnes (1952).

4

Duality Theory

4.1 Introduction

Every linear program has associated with it another linear program called the dual. A given linear program and its dual will be related in important ways. For example, a feasible solution for one will provide a bound on the optimal objective function value of the other. Also, if one has an optimal solution, then the other will have an optimal solution as well and the objective function values of both will be the same. In particular, if one problem has an optimal solution, then a "certificate" of optimality can be obtained from the corresponding dual problem verifying the optimality.

The theory related to the relationship between a linear program and its dual is called duality theory, and has important consequences for optimization and is not only of theoretical, but of practical importance as well. This chapter will develop and explore the implications and economic interpretations of duality theory and its role in optimal algorithm design and sensitivity analysis. With the development of duality theory, a variant of the simplex method called the dual simplex method is developed which can enhance the computation of optimal solutions of linear programs that are modifications of an existing problem.

4.2 Motivation for Duality

Consider the following linear program, which is referred to as the canonical primal problem P

$$\begin{array}{lll} \textit{maximize} & 6x_1 + x_2 + 2x_3 & \\ \textit{subject to} & x_1 + 2x_2 - x_3 & \leq 20 \\ & 2x_1 - x_2 + 2x_3 & \leq 30 \\ & x_1 \geq 0, x_2 \geq 0, x_3 \geq 0, & \end{array}$$

which is of the following matrix form:

$$\begin{array}{ll} \textit{maximize} & c^T x \\ \textit{subject to} & Ax \leq b \\ & x \geq 0. \end{array}$$

The dual (D) of the above canonical primal linear program is

$$\begin{array}{lll} \textit{minimize} & 20\pi_1 + 30\pi_2 & \\ \textit{subject to} & \pi_1 + 2\pi_2 & \geq 6 \\ & 2\pi_1 - \pi_2 & \geq 1 \\ & -\pi_1 + 2\pi_2 & \geq 2 \\ & \pi_1 \geq 0, \pi_2 \geq 0, & \end{array}$$

which is of the form

$$\begin{array}{ll} \textit{minimize} & b^T \pi \\ \textit{subject to} & A^T \pi \geq c \\ & \pi \geq 0. \end{array}$$

We call problems P and D above a primal-dual pair of linear programs. Why does the dual problem take on the form above? One way to motivate this is to consider obtaining bounds on the objective function value for the primal problem.

Consider the following feasible solution for the primal problem P, $x_1 = 10, x_2 = 2$, and $x_3 = 2$. The corresponding objective function value is $z = 66$. Let $z^* =$ the optimal objective function value of the primal problem P which is currently not known, then clearly $z = 66 \leq z^*$, since the feasible solution may not be the optimal solution. In other words, the optimal objective function value of the primal linear program is greater than or equal to 66 and thus 66 is a lower bound on the optimal objective function value. In general, any feasible solution x for a maximization (minimization) linear program will provide a lower (upper) bound on the optimal objective function value, i.e., $c^T x \leq z^*$ ($c^T x \geq z^*$).

A very important observation is that it is possible to obtain upper bounds on the objective function value for P using the constraints of P. For example, consider multiplying the first and second constraints in P by 2 and then adding them together, i.e.,

$$\begin{array}{rll} & (2) * (x_1 + 2x_2 - x_3) & \leq (2) * 20 \\ + & (2) * (2x_1 - x_2 + 2x_3) & \leq (2) * 30 \\ \hline = & 6x_1 + 2x_2 + 2x_3 & \leq 100. \end{array}$$

Thus, we get that $6x_1 + 2x_2 + 2x_3 \leq 100$ for all x_1, x_2, and x_3 non-negative and this will hold for all feasible solutions x of P. Now $6x_1 + x_2 + 2x_3 \leq 6x_1 + 2x_2 + 2x_3$ where the left-hand side of the inequality is the objective function of P, and so the objective function value associated with any feasible solution will always be less than or equal to 100. In particular, for an optimal solution x^*, the objective value $c^T x^*$ is bounded above by 100, i.e., $c^T x^* \leq 100$.

Then, combining the lower and upper bounds it is deduced that $66 \leq z^* \leq 100$. This means that the optimal objective function value is somewhere between 66 and 100. If we can find a smaller upper bound and a larger lower bound, then we can bound the objective function value in a tighter manner and then the bounds would be more informative of what the optimal objective function value could be. Ideally, if we can find a lower bound that is equal to the upper bound then the optimal objective function value would be determined and the corresponding feasible solution would be an optimal solution.

To find the smallest possible upper bound on the objective function value we can generalize the idea above in taking multiples of the primal constraints and adding them together to form an expression involving the variables to bound the objective function value of the primal. Instead of selecting as a multiplier 2 for each of the constraints, we let these quantities be represented by the non-negative variables π_1 for the multiplier for the first primal constraint and π_2 for the multiplier for the second primal constraints. Then, we add the constraints after multiplying each of the constraints by their respective multipliers to get

$$\begin{array}{rll} & (\pi_1)(x_1 + 2x_2 - x_3) & \leq (\pi_1)20 \\ + & (\pi_2)(2x_1 - x_2 + 2x_3) & \leq (\pi_2)30 \\ \hline = & (\pi_1 + 2\pi_2)x_1 + (2\pi_1 - \pi_2)x_2 + (-\pi_1 + 2\pi_2)x_3 & \leq 20\pi_1 + 30\pi_2. \end{array}$$

In order for the quantity $(\pi_1 + 2\pi_2)x_1 + (2\pi_1 - \pi_2)x_2 + (-\pi_1 + 2\pi_2)x_3$ to be an upper bound for the objective function of P we require that

$$(\pi_1 + 2\pi_2) \geq 6, (2\pi_1 - \pi_2) \geq 1, \text{ and } (-\pi_1 + 2\pi_2) \geq 2. \quad (4.1)$$

Then, $6x_1+x_2+2x_3 \leq (\pi_1+2\pi_2)x_1+(2\pi_1-\pi_2)+(-\pi_1+2\pi_2) \leq 20\pi_1+30\pi_2$ for all $x_1, x_2, x_3 \geq 0$ and $\pi_1, \pi_2 \geq 0$.

We can reduce the upper bound on the objective function by choosing values for the multipliers $\pi_1 \geq 0$ and $\pi_2 \geq 0$ so that $20\pi_1 + 30\pi_2$ is as small as possible subject to the conditions in (4.1). But observe that this can be expressed directly as an optimization problem in the form of the dual problem of P. That is, the problem of finding the lowest possible bound for the objective function value of P is equivalent to its dual problem since the requirements on π_1 and π_2 embodied in (4.1) and the non-negativity requirements are exactly the constraints of the dual problem, and the goal is to select π_1 and π_2 to minimize $20\pi_1 + 30\pi_2$, which is the same objective in the dual problem D. Thus, we see that the dual problem is intimately related to the primal problem and not an ad hoc construction.

4.3 Forming the Dual Problem for General Linear Programs

For the canonical primal and dual problems, observe that the dual objective is a minimization, whereas for the primal it is a maximization problem. Both the primal and the dual use the same data c, A, and b, except that for the dual the vector b is the objective function coefficient vector, c is the right-hand side vector, and A^T is the matrix of constraint coefficients. Also, the constraints in D are inequalities of the ($\geq$) type where the primal constraints are of the $\leq$ type. Thus, where the primal has m constraints and n variables, the dual has n constraints and m variables. In fact, the *ith* variable in one problem corresponds to the *ith* constraint in the other problem.

However, most linear programs are not naturally expressed in a form where all constraints are uniformly of one type such as $\geq$ and where all variables are non-negative. But we know that any linear program can be converted into canonical form, for which the dual is known, by using the transformation rules from Chapter 1. It turns out that the dual of a linear program in any form can be obtained without the need to first transform the linear program into canonical form.

Example 4.1
Consider the following linear program as a primal problem

$$\begin{array}{lll} \textit{minimize} & c_1x_1 + c_2x_2 + c_3x_3 & \\ \textit{subject to} & a_{11}x_1 + a_{12}x_2 + a_{13}x_3 & \leq b_1 \\ & a_{21}x_1 + a_{22}x_2 + a_{23}x_3 & \geq b_2 \\ & a_{31}x_1 + a_{32}x_2 + a_{33}x_3 & = b_3 \\ & x_1 \geq 0, x_2 \text{ unrestricted}, x_3 \leq 0. & \end{array}$$

This problem can be converted into the canonical form by letting $x_2 = x_2^+ - x_2^-$ and $x_3 = -x_3^{'}$ and splitting the third constraint into inequalities and then multiplying all constraints of the $\leq$ type by -1 to get

$$\begin{array}{lll} \textit{minimize} & c_1x_1 + c_2x_2^+ - c_2x_2^- - c_3x_3^{'} & \\ \textit{subject to} & -a_{11}x_1 - a_{12}x_2^+ + a_{12}x_2^- + a_{13}x_3^{'} & \geq -b_1 \\ & a_{21}x_1 + a_{22}x_2^+ - a_{22}x_2^- - a_{23}x_3^{'} & \geq b_2 \\ & a_{31}x_1 + a_{32}x_2^+ - a_{32}x_2^- - a_{33}x_3^{'} & \geq b_3 \\ & -a_{31}x_1 - a_{32}x_2^+ + a_{32}x_2^- + a_{33}x_3^{'} & \geq -b_3 \\ & x_1 \geq 0, x_2^+ \geq 0, x_2^- \geq 0, x_3^{'} \geq 0. & \end{array}$$

Then by letting $\pi_1^{'} \geq 0$ be the dual variable corresponding to the first constraint above, $\pi_2 \geq 0$ be the dual variable corresponding to the second constraint above, and π_3^+ and π_3^- be the dual variables corresponding to the third and fourth constraints, the dual problem becomes

$$
\begin{array}{lll}
\textit{maximize} & -b_1\pi_1^{'} + b_2\pi_2 + b_3\pi_3^{+} - b_3\pi_3^{-} & \\
\textit{subject to} & -a_{11}\pi_1^{'} + a_{21}\pi_2 + a_{31}\pi_3^{+} - a_{31}\pi_3^{-} & \leq c_1 \\
& -a_{12}\pi_1^{'} + a_{22}\pi_2 + a_{32}\pi_3^{+} - a_{32}\pi_3^{-} & \leq c_2 \\
& a_{12}\pi_1^{'} - a_{22}\pi_2 - a_{32}\pi_3^{+} + a_{32}\pi_3^{-} & \leq -c_2 \\
& a_{13}\pi_1^{'} - a_{23}\pi_2 - a_{33}\pi_3^{+} + a_{33}\pi_3^{-} & \leq -c_3 \\
& \pi_1^{'} \geq 0, \pi_2 \geq 0, \pi_3^{+} \geq 0, \pi_3^{-} \geq 0, &
\end{array}
$$

which by setting $\pi_1 = -\pi_1^{'}$ and $\pi_3 = \pi_3^{+} - \pi_3^{-}$ is equivalent to

$$
\begin{array}{lll}
\textit{maximize} & b_1\pi_1 + b_2\pi_2 + b_3\pi_3 & \\
\textit{subject to} & a_{11}\pi_1 + a_{21}\pi_2 + a_{31}\pi_3 & \leq c_1 \\
& a_{12}\pi_1 + a_{22}\pi_2 + a_{32}\pi_3 & = c_2 \\
& a_{13}\pi_1 + a_{23}\pi_2 + a_{33}\pi_3 & \geq c_3 \\
& \pi_1 \leq 0, \pi_2 \geq 0, \pi_3 \text{ unrestricted.} &
\end{array}
$$

The significance of the above example is that it is possible to take the dual directly from a linear program in its original form. In particular, associated with a primal constraint of the type $\leq$ will be a dual variable $\pi \leq 0$. For a primal constraint of the $\geq$ type, the corresponding dual variable will be nonnegative $\pi \geq 0$ (as demonstrated before in the canonical case). Finally, for an equality constraint, the corresponding dual variable π is unrestricted.

The following tables give the relationship between a primal-dual pair of linear programs.

Table 4.1 Primal and dual data correspondence

Constraint matrix A	$\Longleftrightarrow$	Constraint matrix A^T
Cost vector c	$\Longleftrightarrow$	Cost vector b
Right-Hand Side vector b	$\Longleftrightarrow$	Right-Hand Side vector c

Table 4.2 Primal dual constraint variable correspondence

Maximization	$\Longleftrightarrow$	Minimization
variables		constraints
≥ 0	$\Longleftrightarrow$	$\geq$
≤ 0	$\Longleftrightarrow$	$\leq$
unrestricted	$\Longleftrightarrow$	$=$
constraints		variables
$\geq$	$\Longleftrightarrow$	≤ 0
$\leq$	$\Longleftrightarrow$	≥ 0
$=$	$\Longleftrightarrow$	unrestricted

Example 4.2

Find the dual of the linear program

$$\begin{array}{lll} \textit{maximize} & 5x_1 - 3x_2 + 2x_3 & \\ \textit{subject to} & x_1 + 4x_2 - 2x_3 & \geq 1 \\ & 2x_1 - x_2 + x_3 & = -5 \\ & x_1 + x_2 - x_3 & \leq 2 \\ & x_1 \leq 0, x_2 \geq 0, x_3 \text{unrestricted.} & \end{array}$$

Solution:
Now the primal data is such that

$$c = \begin{bmatrix} 5 \\ -3 \\ 2 \end{bmatrix}, b = \begin{bmatrix} 1 \\ -5 \\ 2 \end{bmatrix}, \text{ and}$$

$$A = \begin{bmatrix} 1 & 4 & -2 \\ 2 & -1 & 1 \\ 1 & 1 & -1 \end{bmatrix}.$$

Thus, the dual cost vector is b, the right-hand side vector is c, and the constraint coefficient matrix is

$$A^T = \begin{bmatrix} 1 & 2 & 1 \\ 4 & -1 & 1 \\ -2 & 1 & -1 \end{bmatrix}.$$

There are three dual variables since there are three primal constraints (excluding the restrictions, if any, on the primal variables). The first dual variable π_1 corresponds to the first primal constraint, which is of the $\geq$ type, so $\pi_1 \leq 0$, the second dual variable corresponds to the second primal constaint which is an equality, so π_2 is unrestricted, and the third dual variable π_3 corresponds to the last constraint, which is of the $\leq$ type, so $\pi_3 \geq 0$.

Now there are three dual constraints since there are three primal variables. The first dual constraint will be of type $\leq$ since the first primal variable $x_1 \leq 0$, the second dual constraint will be of type $\geq$ since the second primal variable $x_2 \geq 0$, and the third dual constraint will be an equality since the third primal variable x_3 is unrestricted. Thus, the dual is

$$\begin{array}{lll} \textit{minimize} & \pi_1 - 5\pi_2 + 2\pi_3 & \\ \textit{subject to} & \pi_1 + 2\pi_2 + \pi_3 & \leq 5 \\ & 4\pi_1 - \pi_2 + \pi_3 & \geq -3 \\ & -2\pi_1 + \pi_2 - \pi_3 & = 2 \\ & \pi_1 \leq 0, \pi_2 \text{ unrestricted}, \pi_3 \geq 0. & \end{array}$$

4.4 Weak and Strong Duality Theory

For the development of duality theory, we will assume that our primal problem (P) is of the form

$$\begin{array}{lll} \textit{maximize} & \sum_{i=1}^{n} c_i x_i & \\ \textit{subject to} & \sum_{j=1}^{n} a_{ij} x_j \leq b_i & i = 1, ..., m \\ & x_j \geq 0 & j = 1, ..., n, \end{array}$$

and so the corresponding dual problem (D) is

$$\begin{array}{lll} \textit{minimize} & \sum_{i=1}^{m} b_i \pi_i & \\ \textit{subject to} & \sum_{i=1}^{m} a_{ij} \pi_i \geq c_j & j = 1, ..., n \\ & \pi_j \geq 0 & j = 1, ..., m. \end{array}$$

Our first result shows that it does not matter which linear program in a primal-dual pair is called the primal.

Theorem 4.3
The dual of the dual of P is P.
Proof: The dual of problem P is

$$\begin{array}{lll} \textit{minimize} & \sum_{i=1}^{m} b_i \pi_i & \\ \textit{subject to} & \sum_{i=1}^{m} a_{ij} \pi_i \geq c_j & j = 1, ..., n \\ & \pi_i \geq 0 & i = 1, ..., m. \end{array}$$

The objective function can be converted to a maximization problem via maximizing the negation of the objective function, i.e., minimize $\sum_{i=1}^{m} b_i \pi_i =$ $-$maximizing $-\sum_{i=1}^{m} b_i \pi_i$, and the constraints can be converted to less than or equal to type by multiplying the constraints by -1. Then, the dual of P can be written as

$$\begin{array}{lll} -\textit{maximize} & \sum_{i=1}^{m} (-b_i) \pi_i & \\ \textit{subject to} & \sum_{i=1}^{m} (-a_{ij}) \pi_i \leq (-c_j) & j = 1, ..., n \\ & \pi_j \geq 0 & j = 1, ..., m. \end{array}$$

If we let x_i be the dual variable corresponding to the *ith* constraint of the dual of P, then the dual of the dual is

$$\begin{array}{lll} -\textit{minimize} & \sum_{i=1}^{n} (-c_i) x_i & \\ \textit{subject to} & \sum_{i=1}^{n} (-a_{ij}) x_i \leq (-b_j) & i = 1, ..., m \\ & x_i \geq 0 & i = 1, ..., n, \end{array}$$

which is equivalent to the primal problem P. ■

Recall that we motivated the dual of a linear program P in terms of obtaining bounds on the objective function value of P. In particular, the dual provided upper bounds on P. In fact, by construction of the dual any feasible solution π of the dual will provide an upper bound for P. This means that even for an optimal solution x^* of P, any dual feasible solution π is an upper bound of P, i.e., $c^T x^* \leq b^T \pi$. This result is called weak duality.

Theorem 4.4 (Weak Duality)
Let $x = (x_1, ..., x_n)^T$ be a feasible solution for the primal problem P and $\pi = (\pi_1, ..., \pi_m)^T$ be a feasible solution for the dual problem D. Then,

$$\sum_{j=1}^{n} c_j x_j \leq \sum_{i=1}^{m} b_j \pi_j.$$

Proof: We have

$$\begin{aligned}\sum_{j=1}^{n} c_j x_j &\leq \sum_{j=1}^{n} (\sum_{i=1}^{m} a_{ij}\pi_i) x_j \\ &= \sum_{i=1}^{m} (\sum_{j=1}^{n} a_{ij} x_j)\pi_j \leq \sum_{j=1}^{m} b_j \pi_j\end{aligned}$$

where the first inequality holds due the feasibility of the dual solution π and the non-negativity of the primal solution x and the second inequality holds due to the feasibility of the primal solution x and the non-negativity of the dual solution π. ■

Example 4.5
Consider again the linear program

$$\begin{array}{lll} \textit{maximize} & 6x_1 + x_2 + 2x_3 & \\ \textit{subject to} & x_1 + 2x_2 - x_3 & \leq 20 \\ & 2x_1 - x_2 + 2x_3 & \leq 30 \\ & x_1 \geq 0, x_2 \geq 0, x_3 \geq 0 & \end{array}$$

and its dual

$$\begin{array}{lll} \textit{minimize} & 20\pi_1 + 30\pi_2 & \\ \textit{subject to} & \pi_1 + 2\pi_2 & \geq 6 \\ & 2\pi_1 - \pi_2 & \geq 1 \\ & -\pi_1 + 2\pi_2 & \geq 2 \\ & \pi_1 \geq 0, \pi_2 \geq 0. & \end{array}$$

A feasible solution for the dual is $\pi_1 = 2$ and $\pi_2 = 2$ with objective function 100. Consider the feasible solution $x_1 = 15, x_2 = 2.5, x_3 = 0$ for the primal problem with objective function 92.5, which is less than 100. The difference between the dual objective function value and the primal objective function is $100-92.5 = 7.5$. This difference is called the duality gap and weak

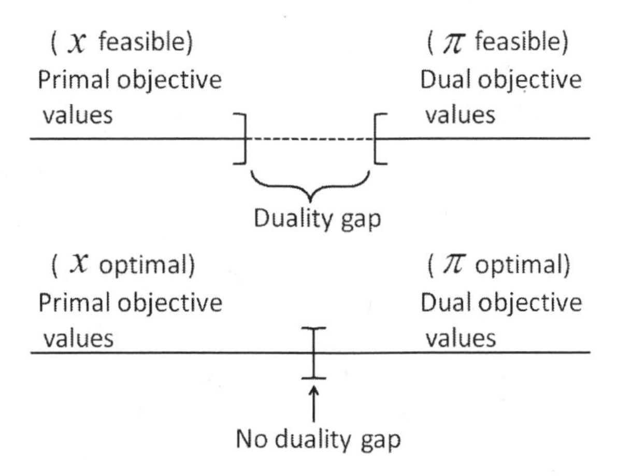

FIGURE 4.1
Duality gap for linear programs where primal is a maximization.

duality states that for a pair of primal and dual feasible solutions this gap is non-negative; see Figure 4.1.

Consider another solution for the primal problem $x_1 = 16, x_2 = 2, x_3 = 0$ with an objective value 98, which is higher than 92.5, but still less than 100, i.e., there is a duality gap of 2. Now consider the dual solution $\pi_1 = 1.6$ and $\pi_2 = 2.2$ with an objective function value of 98. Observe that the objective function values for both the primal and dual are equal at these particular solutions and the duality gap is zero. Thus, these primal and dual solutions must be optimal for their respective problems since the primal solution, which represents a lower bound, attained the value of an upper bound, and the dual solution, which represents an upper bound, attained the value of a lower bound. The next result summarizes this situation.

Corollary 4.6
If $x^* = (x_1^*, ..., x_n^*)^T$ *is a feasible solution for the primal problem* P *and* $\pi^* = (\pi_1^*, ..., \pi_m^*)^T$ *is a feasible solution for the dual problem* D *and* $\sum_{j=1}^{n} c_j x_j^* = \sum_{i=1}^{m} b_j \pi_j^*$*, then* x^* *and* π^* *are optimal solutions for their respective problems.*

Proof: By weak duality, we have that for any feasible solution $x = (x_1, ..., x_n)^T$ for P

$$\sum_{j=1}^{n} c_j x_j \leq \sum_{i=1}^{m} b_j \pi_j^* = \sum_{j=1}^{n} c_j x_j^*.$$

Thus, x^* must be an optimal solution for the primal. Similarly, for any feasible solution $\pi = (\pi_1, ..., \pi_m)^T$ for the dual problem D,

$$\sum_{i=1}^{m} b_j\pi_j^* = \sum_{j=1}^{n} c_j x_j^* \leq \sum_{i=1}^{m} b_j\pi_j.$$

So π^* must be an optimal solution for the dual. ■

Next we consider the implications of weak duality in the case where one of the problems in a primal-dual pair is unbounded.

Corollary 4.7

If the primal problem P is unbounded, then the dual problem D is infeasible.

Proof: If the dual problem P had a feasible solution π, then by weak duality $b^T\pi$ would be a finite upper bound for P, i.e., $c^Tx \leq b^T\pi$ for all x feasible for P, which is a contradiction since by assumption there is an infinite sequence of feasible solutions of P $\{x_{(k)}\}$ such that $c^Tx_{(k)} \longrightarrow \infty$ as $k \longrightarrow \infty$ ■

Similarly, we have the result.

Corollary 4.8

If the dual problem P is unbounded, then the primal problem P is infeasible.

Proof: Proof is similar to the proof of Corollary 4.7. ■

From Corollaries 4.7 and 4.8 above, one can deduce that if both the primal P and dual D have feasible solutions, then both problems should admit finite optimal solutions. It turns out that a stronger claim can be made in the following result, which can be proved via the simplex method by observing that at optimality, the simplex method solves not only the primal but the dual problem as well.

Theorem 4.9 (Strong Duality)

If a linear programming problem has a finite optimal solution, then its dual problem will also have a finite optimal solution, and the optimal objective function values of both problems are the same.

Proof: Without loss of generality assume, the primal problem P is in standard form

$$\begin{array}{ll} \textit{minimize} & c^Tx \\ \textit{subject to} & Ax = b \\ & x \geq 0 \end{array}$$

and has been solved by the simplex method to generate $x^* = \begin{bmatrix} x_B^* \\ x_N^* \end{bmatrix}$ an optimal basic feasible solution. The dual of a linear program in standard form is

$$\begin{array}{ll} \textit{maximize} & b^T\pi \\ \textit{subject to} & A^T\pi \leq c \\ & \pi \text{ unrestricted.} \end{array}$$

Since x^* is optimal, the corresponding reduced costs are non-negative, i.e.,

$$r_N = c_N - c_B^T B^{-1} N \geq 0.$$

Now, let $\pi^* = (c_B^T B^{-1})^T$, then

$$c - A^T\pi^* = \begin{bmatrix} c_B \\ c_N \end{bmatrix} - \begin{bmatrix} B^T \\ N^T \end{bmatrix} \pi^*$$

$$= \begin{bmatrix} c_B \\ c_N \end{bmatrix} - \begin{bmatrix} B^T \\ N^T \end{bmatrix} (c_B^T B^{-1})^T$$

$$= \begin{bmatrix} c_B \\ c_N \end{bmatrix} - \begin{bmatrix} c_B \\ (B^{-1}N)^T c_B \end{bmatrix} = \begin{bmatrix} 0 \\ r_N \end{bmatrix} \geq 0.$$

Thus, π^* is a feasible solution for the dual problem. Furthermore,

$$b^T\pi^* = (\pi^*)^T b = c_B^T B^{-1} b = c_B^T x_B^* = c^T x^*,$$

so by Corollary 4.6, π^* is an optimal solution for the dual. ■

Example 4.10
Consider the linear program

$$\begin{array}{lll} \textit{maximize} & -2x_1 + x_2 & \\ \textit{subject to} & -x_1 + x_2 & \leq 4 \\ & \;\;2x_1 + x_2 & \leq 6 \\ & x_1 \geq 0, x_2 \geq 0. & \end{array}$$

To use the simplex method on the primal, we convert to standard form by adding slack variables x_3 and x_4 and convert the objective to a minimization problem by multiplying the objective function by -1 (we omit the outer negation of the minimization) to get

$$\begin{array}{lll} \textit{minimize} & 2x_1 - x_2 & \\ \textit{subject to} & -x_1 + x_2 + x_3 & = 4 \\ & \;\;2x_1 + x_2 \quad\; + x_4 & = 6 \\ & x_1 \geq 0, x_2 \geq 0, x_3 \geq 0, x_4 \geq 0. & \end{array}$$

The dual is

$$\begin{array}{lll} \textit{maximize} & 4\pi_1 + 6\pi_2 & \\ \textit{subject to} & -\pi_1 + 2\pi_2 & \leq 2 \\ & \pi_1 + \;\; \pi_2 & \leq -1 \\ & \pi_1 & \leq 0 \\ & \qquad\quad \pi_2 & \leq 0 \\ & \pi_1 \text{ unrestricted}, \pi_2 \text{ unrestricted}, & \end{array}$$

which simplifies to

$$\begin{array}{lll} \textit{maximize} & 4\pi_1 + 6\pi_2 & \\ \textit{subject to} & -\pi_1 + 2\pi_2 & \leq 2 \\ & \pi_1 + \quad \pi_2 & \leq -1 \\ & \pi_1 \leq 0, \pi_2 \leq 0. & \end{array}$$

At termination of the simplex method it is found that

$$x^* = \begin{bmatrix} x_B^* \\ x_N^* \end{bmatrix} = \begin{bmatrix} x_2^* \\ x_4^* \\ x_1^* \\ x_3^* \end{bmatrix} = \begin{bmatrix} 4 \\ 2 \\ 0 \\ 0 \end{bmatrix}$$

is an optimal solution for the primal with basis $B = \begin{bmatrix} 1 & 0 \\ 1 & 1 \end{bmatrix}$ and $c_B^T = (-1, 0)$. Now

$$\begin{aligned} \pi^* &= [c_B^T B^{-1}]^T \\ &= \left[(-1,0) \begin{bmatrix} 1 & 0 \\ 1 & 1 \end{bmatrix}^{-1} \right] \\ &= \begin{bmatrix} \pi_1^* \\ \pi_2^* \end{bmatrix} = \begin{bmatrix} -1 \\ 0 \end{bmatrix} \end{aligned}$$

is feasible for the dual problem, and furthermore,

$$\begin{aligned} c^T x^* &= 2x_1^* - x_2^* = 2(0) - 4 = -4 \\ &= b^T \pi^* = 4\pi_1^* + 6\pi_2^* = 4(-1) + 6(0) = -4, \end{aligned}$$

thus, π^* must be an optimal solution to the dual.

Strong duality asserts that if a problem has a finite optimal solution, then there will be no positive duality gap; see Figure 4.1.

4.4.1 Primal-Dual Possibilities

With weak duality and strong duality, it is almost enough to characterize the possibilities a primal has with its dual, e.g., by weak duality if the primal is unbounded above, the dual D is infeasible. Thus, it would be impossible for the dual in this case to ever be feasible. Strong duality says that if one problem has a finite optimal solution, then the other problem also has a finite optimal solution as well and thus could never be unbounded or infeasible.

One possibility that has not been considered thus far is when both problems in a primal-dual pair are infeasible. Consider the linear program

$$\begin{array}{ll} \textit{maximize} & c_1 x_1 + c_2 x_2 \\ \textit{subject to} & x_1 - x_2 \leq -c_1 \\ & -x_1 + x_2 \leq -c_2 \\ & x_1 \geq 0, x_2 \geq 0. \end{array}$$

The dual is

$$\begin{array}{ll} \textit{minimize} & -c_1\pi_1 - c_2\pi_2 \\ \textit{subject to} & \pi_1 - \pi_2 \geq c_1 \\ & -\pi_1 + \pi_2 \geq c_2 \\ & \pi_1 \geq 0, \pi_2 \geq 0. \end{array}$$

Observe that when $c_1 = c_2$, the primal and dual problems are both infeasible. With this in hand, a complete characterization is now attainable; see Table 4.3 below where an entry is either yes or no indicating whether that combination for a primal-dual pair is possible or not.

Table 4.3 Primal-dual possibilities

	Optimal	Infeasible	Unbounded
Optimal	yes	no	no
Infeasible	no	yes	yes
Unbounded	no	yes	no

4.5 Complementary Slackness

In Example 4.10, observe that for the primal problem, the first constraint $-x_1 + x_2 \leq 4$ is tight (active) at the optimal solution x^*, meaning the constraint is satisfied as an equality at x^*. That is, $-x_1^* + x_2^* = 4$ where the slack variable is $x_3^* = 0$. But the second constraint $2x_1 + x_2 \leq 6$ is not active at x^* since $2x_1^* + x_2^* = 4 < 6$ where the slack variable $x_4^* = 2 \neq 0$.

Furthermore, observe that the first dual constraint is not active at the optimal dual solution π^*, that is,

$$-\pi_1^* + 2\pi_2^* = -(-1) + 2(0) = 1 < 2,$$

and so a slack variable π_3 for this constraint would have a non-zero value of 1. The second dual constraint is tight at π^*, that is,

$$\pi_1^* + \pi_2^* = -1 + 0 = -1,$$

and so a slack variable π_4 for this constraint would have a value of 0.

An important observation now is that the product of $x_1^* = 0$ and the first dual slack $\pi_3 = 1$ is 0, that is,

$$x_1^* * \pi_3 = 0 * 1 = 0,$$

and the product of $x_2^* = 4$ and second dual slack $\pi_4 = 0$ is 0 as well.

Now, the product of π_1^* and first primal slack variable x_3^* is also 0, that is,

$$\pi_1^* * x_3^* = -1 * 0 = 0,$$

and similarly

$$\pi_2^* * x_4^* = 0 * 2 = 0.$$

So for this instance we have at optimality that

$$x_i^* * (\text{slack value of } ith \text{ dual constraint}) = 0 \text{ for } i = 1, ..2$$

and

$$\pi_j^* * (\text{slack value of } jth \text{ primal constraint}) = 0 \text{ for } i = 1, ..2.$$

These conditions turn out to be necessary for optimality because, as will be seen below, this implies that the duality gap will be zero indicating that primal and dual objective function values are equal. We now develop the case more generally.

Consider the canonical primal linear program (P)

$$\begin{array}{lll} maximize & c^T x & \\ subject\ to & Ax & \leq b \\ & x \geq 0 & \end{array}$$

and its dual (D)

$$\begin{array}{lll} minimize & b^T \pi & \\ subject\ to & A^T \pi & \geq c \\ & \pi \geq 0. & \end{array}$$

Let the primal slack vector be denoted by

$$x_s = b - Ax$$

and the dual slack vector be denoted by

$$\pi_s = A^T \pi - c\,.$$

Since the matrix A is of dimension $m \times n$, x_s is a vector of m components and π_s, is a vector of n components. Then, for any feasible primal solution x and dual feasible solution π, the primal and dual slack vectors x_s and π_s will both be non-negative, thus we have

$$\begin{aligned} 0 &\leq x^T \pi_s + x_s^T \pi \\ &= (\pi^T A - c^T)x + \pi^T (b - Ax) \\ &= \pi^T b - c^T x. \end{aligned}$$

Thus, the quantity $x^T \pi_s + x_s^T \pi$ is equal to the duality gap between the primal feasible solution x and the dual feasible solution π. Then, since all vectors x, x_s, π, and π_s, are all non-negative, the duality gap will be zero if and only if both terms in $x^T \pi_s + x_s^T \pi$ are zero, that is,

$$x^T \pi_s = 0 \text{ and } x_s^T \pi = 0,$$

in which case both x and π are optimal solutions to their respective problems.

The requirement that $x^T \pi_s = 0$ can be met if for every $i = 1, ..., n$ either $x_i = 0$ or the ith component of π_s is 0, and similarly $x_s^T \pi = 0$ can be satisfied if for every $j = 1, ..., m$ either the jth component of x_s is 0 or $\pi_j = 0$. The requirement that both of these quantities are 0 is called the complementary slackness condition and is summarized in the following theorem.

Theorem 4.11 (Complementary Slackness)
Let x be a feasible solution for the canonical primal problem P and π a feasible solution for its dual D. Then, x and y are optimal solutions if and only if

$$\text{(1) } x_i(A^T\pi - c)_i = 0 \text{ for } i = 1, ..., n$$
and
$$\text{(2) } (b - Ax)_j \; \pi_j = 0 \text{ for } j = 1, ..., m$$

are satisfied. (1) is called the primal complementary slackness condition and (2) the dual complementary slackness condition.

Example 4.12
Consider the following linear program as the primal problem P

$$\begin{array}{lll} \textit{maximize} & 50x_1 + 45x_2 + 30x_3 & \\ \textit{subject to} & 4x_1 + 3x_2 + x_3 & \leq 45 \\ & 2x_1 + 2x_2 + 2x_3 & \leq 25 \\ & 2x_1 + x_2 + 0.5x_3 & \leq 9 \\ & x_1 \geq 0, x_2 \geq 0, x_3 \geq 0. & \end{array}$$

Then the dual problem D is

$$\begin{array}{lll} \textit{minimize} & 45\pi_1 + 25\pi_2 + 9\pi_3 & \\ \textit{subject to} & 4\pi_1 + 2\pi_2 + 2\pi_3 & \geq 50 \\ & 3\pi_1 + 2\pi_2 + \pi_3 & \geq 45 \\ & \pi_1 + 2\pi_2 + 0.5\pi_3 & \geq 30 \\ & \pi_1 \geq 0, \pi_2 \geq 0, \pi_3 \geq 0. & \end{array}$$

The optimal solution for P is $x_1^* = 0, x_2^* = 5.5, x_3^* = 7$ with optimal objective function 457.5. Let $x_s^T = (x_4, x_5, x_6)$ denote the slack variables for the three primal constraints, respectively. Then, the first primal constraint is not tight at optimality and has a slack value $x_4 = 21.5$. The second and third constraints are tight at optimality, so $x_5 = x_6 = 0$.

Now the optimal solution for the dual is $\pi_1^* = 0, \pi_2^* = 7.5, \pi_3^* = 30$ with an objective function value of 457.5. Let $\pi_s^T = (\pi_4, \pi_5, \pi_6)$ denote the dual slack variables. The first dual constraint is not tight at optimality and the slack variable $\pi_4 = 25$. The second and third dual constraints are tight and so the respective slack variables π_5 and π_6 are both zero.

Observe that the following products are all zero (see Tables 4.4 and 4.5), indicating that complementary slackness is satisfied.

Table 4.4 Primal complementary slackness for Example 4.11

Optimal primal	Dual slack	Product
$x_1^* = 0$	$\pi_4 = 25$	0
$x_2^* = 5.5$	$\pi_5 = 0$	0
$x_3^* = 7$	$\pi_6 = 0$	0

Table 4.5 Dual complementary slackness for Example 4.11

Optimal dual	Primal slack	Product
$\pi_1^* = 0$	$x_4 = 21.5$	0
$\pi_2^* = 7.5$	$x_5 = 0$	0
$\pi_3^* = 30$	$x_6 = 0$	0

4.5.1 Complementary Slackness for Standard Form

Now consider the primal problem P in standard form:

$$\begin{array}{ll} \textit{minimize} & c^T x \\ \textit{subject to} & Ax = b \\ & x \geq 0 \end{array}$$

with its dual D

$$\begin{array}{ll} \textit{maximize} & b^T \pi \\ \textit{subject to} & A^T \pi \leq c \\ & \pi \text{ unrestricted.} \end{array}$$

For this case, note that any feasible solution x of the primal problem P will satisfy condition (2) of complementary slackness, i.e., $\pi^T x_s = 0$ since $x_s = b - Ax = 0$ for all feasible x for P. Then, the complementary slackness conditions reduce to condition (1) only, i.e.,

$$x^T \pi_s = x^T (c - A^T \pi) = 0.$$

We can now state the necessary and sufficient conditions for a vector x to be an optimal solution for a linear program in standard form.

Theorem 4.13 (KKT Conditions for Linear Programming)
Suppose a primal problem P is in standard form. Then, a vector x^* is optimal for P if and only if the following conditions all hold:

(1) $Ax^* = b$ and $x^* \geq 0$ (primal feasibility).
(2) There are vectors π^* and π_s such that $A^T \pi^* + \pi_s = c$ and $\pi_s \geq 0$ (dual feasibility).
(3) $x^T \pi_s = 0$ (complementary slackness).

Corollary 4.14

The vector π^* is an optimal solution for the dual of P.

The main implication of Theorem 4.13 and Corollary 4.14 is that if a vector x is claimed to be optimal, then there would be a corresponding dual optimal solution π that could verify the optimality of x, i.e., the optimal dual solution would be a certificate of optimality. This should not be surprising in light of the proof of the strong duality theorem where the optimal dual solution is embedded within the reduced costs. In particular, if x is a non-degenerate optimal basic feasible solution for P in standard form, then it will be possible to uniquely generate an optimal dual solution via the complementary slackness conditions.

Theorem 4.13 is also known as the Karush Kuhn Tucker (KKT) conditions for linear programming see Karush (1939) and Kuhn and Tucker (1951).

Example 4.15

Consider the following primal problem P:

$$\begin{array}{lll} \textit{minimize} & x_1 + x_2 + 3x_3 + x_4 & \\ \textit{subject to} & x_1 + x_2 + x_3 & = 4 \\ & 2x_1 + x_2 \quad\quad + x_4 & = 6 \\ & x_1 \geq 0, x_2 \geq 0, x_3 \geq 0, x_4 \geq 0. & \end{array}$$

Could the solution $x_1 = 2, x_2 = 2, x_3 = x_4 = 0$ be an optimal solution?

Solution:

The solution is a basic feasible solution and so satisfies condition (1) in Theorem 4.13. Furthermore, the primal objective function value is 4. Now the dual problem is

$$\begin{array}{lll} \textit{maximize} & 4\pi_1 + 6\pi_2 & \\ \textit{subject to} & \pi_1 + 2\pi_2 & \leq 1 \\ & \pi_1 + \quad \pi_2 & \leq 1 \\ & \pi_1 & \leq 3 \\ & \quad\quad\quad \pi_2 & \leq 1 \\ & \pi_1 \text{ unrestricted}, \pi_2 \text{ unrestricted}. & \end{array}$$

If the proposed primal feasible solution is optimal then complementary slackness condition (1) must hold, i.e., $x_i(A^T\pi - c)_i = 0$ for $i = 1, ..., 4$. Since $x_3 = x_4 = 0$, the condition holds for $i = 3$ and 4. Now $x_1 = 2 > 0$ and $x_2 = 2 > 0$, so the slacks in the first and second dual constraints must each be 0 in order for the condition to hold for $i = 1$ and 2. This will lead to the following system of equations:

$$\begin{array}{ll} \pi_1 + 2\pi_2 & = 1 \\ \pi_1 + \quad \pi_2 & = 1. \end{array}$$

Solving this system gives the unique solution $\pi_1 = 1$ and $\pi_2 = 0$. This is a feasible dual solution and the dual objective function value at this solution is

4 which is the same as the primal objective function value under the proposed primal solution. Therefore, the solution $x_1 = 2, x_2 = 2, x_3 = x_4 = 0$ is optimal for the primal problem and the certificate of optimality is $\pi_1 = 1$ and $\pi_2 = 0$, which is optimal for the dual problem.

4.6 Duality and the Simplex Method

By construction, the simplex method at each iteration generates a primal feasible solution. We have also seen in the proof of the Strong Duality Theorem that at termination an optimal dual solution is generated as well. In fact, the simplex method maintains complementary slackness for each iteration.

Consider again the linear program from Example 3.6 from Chapter 3:

$$\begin{array}{lllll} \textit{minimize} & -x_1 - x_2 & & & \\ \textit{subject to} & x_1+ & & x_3 & = 1 \\ & & x_2+ & & x_4 = 1 \\ & \multicolumn{4}{l}{x_1 \geq 0, x_2 \geq 0, x_3 \geq 0, x_4 \geq 0.} \end{array}$$

The dual is

$$\begin{array}{llll} \textit{maximize} & \pi_1 + \pi_2 & & \\ \textit{subject to} & \pi_1 & & \leq -1 \\ & & \pi_2 & \leq -1 \\ & \pi_1 & & \leq 0 \\ & & \pi_2 & \leq 0 \\ & \multicolumn{3}{l}{\pi_1 \text{ unrestricted}, \pi_2 \text{ unrestricted.}} \end{array}$$

With dual slacks the dual becomes

$$\begin{array}{lllllll} \textit{maximize} & \pi_1 + \pi_2 & & & & & \\ \textit{subject to} & \pi_1+ & & \pi_5 & & & = -1 \\ & & \pi_2+ & & \pi_6 & & = -1 \\ & \pi_1+ & & & & \pi_7 & = 0 \\ & & \pi_2+ & & & \pi_8 & = 0 \\ & \multicolumn{6}{l}{\pi_1 \text{ unrestricted}, \pi_2 \text{ unrestricted}, \pi_5 \geq 0, \pi_6 \geq 0, \pi_7 \geq 0, \pi_8 \geq 0.} \end{array}$$

It was seen in Example 3.6 from Chapter 3 that the simplex method generated the following sequence of solutions.

Iteration 1: The initial primal solution is

$$x^{(0)} = \begin{bmatrix} x_1 \\ x_2 \\ x_3 \\ x_4 \end{bmatrix} = \begin{bmatrix} 0 \\ 0 \\ 1 \\ 1 \end{bmatrix} \text{ where } x_B = \begin{bmatrix} x_3 \\ x_4 \end{bmatrix},$$

and the initial dual solution is $\pi^{(0)} = \begin{bmatrix} \pi_1 \\ \pi_2 \end{bmatrix} = (c_B^T B^{-1})^T$

$$= \begin{bmatrix} 1 & 0 \\ 0 & 1 \end{bmatrix}^{-1} \begin{bmatrix} 0 \\ 0 \end{bmatrix} = \begin{bmatrix} 0 \\ 0 \end{bmatrix}.$$

The dual slacks $\pi_s^{(0)} = c - A^T \pi^{(0)} = \begin{bmatrix} \pi_5 \\ \pi_6 \\ \pi_7 \\ \pi_8 \end{bmatrix}$

$$= \begin{bmatrix} -1 \\ -1 \\ 0 \\ 0 \end{bmatrix} - \begin{bmatrix} 1 & 0 \\ 0 & 1 \\ 1 & 0 \\ 0 & 1 \end{bmatrix} \begin{bmatrix} 0 \\ 0 \end{bmatrix} = \begin{bmatrix} -1 \\ -1 \\ 0 \\ 0 \end{bmatrix},$$

and the primal slacks $x_s^{(0)} = b - Ax^{(0)} = 0$ by feasibility of $x^{(0)}$.

Observe that $(\pi_s^{(0)})^T x^{(0)} = 0$ and so the complementary slackness conditions are satisfied. However, observe that the dual solution $\pi^{(0)}$ is infeasible since the first two constraints are violated.

Iteration 2:
The primal solution is

$$x^{(1)} = \begin{bmatrix} x_1 \\ x_2 \\ x_3 \\ x_4 \end{bmatrix} = \begin{bmatrix} 1 \\ 0 \\ 0 \\ 1 \end{bmatrix} \text{ where } x_B = \begin{bmatrix} x_1 \\ x_4 \end{bmatrix},$$

and the dual solution is $\pi^{(1)} = \begin{bmatrix} \pi_1 \\ \pi_2 \end{bmatrix} = (c_B^T B^{-1})^T$

$$= \begin{bmatrix} 1 & 0 \\ 0 & 1 \end{bmatrix}^{-1} \begin{bmatrix} 0 \\ -1 \end{bmatrix} = \begin{bmatrix} 0 \\ -1 \end{bmatrix}.$$

The dual slacks $\pi_s^{(1)} = c - A^T \pi^{(1)} = \begin{bmatrix} \pi_5 \\ \pi_6 \\ \pi_7 \\ \pi_8 \end{bmatrix}$

$$= \begin{bmatrix} -1 \\ -1 \\ 0 \\ 0 \end{bmatrix} - \begin{bmatrix} 1 & 0 \\ 0 & 1 \\ 1 & 0 \\ 0 & 1 \end{bmatrix} \begin{bmatrix} 0 \\ -1 \end{bmatrix} = \begin{bmatrix} -1 \\ 0 \\ 0 \\ 1 \end{bmatrix},$$

and the primal slacks $x_s^{(1)} = b - Ax^{(1)} = 0$ by feasibility of $x^{(1)}$.

Observe that $(\pi_s^{(1)})^T x^{(1)} = 0$, and so the complementary slackness conditions are satisfied. The dual solution $\pi^{(1)}$ is infeasible since the first constraint is violated.

Iteration 3

The primal solution is

$$x^{(2)} = \begin{bmatrix} x_1 \\ x_2 \\ x_3 \\ x_4 \end{bmatrix} = \begin{bmatrix} 1 \\ 1 \\ 0 \\ 0 \end{bmatrix} \text{ where } x_B = \begin{bmatrix} x_1 \\ x_2 \end{bmatrix}$$

and the dual solution is $\pi^{(2)} = \begin{bmatrix} \pi_1 \\ \pi_2 \end{bmatrix} = (c_B^T B^{-1})^T$

$$= \begin{bmatrix} 1 & 0 \\ 0 & 1 \end{bmatrix}^{-1} \begin{bmatrix} -1 \\ -1 \end{bmatrix} = \begin{bmatrix} -1 \\ -1 \end{bmatrix}.$$

The dual slacks $\pi_s^{(2)} = c - A^T \pi^{(2)} = \begin{bmatrix} \pi_5 \\ \pi_6 \\ \pi_7 \\ \pi_8 \end{bmatrix}$

$$= \begin{bmatrix} -1 \\ -1 \\ 0 \\ 0 \end{bmatrix} - \begin{bmatrix} 1 & 0 \\ 0 & 1 \\ 1 & 0 \\ 0 & 1 \end{bmatrix} \begin{bmatrix} -1 \\ -1 \end{bmatrix} = \begin{bmatrix} 0 \\ 0 \\ 1 \\ 1 \end{bmatrix}$$

and the primal slacks $x_s^{(2)} = b - Ax^{(2)} = 0$ by feasibility of $x^{(2)}$.

Observe that $(\pi_s^{(2)})^T x^{(2)} = 0$, and so the complementary slackness conditions are satisfied. Now the dual solution $\pi^{(2)}$ is feasible and thus both $x^{(2)}$ and $\pi^{(2)}$ are optimal solutions for the primal and dual problems, respectively.

In summary, the strategy of the simplex method is to start with a feasible primal solution and strive for a dual feasible solution while always maintaining complementary slackness. The optimality condition for the primal problem that the reduced costs are non-negative is equivalent to dual feasibility. Before the development of duality theory and its implications for optimality conditions for linear programming, it was difficult or impossible to see this equivalence. This is not the only strategy for solving linear programs to optimality, as will be seen in the development of the dual simplex method below and interior point methods in Chapter 6. However, all algorithmic strategies that terminate with an optimal solution must satisfy (1) primal feasibility (2) dual feasibility, and (3) complementary slackness, i.e., the KKT conditions for linear programming.

4.6.1 Dual Simplex Method

We now develop a variant of the simplex method called the dual simplex method. The strategy in this approach relaxes primal feasibility, but maintains dual feasibility and complementary slackness for all iterations and at termination, satisfies primal feasibility. The dual simplex method is suitable when it may be difficult to obtain an initial primal basic feasible solution, but a dual feasible solution is readily available. However, its more substantial value may be seen in its use in sensitivity analysis, which will be covered later in this chapter.

The dual simplex method starts with a primal basic solution such that the reduced costs are non-negative or equivalently $\pi = (c_B^T B^{-1})^T$ is feasible for the dual problem. A *primal basic solution* for a linear program in standard form is a vector $x \in R^n$ such $Ax = b$ and the variables that are deemed basic have an associated basis matrix B that is invertible. $x_B = B^{-1}b$ may have negative components, in which case, the basic solution is not feasible. Of course, if the primal basic solution is feasible as well and the reduced costs are non-negative, then the solution is an optimal basic feasible solution for the primal problem.

Consider the following LP:

$$\begin{array}{lll} \textit{minimize} & 2x_1 + 3x_2 & \\ \textit{subject to} & 4x_1 - 3x_2 & \geq 5 \\ & x_1 + 2x_2 & \geq 4 \\ & x_1 \geq 0, x_2 \geq 0. & \end{array}$$

Adding the surplus variables gives

$$\begin{array}{llllll} \textit{minimize} & 2x_1 + 3x_2 & & & \\ \textit{subject to} & 4x_1 - 3x_2 & -x_3 & & = 5 \\ & x_1 + 2x_2 & & -x_4 & = 4 \\ & \multicolumn{4}{l}{x_1 \geq 0, x_2 \geq 0, x_3 \geq 0, x_4 \geq 0.} \end{array}$$

Multiplying each constraint by -1 gives

$$\begin{array}{lllll} \textit{minimize} & 2x_1 + 3x_2 & & & \\ \textit{subject to} & -4x_1 + 3x_2 & +x_3 & & = -5 \\ & -x_1 - 2x_2 & & +x_4 & = -4 \\ & \multicolumn{4}{l}{x_1 \geq 0, x_2 \geq 0, x_3 \geq 0, x_4 \geq 0.} \end{array}$$

Consider the basis $B = \begin{bmatrix} 1 & 0 \\ 0 & 1 \end{bmatrix}$ corresponding to x_3 and x_4, then

$$x_B = \begin{bmatrix} x_3 \\ x_4 \end{bmatrix} = B^{-1}b = \begin{bmatrix} -5 \\ -4 \end{bmatrix}.$$

The non-basic variables are $x_N = \begin{bmatrix} x_1 \\ x_2 \end{bmatrix}$. Thus, the primal basic solution is infeasible. However, $\pi = B^{-1}c_B = \begin{bmatrix} \pi_1 \\ \pi_2 \end{bmatrix} = \begin{bmatrix} 0 \\ 0 \end{bmatrix}$, and so the reduced costs $r_1 = 2$ and $r_2 = 3$ and so the primal solution satisfies the primal optimality conditions (reduced costs are non-negative) or equivalently π is dual feasible.

The idea in the dual simplex method is to start by choosing a current infeasible (negative) basic variable and have it exit the basis (typically, if there is more than one negative basic variable, then the most negative variable is selected).

For example, the basic variable $x_3 = -5$ is selected to leave. Then, a non-basic variable is chosen to enter the basis. Consider the constraint associated with x_3

$$-4x_1 + 3x_2 \qquad +x_3 \qquad = -5.$$

The non-basic variables are x_1 and x_2. Only variable x_1 could enter since x_1 would equal 5/4 in this case, while all other non-basic variables in the constraint remain at 0. Also, x_1 is positive and this occurs since x_1 has a negative coefficient.

In general, suppose that the leaving variable is the *pth* element of the current primal infeasible basis x_B denoted as $(x_B)_p$ and $\bar{b}$ is the *pth* element of $\bar{b} = B^{-1}\, b$. Then, since

$$x_B + B^{-1}Nx_N = B^{-1}b = \bar{b},$$

the *pth* row is

$$(x_B)_p + \sum_{l \in \bar{N}} \tilde{a}_{p,l}x_l = \bar{b}_p < 0.$$

A non-basic variable x_j where $j \in \bar{N}$ must be selected to enter the basis such that $\tilde{a}_{p,\,j} < 0$. In addition, the selection of the non-basic variable must ensure that dual feasibility is maintained (i.e., the reduced costs of the new primal basic solution must remain non-negative).

Suppose that a non-basic variable x_j with $\tilde{a}_{p,\,j} < 0$ will enter the primal basis. To shed light on the requirements for the entering non-basic variable x_j to maintain the non-negativity of the reduced costs, we start with the objective function at the current primal basic solution x,

$$\begin{aligned} z &= c^T x \\ &= c_B^T x_B + c_N^T x_N \\ &= c_B^T(B^{-1}b - B^{-1}Nx_N) + c_N^T x_N \end{aligned}$$

$$= c_B^T B^{-1} b + (c_N^T - c_B^T B^{-1} N) x_N$$
$$= z^* + r_N x_N$$
$$= z^* + \sum_{l \in \bar{N}} r_l x_l$$
$$= z^* + r_j x_j + \sum_{\substack{l \in \bar{N} \\ l \neq j}} r_l x_l, \qquad (4.2)$$

where $z^* = c_B^T B^{-1} b$ and $r_N = (c_N - (c_B^T B^{-1} N)^T)$, the vector of reduced costs for the current basis.

Now suppose that the infeasible basic variable selected to leave is x_i and is in the pth row of $x_B + B^{-1} N x_N = B^{-1} b = \bar{b}$, and so

$$x_i + \tilde{a}_{p,j} x_j + \sum_{\substack{l \in \bar{N} \\ l \neq j}} \tilde{a}_{p,l} x_l = \bar{b}_p < 0.$$

Expressing the entering variable x_j in terms of the non-basic variables in this equation gives

$$x_j = (\bar{b}_p - x_i - \sum_{\substack{l \in \bar{N} \\ l \neq j}} \tilde{a}_{p,l} x_l) / \tilde{a}_{p,\,j},$$

Now, substituting this expression for x_j in (4.2) gives

$$z = z^* + r_j((\bar{b}_p - x_i - \sum_{\substack{l \in \bar{N} \\ l \neq j}} \tilde{a}_{p,l} x_l) / \tilde{a}_{p,\,j}) + \sum_{\substack{l \in \bar{N} \\ l \neq j}} r_l x_l,$$

which after some rearrangement becomes

$$z = z^* + \bar{b}_p (r_j / \tilde{a}_{p,\,j}) - (r_j / \tilde{a}_{p,\,j}) x_i + \sum_{\substack{l \in \bar{N} \\ l \neq j}} (r_l - \tilde{a}_{p,l}(r_j / \tilde{a}_{p,\,j})) x_l.$$

Thus, we see that by entering x_j into the primal basis and exiting x_i that the reduced cost of x_i is $-(r_j / \tilde{a}_{p,\,j})$ and the updated reduced cost for x_l for $l \in \bar{N}$ and $l \neq j$ is $r_l = (r_l - \tilde{a}_{p,l}(r_j / \tilde{a}_{p,\,j}))$.

Now we are in a position to address the issue of ensuring that the reduced costs of the non-basic variables are non-negative. First, observe that $-(r_j / \tilde{a}_{p,\,j})$ is non-negative, since r_j is non-negative, since it was the reduced cost of x_j prior to becoming a basic variable and $\tilde{a}_{p,\,j} < 0$.

Finally, we need the updated reduced costs $(r_l - \tilde{a}_{p,l}(r_j / \tilde{a}_{p,\,j}))$ to be non-negative and this will be ensured if we select the entering variable x_j such that $\tilde{a}_{p,\,j} < 0$ and minimizes the quantity $\left|\frac{r_j}{\tilde{a}_{p,\,j}}\right|$. This is the corresponding minimum ratio test for the dual simplex method and is necessary and sufficient for ensuring the non-negativity of the reduced costs.

In performing the minimum ratio test, the dual simplex method will require the values of $\tilde{a}_{p,l}$ from the pth row of $x_B + B^{-1}Nx_N = \bar{b}$ such that the pth component of x_B is the entering variable x_i. These values are in the pth row of $B^{-1}N$ and can be represented by $e_p^T B^{-1}N$ where e_p is a vector whose entries are all 0 except with the value of 1 at the pth position. Then, $\tilde{a}_{p,l} = e_p^T B^{-1}N_l$ and this quantity can be computed in two steps. First, solve $tB = e_p^T$, then let $w = tN$.

In the case where there does not exist an $\tilde{a}_{p,\,j} < 0$, then $\tilde{a}_{p,l} \geq 0$ for all l in

$$x_i + \tilde{a}_{p,j}x_j + \sum_{\substack{l \in \bar{N} \\ l \neq j}} \tilde{a}_{p,l}x_l = \bar{b}_p$$

and we get

$$x_i = \bar{b}_p - \sum_{l \in \bar{N}} \tilde{a}_{p,l}x_l.$$

So x_i will always be infeasible for any non-negative x_l with $l \in \bar{N}$ since $\bar{b}_i < 0$. Thus, the primal problem is infeasible and therefore the dual unbounded and the dual simplex method will stop.

If there is a suitable x_j with a negative $\tilde{a}_{i,j}$ and once such an x_j has been determined, the entering column N_j is used to compute the improving direction analogously as in the simplex method, that is, the system $Bd = N_j$ is solved for d to get $d = B^{-1}N_j$.

What then are the values of the new basic variables? Clearly, the new basic (entering) variable $x_j = \frac{\bar{b}_p}{\tilde{a}_{p,\,j}}$ and the other basic variables are adjusted according to the following update, which is analogous to the simplex method where the next iterate is $x_B = x_B + \alpha d$ where $\alpha = -\frac{\bar{b}_p}{\tilde{a}_{p,\,j}}$. Note that the leaving variable x_i will be set to 0 in this update and will no longer be a basic variable. Finally, the basis is updated as well as the reduced costs and a new iteration begins.

We now describe the dual simplex method below.

Dual Simplex Method
Step 0: (Initialization)

Start with a basic solution $x^{(0)} = \begin{bmatrix} x_B^{(0)} \\ x_N^{(0)} \end{bmatrix}$ whose reduced costs are non-negative, i.e., $r_l = c_l - c_B^T B^{-1} N_l \geq 0$ for all $q \in \bar{N}$ where B is the basis matrix, N the non-basis matrix with corresponding partition $c = (c_B, c_N)^T$ and $\bar{B}$ and $\bar{N}$ are the index sets of the basic and non-basic variables. Let $k = 0$ and go to Step 1.

Step 1: (Optimality Check)

If $x^{(k)} \geq 0$, then the solution is optimal STOP, else select a negative component $x_i^{(k)} < 0$ and let $p =$ the position in $x_B^{(k)}$ of x_i. Go to Step 2.

Step 2: (Minimum Ratio Test)

Solve $tB = e_p^T$, then let $w = tN$. Let N^* be the set of non-basic variables $x_l^{(k)}$ such that $w_l < 0$. If $N^* = \varnothing$, then STOP; the primal problem is infeasible, else select an entering variable $x_j^{(k)} \in N^*$ such that $\left| \frac{r_j}{\bar{a}_{p,\ j}} \right|$ is minimized. Go to Step 3.

Step 3: (Direction Generation)

Let N_j be the column associated with $x_j^{(k)}$. Then solve the system $Bd^{(k)} = N_j$ for $d^{(k)}$. Go to Step 4.

Step 4: (Basic Variable Updates)

Now $x_j^{(k+1)} = \frac{\bar{b}_p}{\bar{a}_{p,\ j}}$ and let $x_B^{(k+1)} = x_B^{(k)} + \alpha d^{(k)}$ where $\alpha = -x_j^{(k+1)}$. Replace the leaving variable $x_i^{(k+1)}$ with the entering variable $x_j^{(k+1)}$ in the basic variable set. Go to Step 5.

Step 5: (Basis and Reduced Cost Update)

Let B_i be the column in B associated with the leaving basic variable x_i. Update the basis matrix B by removing column B_i and adding column N_j, thus $\bar{B} = \bar{B} - \{i\} \cup \{j\}$. Update the non-basis matrix N by the removing column N_j and adding B_i, thus $\bar{N} = \bar{N} - \{j\} \cup \{i\}$. Update the reduced costs where $r_i = -(r_j/\tilde{a}_{p,\ j})$ and $r_l = (r_l - \tilde{a}_{p,l}(r_j/\tilde{a}_{p,\ j}))$ for $l \in \bar{N}$ and $l \neq i$. Let $k = k + 1$ and go to Step 1.

Example 4.16

We continue with the linear program above and solve using the dual simplex method

$$\begin{array}{llllll} \textit{minimize} & 2x_1 + 3x_2 & & & & \\ \textit{subject to} & -4x_1 + 3x_2 & +x_3 & & = -5 \\ & -x_1 - 2x_2 & & +x_4 & = -4 \\ & x_1 \geq 0, x_2 \geq 0, x_3 \geq 0, x_4 \geq 0. & & & \end{array}$$

Step 0: Recall that the initial basic solution has as basic variables $x_B^{(0)} = \begin{bmatrix} x_3^{(0)} \\ x_4^{(0)} \end{bmatrix}$ and so $x_N^{(0)} = \begin{bmatrix} x_1^{(0)} \\ x_2^{(0)} \end{bmatrix}$, $B = \begin{bmatrix} 1 & 0 \\ 0 & 1 \end{bmatrix}$ and $N = \begin{bmatrix} -4 & 3 \\ -1 & -2 \end{bmatrix}$ with $\bar{B} = \{3, 4\}$ and $\bar{N} = \{1, 2\}$.

Now

$x_B^{(0)} = B^{-1}b = \begin{bmatrix} -5 \\ -4 \end{bmatrix}$ and so the initial solution is primal infeasible, but $\pi^{(0)} = B^{-1}c_B = \begin{bmatrix} \pi_1^{(0)} \\ \pi_2^{(0)} \end{bmatrix} = \begin{bmatrix} 0 \\ 0 \end{bmatrix}$ and so the reduced costs are $r_1 = 2$ and $r_2 = 3$, and the primal solution satisfies primal optimality conditions (but not primal feasibility). Let $k = 0$ and go to Step 1.

Iteration 1

Step 1: Both basic variables $x_B^{(0)} = \begin{bmatrix} x_3^{(0)} \\ x_4^{(0)} \end{bmatrix} = \begin{bmatrix} -5 \\ -4 \end{bmatrix}$ are negative, so $x^{(0)}$ is primal infeasible, we select $x_3^{(0)}$ to leave the basis (we select the basic variable that is the most negative, but selecting $x_4^{(0)}$ instead would be equally valid). $x_3^{(0)}$ occurs in the first position in $x_B^{(0)}$ so $p = 1$.

Step 2: Solve for t in $\begin{bmatrix} t_1 & t_2 \end{bmatrix} \begin{bmatrix} 1 & 0 \\ 0 & 1 \end{bmatrix} = \begin{bmatrix} 1 & 0 \end{bmatrix}$ to get $t = \begin{bmatrix} 1 & 0 \end{bmatrix}$ and then $w = \begin{bmatrix} 1 & 0 \end{bmatrix} \begin{bmatrix} -4 & 3 \\ -1 & -2 \end{bmatrix} = \begin{bmatrix} -4 & 3 \end{bmatrix} = \begin{bmatrix} \tilde{a}_{1,1} & \tilde{a}_{1,2} \end{bmatrix}$ and so $x_1^{(0)}$ is the only non-basic variable $x_j^{(0)}$ with a negative coefficient $\tilde{a}_{p,\,j}$ in row $p = 1$, i.e., $\tilde{a}_{1,1} = -4 < 0$, thus $x_1^{(0)}$ is the entering variable. Go to Step 3.

Step 3:

The entering column is $N_1 = \begin{bmatrix} -4 \\ -1 \end{bmatrix}$ and solve for $d^{(0)}$ in

$$\begin{bmatrix} 1 & 0 \\ 0 & 1 \end{bmatrix} \begin{bmatrix} d_1^{(0)} \\ d_2^{(0)} \end{bmatrix} = \begin{bmatrix} -4 \\ -1 \end{bmatrix}$$

to get $d_1^{(0)} = -4$ and $d_2^{(0)} = -1$. Go to Step 4.

Step 4: $x_1^{(1)} = -5/(-4) = 5/4$ and the new values of the variables in $x_B^{(0)}$ are

$x_B^{(1)} = \begin{bmatrix} x_3^{(1)} \\ x_4^{(1)} \end{bmatrix} = \begin{bmatrix} -5 \\ -4 \end{bmatrix} + (-5/4)\begin{bmatrix} -4 \\ -1 \end{bmatrix} = \begin{bmatrix} 0 \\ -11/4 \end{bmatrix}$. $x_1^{(1)}$ enters as a basic variable and $x_3^{(1)}$ becomes non-basic, so the new basic variable set $x_B^{(1)} = \begin{bmatrix} x_1^{(1)} \\ x_4^{(1)} \end{bmatrix} = \begin{bmatrix} 5/4 \\ -11/4 \end{bmatrix}$. Go to Step 5.

Step 5: The new basis is $B = \begin{bmatrix} -4 & 0 \\ -1 & 1 \end{bmatrix}$ and new non-basis matrix is $N = \begin{bmatrix} 1 & 3 \\ 0 & -2 \end{bmatrix}$, $\bar{B} = \{1, 4\}$, and $\bar{N} = \{3, 2\}$. The updated reduced costs are

$$r_3 = -r_1/\tilde{a}_{1,1} = -2/-4 = 1/2$$

and

$$r_2 = r_2 - \tilde{a}_{1,2}(r_1/\tilde{a}_{1,1}) = 3 - 3(2/-4) = 9/2.\ k = 1. \text{ Go to Step 1.}$$

Iteration 2

Step 1: $x_B^{(1)} = \begin{bmatrix} x_1^{(1)} \\ x_4^{(1)} \end{bmatrix} = \begin{bmatrix} -4 & 0 \\ -1 & 1 \end{bmatrix}^{-1} \begin{bmatrix} -5 \\ -4 \end{bmatrix} = \begin{bmatrix} 5/4 \\ -11/4 \end{bmatrix}$ we must select $x_4^{(1)}$ to leave the basis so $p = 2$. Go to Step 2.

Step 2:

Solve for t in $\begin{bmatrix} t_1 & t_2 \end{bmatrix} \begin{bmatrix} -4 & 0 \\ -1 & 1 \end{bmatrix} = \begin{bmatrix} 0 & 1 \end{bmatrix}$ to get $t = \begin{bmatrix} -1/4 & 1 \end{bmatrix}$ and then $w = \begin{bmatrix} -1/4 & 1 \end{bmatrix} \begin{bmatrix} 1 & 3 \\ 0 & -2 \end{bmatrix} = \begin{bmatrix} -1/4 & -11/4 \end{bmatrix} = \begin{bmatrix} \tilde{a}_{2,3} & \tilde{a}_{2,2} \end{bmatrix}$. $x_3^{(1)}$ and $x_2^{(1)}$ both have negative coefficients $\tilde{a}_{2,j}$ in row $p = 2$, then the minimum ratio test gives $\min\{\left|\frac{r_3}{\tilde{a}_{2,3}}\right|, \left|\frac{r_2}{\tilde{a}_{2,2}}\right|\} = \min\{2, 18/11\}$ and so $x_2^{(1)}$ enters the basis. Go to Step 3.

Step 3:

The entering column is $N_2 = \begin{bmatrix} 3 \\ -2 \end{bmatrix}$ and solve for $d^{(1)}$ in

$$\begin{bmatrix} -4 & 0 \\ -1 & 1 \end{bmatrix} \begin{bmatrix} d_1^{(1)} \\ d_2^{(1)} \end{bmatrix} = \begin{bmatrix} 3 \\ -2 \end{bmatrix}$$

to get $d_1^{(1)} = -3/4$ and $d_2^{(1)} = -11/4$. Go to Step 4.

Step 4: $x_2^{(2)} = \frac{-11/4}{-11/4} = 1$ and the new values of the variables in $x_B^{(1)}$ are

$$x_B^{(2)} = \begin{bmatrix} x_1^{(2)} \\ x_4^{(2)} \end{bmatrix} = \begin{bmatrix} 5/4 \\ -11/4 \end{bmatrix} + (-1) \begin{bmatrix} -3/4 \\ -11/4 \end{bmatrix} = \begin{bmatrix} 2 \\ 0 \end{bmatrix}.$$

$x_2^{(2)}$ enters as a basic variable and $x_4^{(2)}$ becomes non-basic, so the new basic variable set $x_B^{(2)} = \begin{bmatrix} x_1^{(2)} \\ x_1^{(2)} \end{bmatrix} = \begin{bmatrix} 2 \\ 1 \end{bmatrix}$.

Since all basic variables are now positive the primal solution

$$x^{(2)} = \begin{bmatrix} x_1^{(2)} \\ x_2^{(2)} \\ x_3^{(2)} \\ x_4^{(2)} \end{bmatrix} = \begin{bmatrix} 2 \\ 1 \\ 0 \\ 0 \end{bmatrix} \text{ is optimal.}$$

Note that there is no need to update the reduced costs at this point.

It turns out that the dual simplex method applied to a linear program P in standard form is equivalent to applying the simplex method on the dual of P. The action of finding a leaving variable x_i in the dual simplex method is equivalent to finding an entering dual variable π_i and entering a variable x_j is equivalent to exiting a dual variable π_j. However, note that the dual simplex method works directly on the primal basis B and not on the dual basis. Not surprisingly, the dual simplex method requires the same amount of computational effort as the simplex method and is usually not the principal method to solve linear programs in practice. However, the dual simplex method is a valuable tool for sensitivity analysis which is covered at the end of this chapter and in situations where a linear program has been solved and then a new constraint is added. The dual simplex method can be used to solve the linear problem with the extra constraint without needing to re-solve the new problem from scratch.

4.7 Economic Interpretation of the Dual

Both a primal and a dual problem have a intimate relationship as they both use the same data and one problem bounds the other. Furthermore, the concept of duality has enabled a characterization (necessary and sufficient conditions) of optimality of the primal and dual problems. In this section, we explore possible economic meanings for dual linear programs and dual variables in relation to primal problems and primal variables.

4.7.1 Dual Variables and Marginal Values

Consider primal linear programs in canonical form

$$\begin{array}{lll} \textit{maximize} & \sum_{j=1}^{n} c_j x_j & \\ \textit{subject to} & \sum_{j=1}^{n} a_{ij} x_j \leq b_i & i = 1, ..., m \\ & x_j \geq 0 & j = 1, ..., n. \end{array}$$

This can represent the case of a manufacturer that produces n products obtaining a market price of c_j for a unit of product j. A unit of each of the n products require m resources where the total amount of the ith resource available for all products is b_i. The constraint coefficient a_{ij} gives the amount of resource i needed for product j. Thus, the jth column of the constraint matrix gives the total resource requirements for the manufacture of product j. The variable x_i is the amount of product i to produce. Then, the manufacturer solves the linear program to determine the optimal production plan x_i for each product i.

Example 4.17

Suppose that you manufacture two products: tables and chairs. Each table sold will generate a revenue of \$12 and each chair sold will generate \$7 of revenue. Manufacturing chairs and tables requires wood, cutting, and finishing. The amount of wood available is 1200 sq meters and there are 1000 hours available for cutting, and 500 hours available for finishing. Each unit of a product requires the following resources.

Resources (units)	Table	Chair
Wood (sq meters)	6	4
Cutting (hrs)	8	2
Finishing (hrs)	2	1

The manufacturer wishes to find the quantities of tables and chairs to produce to maximize revenue. Let x_1 = number of tables to produce and x_2 = number of chairs to produce, then the following linear program will maximize revenue from producing chairs and tables subject to resource limitations.

$$\begin{array}{lll} \textit{maximize} & 12x_1 + 7x_2 & \\ \textit{subject to} & 6x_1 + 4x_2 & \leq 1200 \\ & 8x_1 + 2x_2 & \leq 1000 \\ & 2x_1 + x_2 & \leq 500 \\ & x_1 \geq 0, x_2 \geq 0 & \end{array}$$

The optimal solution is $x_1^* = 80$ and $x_2^* = 180$, i.e., the maximizing production plan produces 80 tables and 180 chairs and generates a revenue of \$2220.

Now the dual of the production planning model is

$$\begin{array}{lll} \textit{minimize} & 1200\pi_1 + 1000\pi_2 + 500\pi_3 & \\ \textit{subject to} & 6\pi_1 + 8\pi_2 + 2\pi_3 & \geq 12 \\ & 4\pi_1 + 2\pi_2 + \pi_3 & \geq 7 \\ & \pi_1 \geq 0, \pi_2 \geq 0, \pi_3 \geq 0. & \end{array}$$

The optimal dual solution is $\pi_1^* = 1.6, \pi_2^* = 0.30, \pi_3^* = 0$ with an optimal objective function value of \$2220. What economic interpretation might be given to the dual problem?

Consider that the manufacturer obtains the resources (wood, cutting services, and finishing services) from a supplier. The manufacturer wishes to negotiate for the price per unit, π_i, of each resource for $i = 1, ..., 3$. Thus, the manufacturer wishes to minimize the total cost of obtaining these three resources, which becomes the objective of the dual problem.

Now assuming that the production requirements for producing tables and chairs are fully known to the public and in particular to the supplier, i.e., the supplier knows how much revenue the manufacturer can get for each table and chair based on its production requirements a_{ij}. Thus, the manufacturer will assume that the prices should be reasonable so that a supplier would accept them. So each dual constraint ensures that prices of resources are such that the total cost to produce a table (chair) is at least the revenue obtained from a table, i.e., 12 (7 for a chair). Note that if a supplier demanded prices for resources such that the cost of producing is more than the market price for each product, then by complementary slackness the corresponding primal variable $x_i = 0$, i.e., there would be no production of that product.

Observe that the first two primal constraints are tight at the optimal primal solution. One interpretation of these results is that both the wood and cutting resources are exhausted at the optimal revenue-maximizing production plan. Now the optimal dual variables that are associated with these constraints are $\pi_1^* = 1.6$ and $\pi_2^* = 0.30$, respectively. The primal and dual objectives are the same at optimality and the dual objective function is

$$1200\pi_1 + 1000\pi_2 + 500\pi_3$$

, and thus an additional unit of wood will increase the revenue by \$1.6 dollars. Similarly, an extra unit of cutting time will increase revenue by 30 cents.

The third primal constraint is not tight at the optimal primal solution meaning that at the optimal production plan, not all of the finishing capacity was used. So this constraint has positive slack and consequently by complementary slackness $\pi_3^* = 0$. The value of adding an extra unit of finishing capacity does not increase the revenue. Thus, we see that the optimal dual prices can be interpreted as marginal prices or shadow prices indicating the value of an extra unit of each resource.

4.8 Sensitivity Analysis

One of the major assumptions behind linear programming is that the data in the form of cost coefficients c, right-hand side vector b, and constraint coefficients A are known. In practice, the data can represent quantities such as product demand, financial security prices, resource requirements, and resource costs that are difficult to estimate and in reality can fluctuate considerably. For example, consider a linear program that constructs a financial portfolio, e.g., the MAD model from Chapter 1, to be held for a period of time and requires the future performance of the financial assets like the expected return over a particular time duration. This performance data will significantly influence the optimal portfolio construction, and if the actual performance of assets over time are different than what was estimated, then one could be holding a portfolio that may be misleading. For example, suppose that an estimation for the expected return for stock i used in the model is 12% but in reality it may be as low as 5%, then an optimal portfolio constructed from using the estimation of 12% can overconcentrate (invest) in stock i since the model expects good performance, but in reality a much lower performance may be realized in which a smaller investment in stock i would be optimal.

Thus, it is natural to consider how an optimal solution of a linear program may be affected by changes in the data or problem itself. In addition, an existing linear program will often be modified. A new variable could be added representing another financial asset, or a new managerial constraint on investment could be added. How will these changes affect the LP? The analysis of such changes to a linear program is called sensitivity analysis. In this section, we explore cases of data changes and problem modifications where a definitive answer can be obtained concerning the impact of the changes.

We assume that the original linear program is in standard form

$$\begin{array}{ll} \textit{minimize} & c^T x \\ \textit{subject to} & Ax = b \\ & x \geq 0 \end{array}$$

and has been solved to optimality. In each case, the central question is how does the change affect the optimality and feasibility of the existing optimal feasible solution? Will the optimal solution continue to be optimal or feasible for the modified problem? For what range of data will the current optimal solution remain optimal? For what ranges of data is the current feasible solution still feasible?

To answer these questions, the starting point is the feasibility and optimality conditions for a linear program. Given a basis B, recall that it is feasible when $B^{-1}b \geq 0$ and optimal when $r_N = c_N^T - c_B^T B^{-1} N \geq 0$. The general strategy for sensitivity analysis is to investigate whether a change in data or modification to the LP will affect these conditions

4.8.1 Changes in the Right-Hand Side Coefficients

We first consider the case when the right-hand side vector b is perturbed (changed) by an amount $\triangle b$. Let $\tilde{b} = b + \triangle b$ and let $x^* = \begin{bmatrix} x_B^* \\ x_N^* \end{bmatrix}$ be an optimal basic feasible solution to the original LP, and B and N the corresponding basis and non-basis matrices.

Consider the perturbed problem

$$\begin{array}{ll} \textit{minimize} & z = c^T x \\ \textit{subject to} & Ax = b + \Delta b = \tilde{b} \\ & x \geq 0. \end{array}$$

Observe that the change Δb in the right-hand side does not affect the optimality conditions of the original problem since the right-hand side vector does not appear in $r_N = c_N^T - c_B^T B^{-1} N \geq 0$. However, the right-hand side vector is involved in the feasibility condition $B^{-1} b \geq 0$. In particular, the original basis B will be feasible for the perturbed problem if it satisfies the feasibility conditions $B^{-1}\tilde{b} = B^{-1}(b + \Delta b) \geq 0$. This condition is satisfied if

$$B^{-1} b \geq -B^{-1} \Delta b.$$

Note that the new objective function value is $\tilde{z} = c_B^T B^{-1} \tilde{b} = \pi^T (b + \Delta b) = z + \pi^T \Delta b$ where $\pi^T = c_B^T B^{-1}$.

Example 4.18
Consider the production planning linear program

$$\begin{array}{lll} -\textit{minimize} & -12x_1 - 7x_2 & \\ \textit{subject to} & 6x_1 + 4x_2 & \leq 1200 \\ & 8x_1 + 2x_2 & \leq 1000 \\ & 2x_1 + x_2 & \leq 500 \\ & x_1 \geq 0, x_2 \geq 0. & \end{array}$$

Note that the maximization has been converted to a minimization. The optimal solution is $x_1^* = 80$ and $x_2^* = 180$, i.e., the maximizing production plan is to produce 80 tables and 180 chairs and generates a revenue of \$2220. The optimal basic feasible solution for the standard form version of the problem

$$\begin{array}{lllllll} -\textit{minimize} & -12x_1 - 7x_2 & & & & \\ \textit{subject to} & 6x_1 + 4x_2 & +x_3 & & & = 1200 \\ & 8x_1 + 2x_2 & & +x_4 & & = 1000 \\ & 2x_1 + x_2 & & & +x_5 & = 500 \\ & \multicolumn{5}{l}{x_1 \geq 0, x_2 \geq 0, x_3 \geq 0, x_4 \geq 0, x_5 \geq 0} \end{array}$$

is $x_1^* = 80$, $x_2^* = 180$, $x_3^* = 0, x_4^* = 0, x_5^* = 160$ where the latter three variables are the optimal slack values and the optimal basic variables are x_1^*, x_2^*, x_5^*. Thus, the basis matrix is

$$B = \begin{bmatrix} 6 & 4 & 0 \\ 8 & 2 & 0 \\ 2 & 1 & 1 \end{bmatrix}.$$

Now suppose that only the right-hand side of the second constraint is perturbed (the cutting capacity is changed). More formally, we have $\tilde{b}_2 = b_2 + \triangle b_2$ where $\triangle b_2 \neq 0$, then

$$\triangle b = \begin{bmatrix} 0 \\ \triangle b_2 \\ 0 \end{bmatrix}.$$

In order for the current basis B to continue to be feasible for the perturbed problem we need

$$B^{-1}b = \begin{bmatrix} 80 \\ 180 \\ 160 \end{bmatrix} \geq B^{-1}\triangle b = \begin{bmatrix} 0.2000\triangle b_2 \\ -0.3000\triangle b_2 \\ -0.1000\triangle b_2 \end{bmatrix}.$$

Thus, the admissible range for $\triangle b_2$ is $-600 \leq \triangle b_2 \leq 400$.

Suppose that $\triangle b_2$ is 100, that is, we increase the cutting capacity by 100 units, then the basic variable set is now

$$x_B^{new} = \begin{bmatrix} x_1 \\ x_2 \\ x_5 \end{bmatrix} = B^{-1}\tilde{b} = B^{-1}b + B^{-1}\triangle b$$

$$= \begin{bmatrix} 80 \\ 180 \\ 160 \end{bmatrix} + \begin{bmatrix} 0.2000(100) \\ -0.3000(100) \\ -0.1000(100) \end{bmatrix} = \begin{bmatrix} 100 \\ 150 \\ 150 \end{bmatrix}.$$

The corresponding basic feasible solution is feasible for the perturbed problem, and so the original basis B remains feasible (and optimal) for the perturbed problem. The new optimal objective function value is $-\tilde{z} = -c_B^T B^{-1}\tilde{b}$ $= -\pi^T(b + \triangle b) = 2250$. The addition of 100 units of cutting capacity resulted in a revenue increase of \$30.

Suppose that $\triangle b_2$ is 650, then the basic variable set is now

$$x_B^{new} = \begin{bmatrix} x_1 \\ x_2 \\ x_5 \end{bmatrix} = B^{-1}\tilde{b} = B^{-1}b + B^{-1}\triangle b$$

$$= \begin{bmatrix} 80 \\ 180 \\ 160 \end{bmatrix} + \begin{bmatrix} 0.2000(650) \\ -0.3000(650) \\ -0.1000(650) \end{bmatrix} = \begin{bmatrix} 210 \\ -15 \\ 95 \end{bmatrix},$$

which is infeasible since the x_2 is negative. Thus, B is infeasible for the perturbed problem.

An important observation here is that although the original basis B is infeasible for the perturbed problem, the reduced costs remain non-negative

($r_3 = 1.6$ and $r_4 = 0.3$), so π is dual feasible for the dual of the perturbed problem. Then, one can use the original basis B with initial basic variables $x_B^{new} = B^{-1}\tilde{b}$ to start the dual simplex method to ultimately get the optimal solution for the perturbed problem $x_1 = 200, x_2 = 0,\ x_3 = 0, x_4 = 50$, and $x_5 = 100$ with an optimal objective function value of 2400. The advantage in using the dual simplex method is that the perturbed problem does have to be solved from scratch using the revised simplex method.

4.8.2 Changes in the Cost (Objective) Coefficients

We now consider changes in the cost coefficients. Let $\bar{c} = \begin{bmatrix} \bar{c_B} \\ \bar{c_N} \end{bmatrix} = c + \triangle c$ $= \begin{bmatrix} c_B \\ c_N \end{bmatrix} + \begin{bmatrix} \triangle c_B \\ \triangle c_N \end{bmatrix} = \begin{bmatrix} c_B + \triangle c_B \\ c_N + \triangle c_N \end{bmatrix}$ represent the perturbed cost vector where $\triangle c_B$ is the perturbation vector for the cost coefficients of the basic variables and $\triangle c_N$ is the perturbation of the vector for the cost coefficients of the non-basic variables. Then, the perturbed problem is

$$\begin{array}{ll} minimize & z = \bar{c}^T x \\ subject\ to & Ax = b \\ & x \geq 0. \end{array}$$

Let B be the optimal basis of the perturbed problem. Observe that perturbation of cost coefficients does not affect the feasibility of the basis B since the cost vector does not appear in the condition $B^{-1}b \geq 0$. Thus, basis B will always be feasible for a perturbed problem where only cost coefficients are changed.

Changes to Cost Coefficients of Basic Variables

We first consider perturbation of cost coefficients of basic variables. In order for the basis B to maintain optimality, the following condition must hold

$$\begin{aligned} c_N^T - \bar{c}_B^T B^{-1} N &= c_N^T - (c_B + \triangle c_B)^T B^{-1} N \\ &= c_N^T - c_B^T B^{-1} N - \triangle c_B{}^T B^{-1} N \\ &= r_N - \triangle c_B{}^T B^{-1} N \geq 0, \end{aligned}$$

or equivalently

$$r_N \geq \triangle c_B{}^T B^{-1} N.$$

Example 4.19

Consider again the linear program in Example 4.18 and its optimal basis B. Suppose that the cost coefficient c_1 of basic variable x_1 is perturbed so that $\triangle c_B = \begin{bmatrix} \triangle c_1 \\ 0 \\ 0 \end{bmatrix}$, then the current basis will remain optimal for the perturbed problem if

$$r_N^T = (r_3, r_4)^T = (1.6, 0.3)^T \geq (\triangle c_1, 0, 0)B^{-1}N$$

$$= (\triangle c_1, 0, 0) \begin{bmatrix} -0.1 & 0.2 \\ 0.4 & -0.3 \\ -0.2 & -0.1 \end{bmatrix}$$

$$= (-0.1\triangle c_1, 0.2\triangle c_1)^T,$$

thus, the admissible range for $\triangle c_1$ is $-16 \leq \triangle c_1 \leq 1.5$.

For instance, let $\triangle c_1 = -1$, then the basis B will remain optimal since -1 is in the admissible range for the perturbed problem and the new reduced costs for the perturbed problem are now

$$(r_3, r_4) = (1.6 + 0.1(-1), 0.3 - 0.2(-1))$$

$$= (1.5, 0.5) \geq (0, 0),$$

and the new objective function is

$$-\bar{z} = -\bar{c}_B^T B^{-1} b = -(-13, -7) \begin{bmatrix} 6 & 4 & 0 \\ 8 & 2 & 0 \\ 2 & 1 & 1 \end{bmatrix}^{-1} \begin{bmatrix} 1200 \\ 1000 \\ 500 \end{bmatrix} = 2300,$$

so an extra \$80 of revenue is generated from the original optimal production plan due to a \$1 increase per unit in the revenue of a table.

Now let $\triangle c_1 = 2$, then the perturbation is not in the admissible range and the reduced costs are

$$(r_3, r_4)^T = (1.6 + 0.1(2), 0.3 - 0.2(2))^T$$

$$= (1.5, -0.1)^T,$$

and $r_4 < 0$. So the basis matrix B is not optimal for the perturbed problem. However, B can be used in the simplex method to solve the perturbed problem to optimality. Recall that B is a feasible basis, so one can select x_4 to enter the basis and proceed with the simplex iterations to ultimately generate the optimal solution for the perturbed problem. In this instance, the basis B is a valid initial basic feasible solution for the perturbed problem.

Changes to Cost Coefficients of Non-basic Variables

We now consider perturbation of cost coefficients of non-basic variables. In order for the basis B to maintain optimality, the following condition must hold

$$\begin{aligned}\bar{c}_N^T - c_B^T B^{-1} N &= (c_N + \triangle c_N)^T - c_B^T B^{-1} N \\ &= c_N^T - c_B^T B^{-1} N + \triangle c_N{}^T \\ &= r_N + \triangle c_N{}^T \geq 0,\end{aligned}$$

or equivalently

$$r_N \geq -\triangle c_N{}^T.$$

Thus, if a perturbation of non-basic variables $\triangle c_N$ violates the above condition, then the reduced costs of the perturbed problem are not all non-negative and the revised simplex method can be used to generate the optimal solution for the perturbed problem.

4.8.3 Changes in the Constraint Matrix

Sensitivity analysis concerning perturbations in the constraint matrix is not easy. A small change in a single coefficient in a current basis can render the basis infeasible or singular. We consider cases where it is tractable to characterize the sensitivity of linear programs in perturbations/modifications to the constraint matrix.

4.8.3.1 Adding a New Variable

We first consider the case where a new variable x_{n+1} is added to a linear program after the original LP is solved to optimality. Let A_{n+1} denote the new column of coefficients associated with x_{n+1} added to the constraint matrix A, and c_{n+1} be the new objective function coefficient associated with x_{n+1} added to the objective coefficients c. We now would like to solve the modified problem

$$\begin{array}{ll} \textit{minimize} & c^T x + c_{n+1} x_{n+1} \\ \textit{subject to} & Ax + A_{n+1} x_{n+1} = b \\ & x \geq 0, x_{n+1} \geq 0. \end{array}$$

Let $x_{new} = \begin{bmatrix} x^* \\ x_{n+1} \end{bmatrix}$ where x^* be the optimal basic feasible solution for the original problem and let $x_{n+1} = 0$, then x_{new} is feasible for the modified problem above. Now, to check optimality of x_{new} it suffices to check the reduced cost of x_{n+1}.

If $r_{n+1} = c_{n+1} - c_B^T B^{-1} A_{n+1} \geq 0$, then x_{new} is optimal for the modified problem, else $r_{n+1} < 0$, so one can use the basis B associated with x^* to start the revised simplex method for the modified problem and select x_{n+1} as an entering variable.

Example 4.20

Consider the production of another product, a wooden shoe box, in addition to tables and chairs in the production planning model in Example 4.18.

Suppose that the revenue from one unit of a shoe box is \$4, and to produce one unit requires 3 sq meters of wood, 4 hours of cutting, and 1 hour of finishing. Let x_6 = number of shoe boxes to produce, and let $c_3 = -4$ (negated for minimization), and $A_6 = (3, 4, 1)^T$. Then, the modified production planning model is

$$
\begin{array}{lllllll}
-minimize & -12x_1 - 7x_2 & & & & -4x_6 & \\
subject\ to & 6x_1 + 4x_2 & +x_3 & & & +3x_6 & = 1200 \\
 & 8x_1 + 2x_2 & & +x_4 & & +4x_6 & = 1000 \\
 & 2x_1 + x_2 & & & +x_5 & & = 500 \\
 & \multicolumn{6}{l}{x_1 \geq 0, x_2 \geq 0, x_3 \geq 0, x_4 \geq 0, x_5 \geq 0.}
\end{array}
$$

Now the reduced cost $r_6 = c_6 - c_B^T B^{-1} A_6$

$$
= -4 - (-12, 7, 0) \begin{bmatrix} 6 & 4 & 0 \\ 8 & 2 & 0 \\ 2 & 1 & 1 \end{bmatrix}^{-1} \begin{bmatrix} 3 \\ 4 \\ 1 \end{bmatrix} = 2.
$$

So the feasible solution $x_{new} = \begin{bmatrix} x^* \\ x_6 \end{bmatrix} = \begin{bmatrix} x^* \\ 0 \end{bmatrix}$ is optimal for the modified problem where x^* is the optimal solution for the original production planning problem. The revenue obtained from a wooden shoe box given its resource requirements is not enough for the production plan to change.

If the revenue from a wooden shoe box is 8, i.e., $c_6 = -8$ then $r_6 = -2 < 0$ and so x_{new} is no longer optimal for the modified problem and the revised simplex method can be applied with x_{new} as an initial basic feasible solution. In this case, the revenue from a wooden shoe box is enough to warrant a change in the production plan.

4.8.3.2 Adding a New Constraint

Now we consider adding a new constraint to a linear program after it has been solved to optimality. We assume that the constraint is an inequality of the form

$$
a_{m+1}^T x \leq b_{m+1}
$$

where a_{m+1} is the vector of coefficients and b_{m+1} is the right-hand side of the constraint. Thus, the new modified linear program is

$$
\begin{array}{lll}
minimize & c^T x & \\
subject\ to & Ax & = b \\
 & a_{m+1}^T x & \leq b_{m+1} \\
 & x \geq 0. &
\end{array}
$$

Let $F = \{x \in R^n | Ax = b, x \geq 0\}$ and $F' = \{x \in R^n | Ax = b, a_{m+1}^T x \leq b_{m+1}, x \geq 0\}$. The concern with adding a constraint to the original LP is that the feasible region F' of the modified problem may be a strict subset of F, i.e., $F' \subset F$ (in general $F' \subseteq F$). In this situation, the optimal solution to the original LP x^* may not be in F', i.e., is infeasible for the modified problem. In any case, the basis B associated with x^* is not suitable to be used for the modified problem since any basis for the new problem will be a square matrix with dimension $(m+1) \times (m+1)$. The idea in proceeding with the new problem is to extend the basis B so that it can be used in the modified problem. To this end, we add a slack variable x_{n+1} for the new constraint and express the modified LP in terms of the partition implied by the basis B from the optimal basic feasible solution of the original LP to get the following version of the modified problem:

$$\begin{array}{lll} minimize & c_B^T x_B + c_N^T x_N & \\ subject\ to & Bx_B + Nx_N & = b \\ & (a_{m+1,B})^T x_B + (a_{m+1,N})^T x_N + x_{n+1} & = b_{m+1} \\ & x_B \geq 0, x_N \geq 0, x_{n+1} \geq 0. & \end{array}$$

where the coefficients a_{m+1} have been partitioned into coefficients $a_{m+1,B}$ corresponding to the variables in basis B and coefficients $a_{m+1,N}$ corresponding to variables in the non-basis matrix N.

Now we can form the extended basis defined as

$$B_E = \begin{bmatrix} B & 0 \\ a_{m+1,B}^T & 1 \end{bmatrix},$$

and it is not hard to show that the inverse of this matrix is

$$B_E^{-1} = \begin{bmatrix} B^{-1} & 0 \\ -a_{m+1,B}^T B^{-1} & 1 \end{bmatrix}.$$

Now let $\bar{b} = \begin{bmatrix} b \\ b_{m+1} \end{bmatrix}$ and let $\bar{x}_B = B_E^{-1}\bar{b}$, then a basic solution for the modified problem is

$$\bar{x} = \begin{bmatrix} \bar{x}_B \\ 0 \end{bmatrix}.$$

We know that since B is an optimal basis for the original problem, then $x_B = B^{-1}b \geq 0$ (strict under non-degeneracy). Thus, it is natural to conjecture that if $\bar{x}_B \geq 0$, then would $\bar{x}$ be optimal for the modified problem? The answer is in the affirmative as summarized in the following result.

Theorem 4.21
Suppose that B is an optimal basis for the original linear programming problem. If $\bar{x}_B \geq 0$, then $\bar{x}$ is an optimal solution to the modified linear program with the extra constraint $a_{m+1}^T x \geq b_{m+1}$.
Proof: Exercise 4.12.

If at least one of the components of $\bar{x}_B$ is negative, then $\bar{x}$ is not feasible for the modified problem. However, the original basis B is dual feasible for the dual of the original linear programming problem, and so if we let

$$\bar{\pi} = \begin{bmatrix} c_B^T B^{-1} \\ 0 \end{bmatrix},$$

then, $\bar{\pi}$ is dual feasible for the modified problem. Thus, the dual simplex method can be initiated with the extended basis B_E to solve the modified problem to optimality.

Example 4.22
Consider the linear program

$$\begin{array}{lll} \textit{minimize} & -3x_1 - 2x_2 & \\ \textit{subject to} & x_1 + x_2 + x_3 & = 3 \\ & x_1 \geq 0, x_2 \geq 0, x_3 \geq 0. & \end{array}$$

An optimal solution x^* can be obtained by inspection to get $x_1^* = 3, x_2^* = 0, x_3^* = 0$. The optimal basis $B = [1]$ and non-basis matrix is $N = [1\ 1]$ with $\bar{B} = \{1\}$ and $\bar{N} = \{2, 3\}$. $c_B = (-3)$ and $c_N^T = (-2, 0)$.

Now suppose that a new constraint $x_1 \leq 2$ is added to the problem to get

$$\begin{array}{lll} \textit{minimize} & -3x_1 - 2x_2 & \\ \textit{subject to} & x_1 + x_2 + x_3 & = 3 \\ & x_1 + \qquad\qquad x_4 & = 2 \\ & x_1 \geq 0, x_2 \geq 0, x_3 \geq 0, x_4 \geq 0 & \end{array}$$

where x_4 is a slack variable for the new constraint. We form the extended basis

$$B_E = \begin{bmatrix} B & 0 \\ a_{2,B}^T & 1 \end{bmatrix} = \begin{bmatrix} 1 & 0 \\ 1 & 1 \end{bmatrix} \text{ where } c_{\bar{B}} = (-3, 0)^T$$

$$\bar{B}_E = \{1, 4\} \text{ and } \bar{N}_E = \{2, 3\}.$$

Now

$$\bar{x}_B = \begin{bmatrix} x_1 \\ x_4 \end{bmatrix} = B_E^{-1}\bar{b} = \begin{bmatrix} 1 & 0 \\ 1 & 1 \end{bmatrix}^{-1} \begin{bmatrix} 3 \\ 2 \end{bmatrix} = \begin{bmatrix} 3 \\ -1 \end{bmatrix},$$

so $\bar{x} = \begin{bmatrix} \bar{x}_B \\ 0 \end{bmatrix}$ is infeasible for the modified problem. Since $\bar{\pi} = \begin{bmatrix} c_B^T B^{-1} \\ 0 \end{bmatrix} = \begin{bmatrix} -3 \\ 0 \end{bmatrix}$ is feasible for the dual of the modified problem (for this we can verify that the reduced costs are positive (non-negative), i.e.,

$$r_2 = -2 - (-3, 0) \begin{bmatrix} 1 & 0 \\ 1 & 1 \end{bmatrix}^{-1} \begin{bmatrix} 1 \\ 0 \end{bmatrix} = 1$$

and

$$r_3 = 0 - (-3, 0) \begin{bmatrix} 1 & 0 \\ 1 & 1 \end{bmatrix}^{-1} \begin{bmatrix} 1 \\ 0 \end{bmatrix} = 3.$$

We can then use the dual simplex method starting with the basis B_E to solve the modified problem to optimality.

4.9 Exercises

Exercise 4.1

Consider the LP

$$\begin{array}{ll} \textit{minimize} & -4x_1 - 3x_2 - 2x_3 \\ \textit{subject to} & 2x_1 + 3x_2 + 2x_3 \leq 6 \\ & -x_1 + x_2 + x_3 \leq 5 \\ & x_1 \geq 0, x_2 \geq 0, x_3 \geq 0. \end{array}$$

(a) Write the dual of this LP.

(b) Solve the primal problem using the simplex method and show that at optimality the dual problem is solved to optimality as well by showing that the primal and dual solutions generated satisfy dual feasibility and complementary slackness.

Exercise 4.2

Consider the following linear program:

$$\begin{array}{ll} \textit{minimize} & 3x_1 + 4x_2 + 5x_3 \\ \textit{subject to} & x_1 + 3x_2 + x_3 \geq 2 \\ & 2x_1 - x_2 + 3x_3 \geq 3 \\ & x_1 \geq 0, x_2 \geq 0, x_3 \geq 0. \end{array}$$

Solve using

(a) the dual simplex method.

(b) the simplex method on the dual of the problem.

Exercise 4.3

Consider the linear program

$$\begin{array}{ll} \textit{minimize} & 3x_1 + 4x_2 + 6x_3 + 7x_4 \\ \textit{subject to} & x_1 + 2x_2 + 3x_3 + x_4 \geq 1 \\ & -x_1 + x_2 - x_3 + 3x_4 \leq -2 \\ & x_1 \geq 0, x_2 \geq 0, x_3 \geq 0, x_4 \geq 0. \end{array}$$

Solve using the dual simplex method.

Exercise 4.4

Consider the following linear program:

$$\begin{array}{ll} \textit{minimize} & -3x_1 - x_2 \\ \textit{subject to} & x_1 + x_2 + x_3 = 2 \\ & x_1 \geq 0, x_2 \geq 0, x_3 \geq 0. \end{array}$$

(a) Find the optimal solution using the simplex method.
(b) Now consider adding the constraint $x_1 + \ x_4 = 1$ to get the LP

$$\begin{array}{ll} \textit{minimize} & -3x_1 - x_2 \\ \textit{subject to} & x_1 + x_2 + x_3 = 2 \\ & x_1 + \qquad\quad x_4 = 1 \\ & x_1 \geq 0, x_2 \geq 0, x_3 \geq 0, x_4 \geq 0. \end{array}$$

(c) Solve the new linear program without re-solving the new model from scratch. (Hint: Use the dual simplex method)

Exercise 4.5
Consider the linear program

$$\begin{array}{ll} \textit{maximize} & 7x_1 + 17x_2 + 17x_3 \\ \textit{subject to} & x_1 + x_2 \leq 8 \\ & x_1 + 4x_2 + 3x_3 \leq 14 \\ & \qquad 3x_2 + 4x_3 \leq 9 \\ & x_1 \geq 0, x_2 \geq 0, x_3 \geq 0. \end{array}$$

(a) What is the optimal solution?
(b) What is the optimal basis?
(c) What are the optimal dual variables?
(d) By how much can the right-hand side of the first constraint be increased or decreased without changing the optimal basis?
(e) By how much can the objective coefficient of x_1 be increased or decreased without changing the optimal basis?
(f) Suppose a new variable x_4 was added with a coefficient value of $c_4 = 6$ and constraint coefficients $A_4 = \begin{bmatrix} 2 & -1 & 4 \end{bmatrix}^T$. Would the optimal basis remain optimal? Why or why not?

Exercise 4.6
Consider the linear program

$$\begin{array}{ll} \textit{maximize} & 101x_1 - 87x_2 - 23x_3 \\ \textit{subject to} & 6x_1 - 13x_2 - 3x_3 \leq 12 \\ & 6x_1 + 11x_2 + 22x_3 \leq 46 \\ & x_1 + 6x_2 + x_3 \leq 13 \\ & x_1 \geq 0, x_2 \geq 0, x_3 \geq 0. \end{array}$$

(a) What is the optimal solution?
(b) What is the optimal basis?
(c) What are the optimal dual variables?
(d) By how much can the right-hand side of the second constraint be increased or decreased without changing the optimal basis?

(e) By how much can the objective coefficient of x_3 be increased or decreased without changing the optimal basis?

(f) Suppose a new variable x_4 was added with a coefficient value of $c_4 = 45$ and constraint coefficients $A_4 = \begin{bmatrix} 10 & -15 & 14 \end{bmatrix}^T$. Would the optimal basis remain optimal? Why or why not?

Exercise 4.7

Find the duals of the following linear program:

$$\begin{array}{ll} \text{(a)}\,\textit{maximize} & 7x_1 + 17x_2 + 17x_3 \\ \quad\textit{subject to} & x_1 + x_2 = 8 \\ & x_1 + 4x_2 + 3x_3 \geq 14 \\ & \quad 3x_2 + 4x_3 \leq 9 \\ & x_1 \leq 0, x_2 \geq 0, x_3 \text{ unrestricted} \end{array}$$

$$\begin{array}{ll} \text{(b)}\,\textit{minimize} & -7x_1 + 22x_2 + 18x_3 \\ \quad\textit{subject to} & x_1 + 5x_2 + x_3 + x_4 \leq 8 \\ & x_1 + x_2 + x_3 \geq 14 \\ & \quad 3x_2 \quad +4x_4 = 9 \\ & x_1 \geq 0, x_2 \leq 0, x_3 \text{ unrestricted}, x_4 \geq 0 \end{array}$$

Exercise 4.8

Consider the standard form of a linear program

$$\begin{array}{ll} \textit{minimize} & c^T x \\ \textit{subject to} & Ax \leq b \\ & x \geq 0. \end{array}$$

(a) Formulate the dual problem.
(b) State and prove the corresponding weak duality theorem.
(c) State the corresponding strong duality theorem.
(d) State the corresponding complementary slackness conditions.

Exercise 4.9

Consider a linear program in standard form

$$\begin{array}{ll} \textit{minimize} & c^T x \\ \textit{subject to} & Ax = b \\ & x \geq 0. \end{array}$$

where A is $m \times n$ and has full row rank. Let this be the primal problem P.

(a) Find the dual of P.

(b) If an optimal solution x^* for the primal is always non-degenerate, then is the optimal solution to the dual always non-degenerate? Explain.

(c) If an optimal solution x^* for the primal is degenerate, then can one use complementary slackness to always find a corresponding optimal dual solution? Explain.

(d) If both the primal and dual problems have feasible solutions, explain why there is no duality gap in this case.

Exercise 4.10

If a linear program in standard form has a constraint matrix A that is symmetric (i.e., $A = A^T$) and $c = b$, then prove that achieving feasibility is the same as achieving optimality.

Exercise 4.11

If a linear program in standard form has a finite optimal solution, show that any new linear program in standard form derived from the original linear program by changing just the original right-hand side vector b to any other vector b^* will always have a finite optimal solution as well.

Exercise 4.12

Prove Theorem 4.21.

Exercise 4.13

Suppose that Torvelo Surfing is a surfboard manufacturer that produces two types of surfboards. The first type of surfboard is a short board and the other a long board. Each surfboard requires molding and then polishing. Suppose that each unit of a short board requires 1 hour of molding and 1 hour of polishing and each unit of the long board requires 1.5 hours of molding and 2 hours of polishing. Each unit of a short board generates a \$100 profit and each unit of a long board generates \$150 profit. The total amount of molding time available is 50 hours and available polishing time is 75 hours. Would it be a good idea to obtain more molding or polishing labor and if so how much should each resource cost per unit?

Exercise 4.14

Consider a linear program in standard form

$$\begin{array}{ll} \textit{minimize} & c^T x \\ \textit{subject to} & Ax = b \\ & x \geq 0. \end{array}$$

Interpret the problem as using n different resources x to meet demands b for each of m different products at minimum cost. The matrix A gives the technology for producing products from resources. Suppose that x^* is a non-degenerate optimal basic feasible solution with basis B. Prove that the corresponding optimal dual values π^* are marginal values (prices) for the resources.

Exercise 4.15

Consider the following two systems

$$\begin{array}{c} \text{(I) } Ax = b, x \geq 0 \\ \text{(II) } A^T\pi \leq 0, b^T\pi > 0. \end{array}$$

(a) Prove that exactly one of the two systems has a solution, but not both. (Hint: Form appropriate linear programs with constraints (I) and (II).)

(b) Use (a) to give an alternative proof of the Strong Duality Theorem.

Notes and References

Duality in linear programming appeared in the work of Gale, Kuhn, and Tucker (1951), although the origins of duality can be seen from the work of Von Neumann (1945). The presentation of linear programming duality has two distinct approaches. In most developments of linear programming, strong duality is proved via the simplex method as in this chapter. Exercise 4.15 maps another approach that does not rely on the simplex method, but only on systems of alternatives; see Farkas (1901), Mangasarian (1969), and Murty (1988). The strong duality and complementary slackness results of linear programming are equivalent to the KKT conditions for a linear program. KKT conditions apply more widely to non-linear programming problems, but in general only provide in most cases necessary conditions for optimality where as the linear programming KKT conditions are both necessary and sufficient; see Avriel (2003) and Bazaraa, Sherali, and Shetty (2006). The dual simplex method was developed by Lemke (1954) and Beale (1954).

5

Dantzig-Wolfe Decomposition

5.1 Introduction

In this chapter, we consider methods for solving linear programs that exhibit special structure. In particular, we consider linear programs that are in block angular form and develop the Dantzig-Wolfe decomposition method for solving such problems. Economic interpretations of the decomposition are discussed.

5.2 Decomposition for Block Angular Linear Programs

Consider a linear program of the form

$$\begin{array}{ll} \textit{minimize} & c^T x \\ \textit{subject to} & Ax \leq b \\ & x \geq 0, \end{array}$$

and where A is an $m \times n$ matrix, and suppose that the constraint matrix is of the form

$$A = \begin{bmatrix} L_1 & L_2 & \cdots & L_K \\ A_1 & & & \\ & A_2 & & \vdots \\ & & \ddots & \\ & & & A_K \end{bmatrix},$$

where L_k is a submatrix with dimension $m_L \times n_k$ for $k = 1, ..., K$ and A_k is a submatrix of dimension $m_k \times n_k$ for $k = 1, ..., K$ such that $\sum_{k=1}^{K} n_k = n$ and $m_L + \sum_{k=1}^{K} m_k = m$. Such a form for A is called block angular.

Let x^k, c^k, and b^k be corresponding vectors such that

$$x = \begin{bmatrix} x^1 \\ \vdots \\ x^K \end{bmatrix}, c = \begin{bmatrix} c^1 \\ \vdots \\ c^K \end{bmatrix} \text{ and } b = \begin{bmatrix} b^1 \\ \vdots \\ b^K \end{bmatrix} \text{ so that } Ax \leq b.$$

Then, the linear program can be written as

$$\begin{array}{lccccccc}
\textit{minimize} & (c^1)^T x^1 + & (c^2)^T x^2 + & \cdots & + (c^K)^T x^K & & \\
\textit{subject to} & L_1 x^1 + & L_2 x^2 + & \cdots & + L_K x^K & \leq & b^0 \\
& A_1 x^1 & & & & \leq & b^1 \\
& & A_2 x^2 & & & \leq & b^2 \\
& & & \ddots & & \vdots & \vdots \\
& & & & A_K x^K & \leq & b^K \\
& x^1 \geq 0 & x^2 \geq 0 & \cdots & x^K \geq 0, & &
\end{array}$$

or more compactly as

$$\begin{array}{llll}
\textit{minimize} & \sum_{k=1}^{K} (c^k)^T x^k & & \\
\textit{subject to} & \sum_{k=1}^{K} L_k x^k & \leq b^0 & \\
& A_k x^k & \leq b^k & k = 1, ..., K \\
& x^k & \geq 0 & k = 1, ..., K.
\end{array}$$

The constraints $\sum_{k=1}^{K} L_k x^k \leq b_0$ are called coupling or linking constraints since without them, the problem decomposes into K independent subproblems where the kth subproblem is

$$\begin{array}{ll}
\textit{minimize} & (c^k)^T x^k \\
\textit{subject to} & A_k x^k \leq b^k \\
& x^k \geq 0,
\end{array}$$

where each sub-problem is a linear program.

Example 5.1

Consider the linear program

$$\begin{array}{lccccc}
\textit{minimize} & -2x_1 & -3x_2 & -5x_3 & -4x_4 & \\
\textit{subject to} & x_1 + & x_2 + & 2x_3 & & \leq 4 \\
& & x_2 + & x_3 + & x_4 & \leq 3 \\
& 2x_1 + & x_2 & & & \leq 4 \\
& x_1 + & x_2 & & & \leq 2 \\
& & & x_3 + & x_4 & \leq 2 \\
& & & 3x_3 + & 2x_4 & \leq 5 \\
& x_1 \geq 0 & x_2 \geq 0 & x_3 \geq 0 & x_4 \geq 0. &
\end{array}$$

The first two constraints

$$\begin{array}{rrrrl} x_1 + & x_2 + & 2x_3 & & \leq 4 \\ & x_2 + & x_3 + & x_4 & \leq 3 \end{array}$$

are the linking constraints and can be represented as $L_1x^1 + L_2x^2 \leq b^0$ where

$$L_1 = \begin{bmatrix} 1 & 1 \\ 0 & 1 \end{bmatrix}, L_2 = \begin{bmatrix} 2 & 0 \\ 1 & 1 \end{bmatrix}, b^0 = \begin{bmatrix} 4 \\ 3 \end{bmatrix}$$

and

$$x^1 = \begin{bmatrix} x_1 \\ x_2 \end{bmatrix}, x^2 = \begin{bmatrix} x_3 \\ x_4 \end{bmatrix}.$$

Without these constraints, the linear program would consist of the two independent linear programming subproblems where the first independent subproblem is

$$\begin{array}{lrrl} \textit{minimize} & -2x_1 & -3x_2 & \\ \textit{subject to} & 2x_1 + & x_2 & \leq 4 \\ & x_1 + & x_2 & \leq 2 \\ & x_1 \geq 0 & x_2 \geq 0 & \end{array}$$

where

$$c^1 = \begin{bmatrix} -2 \\ -3 \end{bmatrix}, A^1 = \begin{bmatrix} 2 & 1 \\ 1 & 1 \end{bmatrix}, \text{ and } b^1 = \begin{bmatrix} 4 \\ 2 \end{bmatrix},$$

and the second independent subproblem is

$$\begin{array}{lrrl} \textit{minimize} & -5x_3 & -4x_4 & \\ \textit{subject to} & & & \\ & x_3 + & x_4 & \leq 2 \\ & 3x_3 + & 2x_4 & \leq 5 \\ & x_3 \geq 0 & x_4 \geq 0 & \end{array}$$

where

$$c^2 = \begin{bmatrix} -5 \\ -4 \end{bmatrix}, A^2 = \begin{bmatrix} 1 & 1 \\ 3 & 2 \end{bmatrix}, \text{ and } b^2 = \begin{bmatrix} 2 \\ 5 \end{bmatrix}.$$

The idea in decomposition is to exploit the structure of such a linear program so that the entire problem does not have to be solved all at once, but instead solve problems that are smaller and usually more tractable. In Dantzig-Wolfe decomposition, the problem will be decomposed into a master problem that will be concerned with the linking constraints only, and into subproblems that result from the decoupling of the linking constraints.

The master problem (MP) is of the form

$$\begin{array}{lll} minimize & \sum\limits_{k=1}^{K} \left(c^k\right)^T x^k & \\ subject\ to & \sum\limits_{k=1}^{K} L_k x^k & \leq b^0. \end{array}$$

5.3 Master Problem Reformulation

It is not enough to have the decomposition into master and subproblems as above. To make the decomposition effective, the master problem needs reformulation. The reformulation will enable the master problem and subproblems to exchange information regarding progress toward an optimal solution for the original linear program while enabling each problem to be solved separately. The key to reformulation is the fact that the subproblems are all linear programs and so the feasible sets are polyhedrons. Recall from the Resolution Theorem in Chapter 2 that any feasible point of a linear program can be represented as a convex combination of the extreme points and a non-negative linear combination of extreme directions of the feasible set.

Let $P_k = \{x^k | A_k x^k \leq b^k, x^k \geq 0\}$ be the feasible set of subproblem SP_k and $v_1^k, v_2^k, ..., v_{N_k}^k$ be the extreme points of P_k and $d_1^k, d_2^k, ..., d_{l_k}^k$ be the extreme directions of P_k. Then, by the Resolution Theorem, any point $x^k \in P_k$ can be expressed as

$$x^k = \sum_{i=1}^{N_k} \lambda_i^k v_i^k + \sum_{j=1}^{l_k} \mu_j^k d_j^k,$$

where $\sum_{i=1}^{N_k} \lambda_i^k = 1$, $\lambda_i^k \geq 0$ for $i = 1, ..., N_k$ and $\mu_j^k \geq 0$ for $j = 1, ..., l_k$. Substituting this representation into the master problem gives

$$
\begin{array}{lll}
\textit{minimize} & \sum_{k=1}^{K} \left(c^k\right)^T \left(\sum_{i=1}^{N_k} \lambda_i^k v_i^k + \sum_{j=1}^{l_k} \mu_j^k d_j^k\right) & \\
\textit{subject to} & \sum_{k=1}^{K} L_k\left(\sum_{i=1}^{N_k} \lambda_i^k v_i^k + \sum_{j=1}^{l_k} \mu_j^k d_j^k\right) \leq b^0 & \\
& \sum_{i=1}^{N_k} \lambda_i^k = 1 & k = 1, ..., K \\
& \lambda_i^k \geq 0 \ , \ i = 1, ..., N_k & k = 1, ..., K \\
& \mu_j^k \geq 0 \ , \ j = 1, ..., l_k & k = 1, ..., K.
\end{array}
$$

After simplification, the master problem becomes

$$
\begin{array}{lll}
\textit{minimize} & \sum_{k=1}^{K} \sum_{i=1}^{N_k} \lambda_i^k \left(c^k\right)^T (v_i^k) + \sum_{k=1}^{K} \sum_{j=1}^{l_k} \mu_j^k \left(c^k\right)^T (d_j^k) & \\
\textit{subject to} & \sum_{k=1}^{K} \sum_{i=1}^{N_k} \lambda_i^k (L_k v_i^k) + \sum_{k=1}^{K} \sum_{j=1}^{l_k} \mu_j^k (L_k d_j^k) \leq b^0 & \\
& \sum_{i=1}^{N_k} \lambda_i^k = 1 & k = 1, ..., K \\
& \lambda_i^k \geq 0 \ , \ i = 1, ..., N_k & k = 1, ..., K \\
& \mu_j^k \geq 0 \ , \ j = 1, ..., l_k & k = 1, ..., K.
\end{array}
$$

The master problem now has as variables λ_i^k and μ_j^k. Corresponding to each extreme point $v_i^k \in P_k$, let $\mathrm{f}_i^k = \left(c^k\right)^T (v_i^k)$ and $q_i^k = L_k v_i^k$ and corresponding to each extreme direction d_j^k of P_k, let $\bar{\mathrm{f}}_j^k = \left(c^k\right)^T (d_j^k)$ and $\bar{q}_j^k = L_k d_j^k$. Then, the master program can be reformulated as

$$
\begin{array}{lll}
\textit{minimize} & \sum_{k=1}^{K} \sum_{i=1}^{N_k} \lambda_i^k \mathrm{f}_i^k + \sum_{k=1}^{K} \sum_{j=1}^{l_k} \mu_j^k \bar{\mathrm{f}}_j^k & \\
\textit{subject to} & \sum_{k=1}^{K} \sum_{i=1}^{N_k} \lambda_i^k q_i^k + \sum_{k=1}^{K} \sum_{j=1}^{l_k} \mu_j^k \bar{q}_j^k \ \leq b^0 & \\
& \sum_{i=1}^{N_k} \lambda_i^k = 1 & k = 1, ..., K \\
& \lambda_i^k \geq 0 \ , \ i = 1, ..., N_k & k = 1, ..., K \\
& \mu_j^k \geq 0 \ , \ j = 1, ..., l_k & k = 1, ..., K.
\end{array}
$$

The constraints of type $\sum_{i=1}^{N_k} \lambda_i^k = 1$ are called convexity constraints associated with subproblem k. The master problem can be more compactly represented as

$$
\begin{array}{ll}
\textit{minimize} & \mathrm{f}_v^T \lambda + \mathrm{f}_d^T \mu \\
\textit{subject to} & Q_v \lambda + Q_d \mu + s = r \\
& \lambda \geq 0, \mu \geq 0, s \geq 0,
\end{array}
$$

where

$$\lambda = (\lambda_1^1, ..., \lambda_{N_1}^1, \lambda_1^2, ..., \lambda_{N_2}^2, ..., \lambda_1^K, ..., \lambda_{N_K}^K)^T$$
$$\mu = (\mu_1^1, ..., \mu_{l_1}^1, \mu_1^2, ..., \mu_{l_2}^2, ..., \mu_1^K, ..., \mu_{l_K}^K)^T$$
$$r^T = ((b^0)^T, e^T) = ((b^0)^T, (1..., 1)^T),$$

where s is a vector of slack variables and e is the vector of dimension K with all components equal to 1. Q_v is a matrix such that the column associated with λ_i^k is

$$\begin{bmatrix} q_i^k \\ e_k \end{bmatrix} = \begin{bmatrix} L_k v_i^k \\ e_k \end{bmatrix},$$

where e_k is the kth unit vector. Q_d is a matrix such that the column associated with μ_j^k is

$$\begin{bmatrix} \bar{q}_j^k \\ 0 \end{bmatrix} = \begin{bmatrix} L_k d_j^k \\ 0 \end{bmatrix}.$$

The number of variables λ_i^k and μ_j^k can be extremely large for even moderately sized problems since feasible sets of subproblems (linear programs) can have an extremely large number of extreme points and extreme directions. In other words, there will be many more columns than rows in the reformulated master problem.

5.4 Restricted Master Problem and the Revised Simplex Method

The key idea to get around this difficulty of handling an extremely large number of variables is to use the revised simplex method to solve the master problem. The major advantage is that it is not necessary to formulate the entire reformulated master problem since the vast majority of the variables will be zero (i.e., non-basic) at an optimal (basic feasible) solution.

This motivates the construction of a smaller version of the master problem called the restricted master problem where only a small subset of the variables λ_i^k and μ_j^k are included corresponding to a current basic feasible solution, and the remaining variables are non-basic, i.e., set to zero. If the reduced costs of the basic feasible solution are non-negative, then the revised simplex method stops with the optimal solution, else some non-basic variable with negative reduced cost is selected to enter the basis.

Given a current basis B for the restricted master problem, let $\pi^T = f_B^T B^{-1}$

where f_B is a vector consisting of the quantities f_i^k and $\bar{f}_j^k$ associated with variables λ_i^k and μ_j^k, which are basic. We assume that the components of π are arranged so that

$$\pi = \begin{bmatrix} \pi^1 \\ \pi_1^2 \\ \vdots \\ \pi_K^2 \end{bmatrix},$$

where π^1 are the dual variables associated with the linking constraints and π_i^2 is the dual variable associated with the convexity constraint of subproblem i in the restricted master problem.

Then, the reduced cost corresponding to a non-basic variable λ_i^k is of the form

$$r_i^k = f_i^k - \pi^T \begin{bmatrix} q_i^k \\ e_k \end{bmatrix} = \left(c^k\right)^T (v_i^k) - (\pi^1)^T L_k v_i^k - \pi_k^2,$$

and the reduced cost for corresponding to a non-basic variable μ_j^k is of the form

$$\bar{r}_j^k = \bar{f}_j^k - \pi^T \begin{bmatrix} \bar{q}_j^k \\ 0 \end{bmatrix} = \left(c^k\right)^T (d_j^k) - (\pi^1)^T L_k d_j^k.$$

There will be a considerable number of non-basic variables, but fortunately one does not have to compute all of the reduced costs. In fact, it will suffice to determine only the minimal reduced cost among all of the non-basic variables. To this end, we let

$$r_{\min} = \min_{k \in \{1,\ldots,K\}} \left\{ \min_{i \in \{1,\ldots,N_k\}} \{r_i^k\} \right\}$$

or

$$r_{\min} = \min_{k \in \{1,\ldots,K\}} \left\{ \min_{i \in \{1,\ldots,N_k\}} \{\left(c^k\right)^T (v_i^k) - (\pi^1)^T L_k v_i^k - \pi_k^2\} \right\}.$$

Let $r_*^k = \min_{i \in \{1,\ldots,N_k\}} \{r_i^k\}$. Then, r_*^K is equivalent to the optimal objective function of the subproblem SP_k

$$\begin{array}{ll} \textit{minimize} & \sigma_k = (\left(c^k\right)^T - (\pi^1)^T L_k) x^k \\ \textit{subject to} & A_k x^k \leq b^k \\ & x^k \geq 0. \end{array}$$

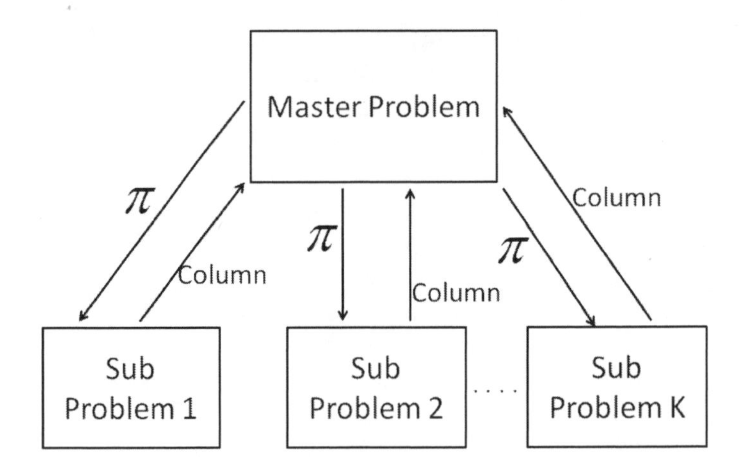

FIGURE 5.1
Dantzig-Wolfe Decomposition.

Since the term π_k^2 is fixed, it can be removed from the objective function. We assume that the revised simplex method will be used to solve the subproblem, so if the subproblem is bounded, an optimal extreme point x_k will be generated and x_k will be one of the extreme points v_i^k.

Let the optimal extreme point of subproblem SP_k be $v_{i^*}^k$ for some index $i^* \in \{1, ..., N_k\}$, and let σ_k^* denote the optimal objective function value of SP_k. Then, $r_*^k = \sigma_k^* - \pi_k^2$.

There are three possibilities in solving the subproblems SP_k in attempting to generate $r_{\min}$.

(1) If all subproblems are bounded and $r_{\min} = \min\limits_{k \in \{1,...,K\}} \{r_*^k\} < 0$, then let t be the index k such that $r_{\min} = r_*^t$. The column $\begin{bmatrix} q_{i^*}^t \\ e_t \end{bmatrix} = \begin{bmatrix} L_t v_{i^*}^t \\ e_t \end{bmatrix}$ associated with the optimal extreme point $v_{i^*}^t$ of subproblem SP_t that achieved $r_{\min} = r_*^t$ is entered into the basis B.

(2) If all subproblems are bounded and $r_{\min} = \min\limits_{k \in \{1,...,K\}} \{r_*^k\} \geq 0$, then the current basis B is optimal.

(3) If there is at least one subproblem that is unbounded, then let s be the index k of such an unbounded subproblem SP_s. The revised simplex method will return an extreme direction $d_{j^*}^s$ for some $j^* \in \{1, ..., l_s\}$ associated with SP_s such that $((c^s)^T - (\pi^1)^T L_s) d_{j^*}^s < 0$, and so the column $\begin{bmatrix} \bar{q}_{j^*}^s \\ 0 \end{bmatrix} = \begin{bmatrix} L_s d_{j^*}^s \\ 0 \end{bmatrix}$ associated with $\mu_{j^*}^s$ (the multiplier of $d_{j^*}^s$) can enter the basis B.

Price-Directed Decomposition The Dantzig-Wolfe decomposition can be interpreted as a price-directed decomposition in that the restricted master problem generates prices in the form of dual values π and these prices are sent to the subproblems SP_k to form the coefficients of the objective function σ_k. The subproblems then compute optimal solutions x_k or generate extreme directions and send back up to the master problem the appropriate column to possibly enter the basis. The process iterates until an optimal solution is found or the problem is declared unbounded. See Figure 5.1.

5.5 Dantzig-Wolfe Decomposition

We now provide the detailed steps of the Dantzig-Wolfe Decomposition.

Dantzig-Wolfe Decomposition Method

Step 0: (Initialization)

Generate an initial basis B for the master problem.
Let x_B be the basic variables and $\bar{B}$ the index set of the basic variables and set all other variables to non-basic (zero) to get the restricted master problem.
Go to Step 1.

Step 1: (Simplex Multiplier Generation)

Solve for π in the linear system $B^T\pi = \mathrm{f}_B$.
Go to Step 2.

Step 2: (Optimality Check)

For each $k = 1, ..., K$ solve SP_k, i.e.,

$$\begin{array}{lll} \textit{minimize} & \sigma_k = \left((c^k)^T - (\pi^1)^T L_k\right)x^k & \\ \textit{subject to} & A_k x^k & \leq b^k \\ & x^k \geq 0 & \end{array}$$

using the revised simplex method. If SP_k is unbounded, then go to Step 3, else let $x^k = v^k_{i^*}$ denote the optimal basic feasible solution and compute $r^k_* = \sigma^*_k - \pi^2_k$.
If $r_{\min} = \min\limits_{k\in\{1,...,K\}} \{r^k_*\} \geq 0$, then STOP; the current basis B is optimal, else go to Step 3.

Step 3: (Column Generation)

If all subproblems SP_k are bounded and $r_{\min} = \min\limits_{k\in\{1,...,K\}} \{r^k_*\} < 0$, then let t

be the index of k in SP_k such that $r_{\min} = r_*^t$. Let $\bar{a} = \begin{bmatrix} q_{i*}^t \\ e_t \end{bmatrix} = \begin{bmatrix} L_t v_{i*}^t \\ e_t \end{bmatrix}$ where v_{i*}^t is the optimal extreme point of SP_t and go to Step 4. Else there is a subproblem SP_s that is unbounded, and so an extreme direction d_{j*}^s will be generated such that $((c^s)^T - (\pi^1)^T L_s)d_{j*}^s < 0$, and so let $\bar{a}$ $= \begin{bmatrix} \bar{q}_{j*}^s \\ 0 \end{bmatrix} = \begin{bmatrix} L_s d_{j*}^s \\ 0 \end{bmatrix}$ and go to Step 4.

Step 4: (Descent Direction Generation)
Solve for d in the linear system $Bd = -\bar{a}$.
If $d \geq 0$, then the linear program is unbounded STOP, else go to Step 5.

Step 5: (Step Length Generation)
Compute the step length $\alpha = \min_{l \in \bar{B}}\{-\frac{x_l}{d_l} | d_l < 0\}$ (the minimum ratio test).
Let l^* be the index of the basic variable that attains the minimum ratio α.
Go to Step 6.

Step 6: (Update Basic Variables and Basis)
Now let $x_B = x_B + \alpha d$. Go to Step 7.

Step 7: (Basis Update)
Let B_{l^*} be the column in B associated with the leaving basic variable x_{l^*}
Update the basis matrix B by removing B_{l^*} and adding the column $\bar{a}$ and update $\bar{B}$.
Go to Step 1.

Example 5.2

We illustrate the Dantzig-Wolfe on the linear program in Example 5.1, which happens to be bounded, with subproblem feasible sets that are also bounded, and so extreme directions will not exist.

The master problem has the form (after adding slack variables $s = (s_1, s_2)^T$)

$$
\begin{array}{lllll}
\textit{minimize} & \sum_{i=1}^{N_1} \lambda_i^1 f_i^1 + \sum_{i=1}^{N_2} \lambda_i^2 f_i^2 & & & \\
\textit{subject to} & \sum_{i=1}^{N_1} \lambda_i^1 L_1 v_i^1 + \sum_{i=1}^{N_2} \lambda_i^2 L_2 v_i^2 + s & & & = b^0 \\
& \sum_{i=1}^{N_1} \lambda_i^1 & & & = 1 \\
& & \sum_{i=1}^{N_2} \lambda_i^2 & & = 1 \\
& \lambda_i^1 \geq 0 & & & i = 1, ..., N_1 \\
& & \lambda_i^2 \geq 0 & & i = 1, ..., N_2 \\
& s \geq 0. & & &
\end{array}
$$

Recall that it is not necessary to know the quantities N_1 and N_2 in advance.

Step 0: Start with as basic variables $s_1 = b_1^0 = 4$, $s_2 = b_2^0 = 3$, $\lambda_1^1 = 1$, and $\lambda_1^2 = 1$, so $x_B = (s_1, s_2, \lambda_1^1, \lambda_1^2)^T$ with initial extreme points $v_1^1 = \begin{bmatrix} x_1 \\ x_2 \end{bmatrix} = \begin{bmatrix} 0 \\ 0 \end{bmatrix}$ and $v_1^2 = \begin{bmatrix} x_3 \\ x_4 \end{bmatrix} = \begin{bmatrix} 0 \\ 0 \end{bmatrix}$.

Then, the restricted master problem is

$$\begin{array}{lllll} \textit{minimize} & \lambda_1^1 \mathrm{f}_1^1 + \lambda_1^2 \mathrm{f}_1^2 & & & \\ \textit{subject to} & \lambda_1^1 L_1 v_1^1 & + \lambda_1^2 L_2 v_1^2 + s & & = b^0 \\ & \lambda_1^1 & & & = 1 \\ & & \lambda_1^2 & & = 1 \\ & \lambda_1^1 \geq 0 & & & \\ & & \lambda_1^2 \geq 0 & & \\ & s \geq 0, & & & \end{array}$$

and since the initial extreme point for each subproblem is the zero vector we get

$$\begin{array}{llllll} \textit{minimize} & 0 & & & & \\ \textit{subject to} & & & s_1 & & = 4 \\ & & & & s_2 & = 3 \\ & \lambda_1^1 & & & & = 1 \\ & & \lambda_1^2 & & & = 1 \\ & \lambda_1^1 \geq 0 & & & & \\ & & \lambda_1^2 \geq 0 & & & \\ & s \geq 0. & & & & \end{array}$$

The basis matrix $B = I$, i.e., the 4 by 4 identity matrix.

Iteration 1

Step 1: $\mathrm{f}_B = \begin{bmatrix} c_{s_1} \\ c_{s_2} \\ \mathrm{f}_1^1 \\ \mathrm{f}_1^2 \end{bmatrix} = \begin{bmatrix} 0 \\ 0 \\ (c^1)^T v_1^1 \\ (c^2)^T v_1^2 \end{bmatrix} = \begin{bmatrix} 0 \\ 0 \\ 0 \\ 0 \end{bmatrix}$, then solving $B^T \pi = \mathrm{f}_B$ gives

$$\pi = \begin{bmatrix} \pi^1 \\ \pi_1^2 \\ \pi_2^2 \end{bmatrix} = \begin{bmatrix} \pi_1^1 \\ \pi_2^1 \\ \pi_1^2 \\ \pi_2^2 \end{bmatrix} = \begin{bmatrix} 0 \\ 0 \\ 0 \\ 0 \end{bmatrix}.$$

Step 2:

Now the objective function of SP_1 is $\sigma_1 = ((c^1)^T - (\pi^1)^T L_1) x^1 = (c^1)^T x^1 = -2x_1 - 3x_2$, so SP_1 is

$$\begin{array}{llll} \textit{minimize} & -2x_1 & -3x_2 & \\ \textit{subject to} & 2x_1 + & x_2 & \leq 4 \\ & x_1 + & x_2 & \leq 2 \\ & x_1 \geq 0 & x_2 \geq 0. & \end{array}$$

The optimal solution is $x^1 = \begin{bmatrix} x_1 \\ x_2 \end{bmatrix} = \begin{bmatrix} 0 \\ 2 \end{bmatrix}$ with objective function value $\sigma_1^* = -6$, and so $r_*^1 = \sigma_1^* - \pi_1^2 = -6 - 0 = -6$. Let $v_2^1 = x^1$.

Now the objective function of SP_2 is $\sigma_2 = \left((c^2)^T - (\pi^1)^T L_2\right)x^2 = (c^2)^T x^2 = -5x_3 - 4x_4$, so SP_2 is

$$\begin{array}{llll} \textit{minimize} & -5x_3 & -4x_4 & \\ \textit{subject to} & & & \\ & x_3 + & x_4 & \leq 2 \\ & 3x_3 + & 2x_4 & \leq 5 \\ & x_3 \geq 0 & x_4 \geq 0. & \end{array}$$

The optimal solution is $x^2 = \begin{bmatrix} x_3 \\ x_4 \end{bmatrix} = \begin{bmatrix} 1 \\ 1 \end{bmatrix}$ with objective function value $\sigma_2^* = -9$, and so $r_*^2 = \sigma_2^* - \pi_2^2 = -9 - 0 = -9$. Let $v_2^2 = x^2$.

Step 3:

$$r_{\min} = r_*^2 = -9 \text{ and so } \bar{a} = \begin{bmatrix} q_2^2 \\ e_2 \end{bmatrix} = \begin{bmatrix} L_2 v_2^2 \\ e_2 \end{bmatrix} = \begin{bmatrix} 2 \\ 2 \\ 0 \\ 1 \end{bmatrix}.$$

Step 4: $Bd = -\bar{a}$ and so $d = -\begin{bmatrix} 2 \\ 2 \\ 0 \\ 1 \end{bmatrix}$.

Step 5: $\alpha = \min\{\frac{4}{2}, \frac{3}{2}, \frac{1}{1}\} = 1$ and so $\lambda_2^2 = \alpha = 1$.

Step 6: $x_B = \begin{bmatrix} s_1 \\ s_2 \\ \lambda_1^1 \\ \lambda_1^2 \end{bmatrix} = \begin{bmatrix} 4 \\ 3 \\ 1 \\ 1 \end{bmatrix} + (1)\begin{bmatrix} -2 \\ -2 \\ 0 \\ -1 \end{bmatrix} = \begin{bmatrix} 2 \\ 1 \\ 1 \\ 0 \end{bmatrix}$, so λ_1^2 leaves the basis and λ_2^2 enters the basis and the updated basic variable set is $x_B = \begin{bmatrix} s_1 \\ s_2 \\ \lambda_1^1 \\ \lambda_2^2 \end{bmatrix} = \begin{bmatrix} 2 \\ 1 \\ 1 \\ 1 \end{bmatrix}$.

Step 7: Column $\bar{a}$ enters the basis and the column $(0,0,1,0)^T$ associated with λ_1^2 leaves the basis and so

$$B = \begin{bmatrix} 1 & 0 & 0 & 2 \\ 0 & 1 & 0 & 2 \\ 0 & 0 & 1 & 0 \\ 0 & 0 & 0 & 1 \end{bmatrix}.$$

Go to Step 1.

Iteration 2

Step 1: $f_B = \begin{bmatrix} c_{s_1} \\ c_{s_2} \\ f_1^1 \\ f_2^2 \end{bmatrix} = \begin{bmatrix} 0 \\ 0 \\ (c^1)^T v_1^1 \\ (c^2)^T v_2^2 \end{bmatrix} = \begin{bmatrix} 0 \\ 0 \\ 0 \\ -9 \end{bmatrix}$, then solving $B^T \pi = f_B$ gives

$$\pi = \begin{bmatrix} \pi^1 \\ \pi_1^2 \\ \pi_2^2 \end{bmatrix} = \begin{bmatrix} \pi_1^1 \\ \pi_2^1 \\ \pi_1^2 \\ \pi_2^2 \end{bmatrix} = \begin{bmatrix} 0 \\ 0 \\ 0 \\ -9 \end{bmatrix}.$$

Step 2:

Now the objective functions of SP_1 and SP_2 remain the same as $\pi^1 = \begin{bmatrix} 0 \\ 0 \end{bmatrix}$, so the subproblems remain the same. Since the optimal solution v_2^2 for SP_2 and its multiplier $\lambda_2^2 = 1$ are already in the restricted master problem and then $r_{\min} = r_*^1 = -6$, so we enter into the basis λ_2^1 of the restricted master problem along with $v_2^1 = \begin{bmatrix} x_1 \\ x_2 \end{bmatrix} = \begin{bmatrix} 0 \\ 2 \end{bmatrix}$ the optimal extreme point of SP_1.

Step 3:

$$\text{So } \bar{a} = \begin{bmatrix} q_2^1 \\ e_1 \end{bmatrix} = \begin{bmatrix} L_1 v_2^1 \\ e_1 \end{bmatrix} = \begin{bmatrix} 2 \\ 2 \\ 1 \\ 0 \end{bmatrix}.$$

Step 4: $Bd = -\bar{a}$ and so $d = -\begin{bmatrix} 2 \\ 2 \\ 1 \\ 0 \end{bmatrix}$.

Step 5: $\alpha = \min\{\frac{2}{2}, \frac{1}{2}, \frac{1}{1}\} = 1$ and so $\lambda_2^1 = \alpha = \frac{1}{2}$.

Step 6: $x_B = \begin{bmatrix} s_1 \\ s_2 \\ \lambda_1^1 \\ \lambda_2^2 \end{bmatrix} = \begin{bmatrix} 2 \\ 1 \\ 1 \\ 1 \end{bmatrix} + (0.5)\begin{bmatrix} -2 \\ -2 \\ -1 \\ 0 \end{bmatrix} = \begin{bmatrix} 1 \\ 0 \\ 0.5 \\ 1 \end{bmatrix}$, so s_2 leaves the basis and λ_2^1 enters the basis and the updated basic variable set is $x_B = \begin{bmatrix} s_1 \\ \lambda_2^1 \\ \lambda_1^1 \\ \lambda_2^2 \end{bmatrix} = \begin{bmatrix} 1 \\ 0.5 \\ 0.5 \\ 1 \end{bmatrix}$.

Step 7: Column $\bar{a}$ enters the basis and the column $(0,1,0,0)^T$ associated with s_2 leaves the basis, and so

$$B = \begin{bmatrix} 1 & 2 & 0 & 2 \\ 0 & 2 & 0 & 2 \\ 0 & 1 & 1 & 0 \\ 0 & 0 & 0 & 1 \end{bmatrix}.$$

$\bar{B}$ is same as before, except λ_2^1 is in the position of the basis that s_2 previously occupied.

Go to Step 1.

Iteration 3

Step 1: $\mathrm{f}_B = \begin{bmatrix} c_{s_1} \\ \mathrm{f}_2^1 \\ \mathrm{f}_1^1 \\ \mathrm{f}_2^2 \end{bmatrix} = \begin{bmatrix} 0 \\ (c^1)^T v_2^1 \\ (c^1)^T v_1^1 \\ (c^2)^T v_2^2 \end{bmatrix} = \begin{bmatrix} 0 \\ -6 \\ 0 \\ -9 \end{bmatrix}$, then solving $B^T\pi = \mathrm{f}_B$ gives

$$\pi = \begin{bmatrix} \pi^1 \\ \pi_1^2 \\ \pi_2^2 \end{bmatrix} = \begin{bmatrix} \pi_1^1 \\ \pi_2^1 \\ \pi_1^2 \\ \pi_2^2 \end{bmatrix} = \begin{bmatrix} 0 \\ -3 \\ 0 \\ -3 \end{bmatrix}.$$

Step 2:

Now the objective functions of SP_1 and SP_2 change since $\pi^1 = \begin{bmatrix} 0 \\ -3 \end{bmatrix}$.

Now the objective function of SP_1 is $\sigma_1 = ((c^1)^T - (\pi^1)^T L_1)x^1 = -2x_1$, so SP_1 is

$$\begin{array}{llll} \textit{minimize} & -2x_1 & & \\ \textit{subject to} & 2x_1 + & x_2 & \leq 4 \\ & x_1 + & x_2 & \leq 2 \\ & x_1 \geq 0 & x_2 \geq 0. & \end{array}$$

The optimal solution is $x^1 = \begin{bmatrix} x_1 \\ x_2 \end{bmatrix} = \begin{bmatrix} 2 \\ 0 \end{bmatrix}$ with objective function value $\sigma_1^* = -4$ and so $r_*^1 = \sigma_1^* - \pi_1^2 = -4 - 0 = -4$. Let $v_3^1 = x^1$.

Now the objective function of SP_2 is $\sigma_2 = ((c^2)^T - (\pi^1)^T L_2)x^2 = -2x_3 - x_4$ so SP_2 is

$$\begin{array}{llllll} \textit{minimize} & -2x_3 & -x_4 & & \\ \textit{subject to} & x_3 + & x_4 & \leq & 2 \\ & 3x_3 + & 2x_4 & \leq & 5 \\ & x_3 \geq 0 & x_4 \geq 0. & & \end{array}$$

The optimal solution is $x^2 = \begin{bmatrix} x_3 \\ x_4 \end{bmatrix} = \begin{bmatrix} \frac{5}{3} \\ 0 \end{bmatrix}$ with objective function value $\sigma_2^* = -\frac{10}{3}$ and so $r_*^2 = \sigma_2^* - \pi_2^2 = -\frac{10}{3} - (-3) = -\frac{1}{3}$.

Step 3:

$$\text{So } \bar{a} = \begin{bmatrix} q_3^1 \\ e_1 \end{bmatrix} = \begin{bmatrix} L_1 v_3^1 \\ e_1 \end{bmatrix} = \begin{bmatrix} 2 \\ 0 \\ 1 \\ 0 \end{bmatrix}.$$

Step 4: $Bd = -\bar{a}$ and so $d = -\begin{bmatrix} 2 \\ 0 \\ 1 \\ 0 \end{bmatrix}$.

Step 5: $\alpha = \min\left\{\frac{1}{2}, \frac{\frac{1}{2}}{1}\right\} = \frac{1}{2}$ and so $\lambda_3^1 = \alpha = \frac{1}{2}$.

Step 6: $x_B = \begin{bmatrix} s_1 \\ \lambda_2^1 \\ \lambda_1^1 \\ \lambda_2^2 \end{bmatrix} = \begin{bmatrix} 1 \\ \frac{1}{2} \\ \frac{1}{2} \\ 1 \end{bmatrix} + (0.5)\begin{bmatrix} -2 \\ 0 \\ -1 \\ 0 \end{bmatrix} = \begin{bmatrix} 0 \\ \frac{1}{2} \\ 0 \\ 1 \end{bmatrix}$, so λ_1^1 leaves the basis, and λ_3^1 enters the basis, and the updated basic variable set is $x_B = \begin{bmatrix} s_1 \\ \lambda_2^1 \\ \lambda_3^1 \\ \lambda_2^2 \end{bmatrix} = \begin{bmatrix} 0 \\ \frac{1}{2} \\ \frac{1}{2} \\ 1 \end{bmatrix}$.

Step 7: Column $\bar{a}$ enters the basis and the column $(0, 0, 1, 0)^T$ associated with λ_1^1 leaves the basis, and so

$$B = \begin{bmatrix} 1 & 2 & 2 & 2 \\ 0 & 2 & 0 & 2 \\ 0 & 1 & 1 & 0 \\ 0 & 0 & 0 & 1 \end{bmatrix}.$$

Go to Step 1.

Iteration 4

Step 1: $f_B = \begin{bmatrix} s_1 \\ f_2^1 \\ f_3^1 \\ f_2^2 \end{bmatrix} = \begin{bmatrix} 0 \\ (c^1)^T v_2^1 \\ (c^1)^T v_3^1 \\ (c^2)^T v_2^2 \end{bmatrix} = \begin{bmatrix} 0 \\ -6 \\ -4 \\ -9 \end{bmatrix}$, then solving $B^T \pi = f_B$ gives

$$\pi = \begin{bmatrix} \pi^1 \\ \pi_1^2 \\ \pi_2^2 \end{bmatrix} = \begin{bmatrix} \pi_1^1 \\ \pi_2^1 \\ \pi_1^2 \\ \pi_2^2 \end{bmatrix} = \begin{bmatrix} 0 \\ -1 \\ -4 \\ -7 \end{bmatrix}.$$

Step 2:

The objective functions of SP_1 and SP_2 change since $\pi^1 = \begin{bmatrix} 0 \\ -1 \end{bmatrix}$.

Now the objective function of SP_1 is $\sigma_1 = ((c^1)^T - (\pi^1)^T L_1)x^1 = -2x_1 - 2x_2$, so SP_1 is

$$\begin{array}{llcl} \textit{minimize} & -2x_1 & -2x_2 & \\ \textit{subject to} & 2x_1 + & x_2 & \leq 4 \\ & x_1 + & x_2 & \leq 2 \\ & x_1 \geq 0 & x_2 \geq 0. & \end{array}$$

The optimal solution is $x^1 = \begin{bmatrix} x_1 \\ x_2 \end{bmatrix} = \begin{bmatrix} 0.0224 \\ 1.9776 \end{bmatrix}$ with objective function value $\sigma_1^* = -4$, and so $r_*^1 = \sigma_1^* - \pi_1^2 = -4 - (-4) = 0$.

Now the objective function of SP_2 is $\sigma_2 = ((c^2)^T - (\pi^1)^T L_2)x^2 = -4x_3 - 3x_4$, so SP_2 is

$$\begin{array}{llcl} \textit{minimize} & -4x_3 & -3x_4 & \\ \textit{subject to} & x_3 + & x_4 & \leq 2 \\ & 3x_3 + & 2x_4 & \leq 5 \\ & x_3 \geq 0 & x_4 \geq 0. & \end{array}$$

The optimal solution is $x^2 = \begin{bmatrix} x_3 \\ x_4 \end{bmatrix} = \begin{bmatrix} 1 \\ 1 \end{bmatrix}$ with objective function value $\sigma_2^* = -7$, and so $r_*^2 = \sigma_2^* - \pi_2^2 = -7 - (-7) = 0$.

Since $r_{\min} = 0$ we STOP, and the current basis B represents an optimal solution. That is, the optimal basic variables to the restricted master problem are

$$x_B = \begin{bmatrix} s_1 \\ \lambda_2^1 \\ \lambda_3^1 \\ \lambda_2^2 \end{bmatrix} = \begin{bmatrix} 0 \\ \frac{1}{2} \\ \frac{1}{2} \\ 1 \end{bmatrix},$$

and so the optimal solution in terms of the original variables can be recovered as

$$x^1 = \begin{bmatrix} x_1 \\ x_2 \end{bmatrix} = \lambda_2^1 v_2^1 + \lambda_3^1 v_3^1 = 0.5 \begin{bmatrix} 0 \\ 2 \end{bmatrix} + 0.5 \begin{bmatrix} 2 \\ 0 \end{bmatrix} = \begin{bmatrix} 1 \\ 1 \end{bmatrix}$$

and

$$x^2 = \begin{bmatrix} x_3 \\ x_4 \end{bmatrix} = \lambda_2^2 v_2^2 = 1 \begin{bmatrix} 1 \\ 1 \end{bmatrix} = \begin{bmatrix} 1 \\ 1 \end{bmatrix}.$$

The optimal objective function value is $z^* = -14$. That is, the optimal solution for the original problem is in this instance a convex combination of extreme points of the subproblems.

Example 5.3 (Unbounded sub-problem case)
Consider the following linear program:

$$\begin{array}{lllllll} \textit{maximize} & 2x_1 & + & 3x_2 & + & 2x_3 & \\ \textit{subject to} & x_1 & + & x_2 & + & x_3 & \leq 12 \\ & -x_1 & + & x_2 & & & \leq 2 \\ & -x_1 & + & 2x_2 & & & \leq 8 \\ & & & & & x_3 & \leq 1 \\ & \multicolumn{6}{l}{x_1 \geq 0, x_2 \geq 0, x_3 \geq 0.} \end{array}$$

The linear program can been seen to exhibit block angular structure with the following partitions

$$x = \begin{bmatrix} x^1 \\ x^2 \end{bmatrix} \text{ where } x^1 = \begin{bmatrix} x_1 \\ x_2 \end{bmatrix} \text{ and } x^2 = [x_3]$$

$$c = \begin{bmatrix} c^1 \\ c^2 \end{bmatrix} \text{ where } c^1 = \begin{bmatrix} c_1 \\ c_2 \end{bmatrix} = \begin{bmatrix} -2 \\ -3 \end{bmatrix} \text{and } c^2 = [c_3] = [-2]$$

$$b^0 = [12],\ b^1 = \begin{bmatrix} 2 \\ 8 \end{bmatrix}, \text{ and } b^2 = [1]$$

$$L_1 = \begin{bmatrix} 1 & 1 \end{bmatrix} \text{ and } L_2 = [1]$$

$$A_1 = \begin{bmatrix} -1 & 1 \\ -1 & 2 \end{bmatrix} \text{ and } A_2 = [1].$$

Initialization

Step 0: Start with as basic variables $s_1 = b_1^0 = 12, \lambda_1^1 = 1$, and $\lambda_1^2 = 1$, so $x_B = (s_1,\ s_2, \lambda_1^1, \lambda_1^2)^T$ with initial extreme points $v_1^1 = \begin{bmatrix} x_1 \\ x_2 \end{bmatrix} = \begin{bmatrix} 0 \\ 0 \end{bmatrix}$ and $v_1^2 = [x_3] = [0]$.

Then, the restricted master problem is

$$\begin{array}{lll} \textit{minimize} & \lambda_1^1 f_1^1 + \lambda_1^2 f_1^2 & \\ \textit{subject to} & \lambda_1^1 L_1 v_1^1 + \lambda_1^2 L_2 v_1^2 + s & = b^0 \\ & \lambda_1^1 & = 1 \\ & \qquad\quad \lambda_1^2 & = 1 \\ & \lambda_1^1 \geq 0, \lambda_1^2 \geq 0,\ s \geq 0 & \end{array}$$

and since the initial extreme point for each subproblem is the zero vector we get

$$\begin{array}{lll} \textit{minimize} & 0 & \\ \textit{subject to} & \qquad\qquad s_1 & = 12 \\ & \lambda_1^1 & = 1 \\ & \qquad \lambda_1^2 & = 1 \\ & \lambda_1^1 \geq 0, \lambda_1^2 \geq 0, s_1 \geq 0. & \end{array}$$

The basis matrix $B = I$, i.e., the 3 by 3 identity matrix.

Iteration 1

Step 1: $f_B = \begin{bmatrix} c_{s_1} \\ f_1^1 \\ f_1^2 \end{bmatrix} = \begin{bmatrix} 0 \\ (c^1)^T v_1^1 \\ (c^2)^T v_1^2 \end{bmatrix} = \begin{bmatrix} 0 \\ 0 \\ 0 \end{bmatrix}$, then solving $B^T \pi = f_B$ gives

$$\pi = \begin{bmatrix} \pi^1 \\ \pi_1^2 \\ \pi_2^2 \end{bmatrix} = \begin{bmatrix} \pi_1^1 \\ \pi_1^2 \\ \pi_2^2 \end{bmatrix} = \begin{bmatrix} 0 \\ 0 \\ 0 \end{bmatrix}.$$

Step 2: (Note that SP_2 is arbitrarily solved first; in general, the ordering of solving subproblems does not affect the decomposition).

Now the objective function of SP_2 is $\sigma_2 = ((c^2)^T - (\pi^1)^T L_2)x^2 = (c^2)^T x^2 = -2x_3$, so SP_2 is

$$\begin{array}{lll} \textit{minimize} & -2x_3 & \\ \textit{subject to} & x_3 & \leq\ 1 \\ & x_3 \geq 0. & \end{array}$$

The optimal solution is $x^2 = [x_3] = [1]$ with objective function value $\sigma_2^* = -2$, and so $r_*^2 = \sigma_2^* - \pi_2^2 = -2 - 0 = -2$. Let $v_2^2 = x^2$.

The objective function of SP_1 is $\sigma_1 = ((c^1)^T - (\pi^1)^T L_1)x^1 = (c^1)^T x^1 = -2x_1 - 3x_2$, so SP_1 is

$$\begin{array}{lrcrcl} \textit{minimize} & -2x_1 & - & 3x_2 & & \\ \textit{subject to} & -x_1 & + & x_2 & \leq & 2 \\ & -x_1 & + & 2x_2 & \leq & 8 \\ & & x_1 \geq 0, & x_2 \geq 0. & & \end{array}$$

SP_1 is unbounded. Go to Step 3.

Step 3:
SP_1 is unbounded, and so

$$\bar{a} = \begin{bmatrix} \bar{q}_1^1 \\ 0 \end{bmatrix} = \begin{bmatrix} L_1 d_1^1 \\ 0 \end{bmatrix} = \begin{bmatrix} \begin{bmatrix} 1 & 1 \end{bmatrix} \begin{bmatrix} 1.85664197e+15 \\ 9.28320987e+14 \end{bmatrix} \\ 0 \\ 0 \end{bmatrix}$$

$$= \begin{bmatrix} 2.78496296e+15 \\ 0 \\ 0 \end{bmatrix}.$$

Step 4: $Bd = -\bar{a}$ and so $d = \begin{bmatrix} -2.78496296e+15 \\ 0 \\ 0 \end{bmatrix}$.

Step 5: $\alpha = \min\left\{-\frac{12}{-2.78496296e+15}\right\} = 4.30885443e-15$ and so $\mu_1^1 = \alpha$.

Step 6: $x_B = \begin{bmatrix} s_1 \\ \lambda_1^1 \\ \lambda_1^2 \end{bmatrix} = \begin{bmatrix} 12 \\ 1 \\ 1 \end{bmatrix} + (4.30885443e-15) \begin{bmatrix} -2.78496296e+15 \\ 0 \\ 0 \end{bmatrix} =$ $\begin{bmatrix} 0 \\ 1 \\ 1 \end{bmatrix}$, so s_1 leaves the basis and μ_1^1 enters the basis and the updated basic variable set is

$$x_B = \begin{bmatrix} \mu_1^1 \\ \lambda_1^1 \\ \lambda_1^2 \end{bmatrix} = \begin{bmatrix} 4.30885443e-15 \\ 1 \\ 1 \end{bmatrix}.$$

Step 7: Column $\bar{a}$ enters the basis and takes the place of column $(1,0,0)^T$ associated with s_1, which exits the basis and so

$$B = \begin{bmatrix} 2.78496296e+15 & 0 & 0 \\ 0 & 1 & 0 \\ 0 & 0 & 1 \end{bmatrix}.$$

Go to Step 1.

Iteration 2

Step 1: $$\mathrm{f}_B = \begin{bmatrix} \mathrm{f}_1^{-1} \\ \mathrm{f}_1^1 \\ \mathrm{f}_1^2 \end{bmatrix} = \begin{bmatrix} (c^1)^T d_1^1 \\ (c^1)^T v_1^1 \\ (c^2)^T v_1^2 \end{bmatrix} = \begin{bmatrix} \begin{bmatrix} -2 & -3 \end{bmatrix} \begin{bmatrix} 1.85664197e+15 \\ 9.28320987e+14 \end{bmatrix} \\ 0 \\ 0 \end{bmatrix}$$

$$= \begin{bmatrix} -6.49824691e+15 \\ 0 \\ 0 \end{bmatrix},$$

then solving $B^T\pi = \mathrm{f}_B$ gives

$$\pi = \begin{bmatrix} \pi^1 \\ \pi_1^2 \\ \pi_2^2 \end{bmatrix} = \begin{bmatrix} \pi_1^1 \\ \pi_1^2 \\ \pi_2^2 \end{bmatrix} = \begin{bmatrix} -2.33333333 \\ 0 \\ 0 \end{bmatrix}.$$

Step 2:
Now the objective functions of SP_1 is $\sigma_1 = ((c^1)^T - (\pi^1)^T L_1)x^1 = .33333333x_1 - .66666667x_2$, and so SP_1 is

$$\begin{array}{lcccl} \textit{minimize} & .33333333x_1 & - & .66666667x_2 & \\ \textit{subject to} & -x_1 & + & x_2 & \leq 2 \\ & -x_1 & + & 2x_2 & \leq 8 \\ & & x_1 \geq 0, & x_2 \geq 0. & \end{array}$$

The optimal solution is $x^1 = \begin{bmatrix} x_1 \\ x_2 \end{bmatrix} = \begin{bmatrix} 152.89068985 \\ 80.44534492 \end{bmatrix}$ with objective function value $\sigma_1^* = -2\frac{2}{3}$, and so $r_*^1 = \sigma_1^* - \pi_1^2 = -2\frac{2}{3} - 0 = -2\frac{2}{3}$. Let $v_2^1 = x^1$.

Now the objective function of SP_2 is $\sigma_2 = ((c^2)^T - (\pi^1)^T L_2)x^2 = (c^2)^T x^2 = .33333333x_3$, so SP_2 is

$$\begin{array}{lcl} \textit{minimize} & .33333333x_3 & \\ \textit{subject to} & x_3 & \leq 1 \\ & x_3 \geq 0. & \end{array}$$

The optimal solution is $x^2 = [x_3] = [0]$ with objective function value $\sigma_2^* = 0$, and so $r_*^2 = \sigma_2^* - \pi_2^2 = 0 - 0 = 0$. Let $v_3^2 = x^2$.

Step 3:
$r_{\min} = r_*^1 = -2\frac{2}{3}$, and so

$$\begin{bmatrix} q_2^1 \\ e_1 \end{bmatrix} = \begin{bmatrix} L_1 v_2^1 \\ e_1 \end{bmatrix} = \bar{a} = \begin{bmatrix} \begin{bmatrix} 1 & 1 \end{bmatrix} \begin{bmatrix} 152.89068985 \\ 80.44534492 \end{bmatrix} \\ 1 \\ 0 \end{bmatrix} =$$

$$\begin{bmatrix} 233.33603477 \\ 1 \\ 0 \end{bmatrix}.$$

Step 4: $Bd = -\bar{a}$ and so $d = \begin{bmatrix} -8.37842506e - 14 \\ -1 \\ 0 \end{bmatrix}$.

Step 5: $\alpha = \min\left\{-\frac{4.30885443e-15}{-8.37842506e-14}, -\frac{1}{-1}\right\} = 0.05142798$ and so $\lambda_2^1 = \alpha$.

Step 6: $x_B = \begin{bmatrix} \mu_1^1 \\ \lambda_1^1 \\ \lambda_1^2 \end{bmatrix} = \begin{bmatrix} 4.30885443e - 15 \\ 1 \\ 1 \end{bmatrix} + (0.05142798) \begin{bmatrix} -8.37842506e - 14 \\ -1 \\ 0 \end{bmatrix}$

$= \begin{bmatrix} 0 \\ 0.94857202 \\ 1 \end{bmatrix}$ so μ_1^1 leaves the basis and λ_2^1 enters the basis and the updated basic variable set is

$$x_B = \begin{bmatrix} \lambda_2^1 \\ \lambda_1^1 \\ \lambda_1^2 \end{bmatrix} = \begin{bmatrix} 0.05142798 \\ 0.94857202 \\ 1 \end{bmatrix}.$$

Step 7: Column $\bar{a}$ enters the basis and the column $(1, 0, 0)^T$ associated with μ_1^1 leaves the basis, and so

$$B = \begin{bmatrix} 233.33603477 & 0 & 0 \\ 1 & 1 & 0 \\ 0 & 0 & 1 \end{bmatrix}.$$

Go to Step 1.

Iteration 3

Step 1: $\mathrm{f}_B = \begin{bmatrix} \mathrm{f}_2^1 \\ \mathrm{f}_1^1 \\ \mathrm{f}_1^2 \end{bmatrix} = \begin{bmatrix} (c^1)^T v_2^1 \\ (c^1)^T v_1^1 \\ (c^2)^T v_1^2 \end{bmatrix} = \begin{bmatrix} \begin{bmatrix} -2 & -3 \end{bmatrix} \begin{bmatrix} 152.89068985 \\ 80.44534492 \end{bmatrix} \\ 0 \\ 0 \end{bmatrix}$

$$= \begin{bmatrix} -547.11741446 \\ 0 \\ 0 \end{bmatrix},$$

then solving $B^T\pi = \mathrm{f}_B$ gives

$$\pi = \begin{bmatrix} \pi^1 \\ \pi_1^2 \\ \pi_2^2 \end{bmatrix} = \begin{bmatrix} \pi_1^1 \\ \pi_1^2 \\ \pi_2^2 \end{bmatrix} = \begin{bmatrix} -2.34476177 \\ 0 \\ 0 \end{bmatrix}.$$

Step 2:

Now the objective functions of SP_1 is $\sigma_1 = ((c^1)^T - (\pi^1)^T L_1)x^1 = .34476177x_1 - .65523823x_2$, and so SP_1 is

$$\begin{array}{lrcrcl} \textit{minimize} & .34476177x_1 & - & .65523823x_2 & & \\ \textit{subject to} & -x_1 & + & x_2 & \leq & 2 \\ & -x_1 & + & 2x_2 & \leq & 8 \\ & & x_1 \geq 0, & x_2 \geq 0. & & \end{array}$$

The optimal solution is $x^1 = \begin{bmatrix} x_1 \\ x_2 \end{bmatrix} = \begin{bmatrix} 4 \\ 6 \end{bmatrix}$ with objective function value $\sigma_1^* = -2.55238227$, and so $r_*^1 = \sigma_1^* - \pi_1^2 = -2.55238227 - 0 = -2.55238227$. Let $v_3^1 = x^1$.

Now the objective function of SP_2 is $\sigma_2 = ((c^2)^T - (\pi^1)^T L_2)x^2 = (c^2)^T x^2 = .34476177x_3$, so SP_2 is

$$\begin{array}{lrcl} \textit{minimize} & .34476177x_3 & & \\ \textit{subject to} & x_3 & \leq & 1 \\ & x_3 \geq 0. & & \end{array}$$

The optimal solution is $x^2 = [x_3] = [0]$ with objective function value $\sigma_2^* = 0$, and so $r_*^2 = \sigma_2^* - \pi_2^2 = 0 - 0 = 0$. Let $v_4^2 = v_3^2 = x^2$.

Step 3:

$r_{\min} = r_*^1 = -2.55238227$, and so

$$\bar{a} = \begin{bmatrix} q_3^1 \\ e_1 \end{bmatrix} = \begin{bmatrix} L_1 v_3^1 \\ e_1 \end{bmatrix} = \begin{bmatrix} \begin{bmatrix} 1 & 1 \end{bmatrix} \begin{bmatrix} 4 \\ 6 \end{bmatrix} \\ 1 \\ 0 \end{bmatrix} = \begin{bmatrix} 10 \\ 1 \\ 0 \end{bmatrix}.$$

Step 4: $Bd = -\bar{a}$ and so $d = \begin{bmatrix} -0.04285665 \\ -0.95714335 \\ 0 \end{bmatrix}$.

Step 5: $\alpha = \min\left\{-\frac{0.05142798}{-0.04285665}, -\frac{0.94857202}{-0.95714335}\right\} = 0.99104489$, and so $\lambda_3^1 = \alpha$.

Step 6: $x_B = \begin{bmatrix} \lambda_2^1 \\ \lambda_1^1 \\ \lambda_1^2 \end{bmatrix} = \begin{bmatrix} 0.05142798 \\ 0.94857202 \\ 1 \end{bmatrix} + (0.99104489) \begin{bmatrix} -0.04285665 \\ -0.95714335 \\ 0 \end{bmatrix} =$
$\begin{bmatrix} 0.00895511 \\ 0 \\ 1 \end{bmatrix}$, so λ_1^1 leaves the basis and λ_3^1 enters the basis and the updated basic variable set is

$$x_B = \begin{bmatrix} \lambda_2^1 \\ \lambda_3^1 \\ \lambda_1^2 \end{bmatrix} = \begin{bmatrix} 0.00895511 \\ 0.99104489 \\ 1 \end{bmatrix}.$$

Step 7: Column $\bar{a}$ enters the basis and the column $(0, 1, 0)^T$ associated with λ_1^1 leaves the basis, and so

$$B = \begin{bmatrix} 233.336035 & 10 & 0 \\ 1 & 1 & 0 \\ 0 & 0 & 1 \end{bmatrix}.$$

Go to Step 1.

Iteration 4

Step 1: $\text{f}_B = \begin{bmatrix} \text{f}_2^1 \\ \text{f}_3^1 \\ \text{f}_1^2 \end{bmatrix} = \begin{bmatrix} (c^1)^T v_2^1 \\ (c^1)^T v_3^1 \\ (c^2)^T v_1^2 \end{bmatrix} = \begin{bmatrix} \begin{bmatrix} -2 & -3 \end{bmatrix} \begin{bmatrix} 152.89068985 \\ 80.44534492 \end{bmatrix} \\ \begin{bmatrix} -2 & -3 \end{bmatrix} \begin{bmatrix} 4 \\ 6 \end{bmatrix} \\ 0 \end{bmatrix}$

$$= \begin{bmatrix} -547.11741446 \\ -26 \\ 0 \end{bmatrix},$$

then solving $B^T \pi = \text{f}_B$ gives

$$\pi = \begin{bmatrix} \pi^1 \\ \pi_1^2 \\ \pi_2^2 \end{bmatrix} = \begin{bmatrix} \pi_1^1 \\ \pi_1^2 \\ \pi_2^2 \end{bmatrix} = \begin{bmatrix} -2.33333333 \\ -2.66666667 \\ 0 \end{bmatrix}.$$

Step 2:

Now the objective functions of SP_1 is $\sigma_1 = ((c^1)^T - (\pi^1)^T L_1)x^1 = .33333333x_1 - .66666667x_2$, and so SP_1 is

$$\begin{array}{llcll} \textit{minimize} & .33333333x_1 & - & .66666667x_2 & \\ \textit{subject to} & -x_1 & + & x_2 & \leq 2 \\ & -x_1 & + & 2x_2 & \leq 8 \\ & & x_1 \geq 0, & x_2 \geq 0. & \end{array}$$

The optimal solution is $x^1 = \begin{bmatrix} x_1 \\ x_2 \end{bmatrix} = \begin{bmatrix} 152.50614560 \\ 80.25307280 \end{bmatrix}$ with objective function value $\sigma_1^* = -2.66666667$, and so $r_*^1 = \sigma_1^* - \pi_1^2 = -2.66666667 - (-2.66666667) = 0$. Let $v_4^1 = x^1$.

Now the objective function of SP_2 is $\sigma_2 = ((c^2)^T - (\pi^1)^T L_2)x^2 = (c^2)^T x^2 = .33333333x_3$, so SP_2 is

$$\begin{array}{lll} \textit{minimize} & .33333333x_3 & \\ \textit{subject to} & x_3 & \leq 1 \\ & x_3 \geq 0. & \end{array}$$

The optimal solution is $x^2 = [x_3] = [0]$ with objective function value $\sigma_2^* = 0$, and so $r_*^2 = \sigma_2^* - \pi_2^2 = 0 - 0 = 0$. Let $v_5^2 = v_4^2 = v_3^2 = x^2$.

Step 3:

Since $r_{\min} = 0$, STOP the optimal solution to the restricted master problem is

$$x_B = \begin{bmatrix} \lambda_2^1 \\ \lambda_3^1 \\ \lambda_1^2 \end{bmatrix} = \begin{bmatrix} 0.00895511 \\ 0.99104489 \\ 1 \end{bmatrix}.$$

Thus, the optimal solution in the original variables are recovered as

$$x^1 = \begin{bmatrix} x_1 \\ x_2 \end{bmatrix} = \lambda_2^1 v_2^1 + \lambda_3^1 v_3^1$$

$$= 0.00895511 * \begin{bmatrix} 152.50614560 \\ 80.25307280 \end{bmatrix} + 0.99104489 * \begin{bmatrix} 4 \\ 6 \end{bmatrix}$$

$$= \begin{bmatrix} 5.33333251 \\ 6.66666626 \end{bmatrix}$$

and

$$x^2 = [x_3] = \lambda_1^2 v_1^2 = 1 * 0 = 0.$$

(Note: the exact optimal solution is $x_1 = 5\frac{1}{3}, x_2 = 6\frac{2}{3}, x_3 = 0$ and the answer provided above by the decomposition is approximate due to rounding estimates.)

5.5.1 Economic Interpretation

The Dantzig Wolfe decomposition algorithm has the following economic interpretation, which supports the view that it is a price-directed decomposition as described earlier. The original linear program with block angular structure represents a company with K subdivisions (subproblems) where subdivision k independently produces its own set of products according to its set of constraints $A_k x^k \leq b^k$. But all subdivisions require the use of a limited set of common resources, which give rise to the linking constraints. The company wishes to minimize the cost of production over all products from the K subdivisions. The master problem represents a company-wide supervisor that manages the use of the common resources. At an iteration of the Dantzig-Wolfe decomposition, the restricted master problem generates a master production plan represented by the basis B based on the production plan (proposals) sent by the subdivisions. The basis represents that fraction or weight of each proposal in the master plan.

Then, the supervisor is responsible for the computation of the vector π^1, which represents the calculation of the prices for the common resources where $-\pi_i^1$ is the price for consuming a unit of common resource i. These (marginal) prices reflect demand for the common resources by the master plan, are announced to all of the subdivisions, and are used by each subdivision in the coefficients of its objective function in constructing its optimal production plan (proposal). Recall that the objective of subdivision k is to minimize

$$\sigma_k = ((c^k)^T - (\pi^1)^T L_k)x^k,$$

where $(c^k)^T x^k$ is the original objective function of the subdivision and represents its own specific (i.e., not related to other subdivisions) costs, and $L_k x^k$, is the quantity of common resources consumed by the production plan (proposal) x^k. Thus, the total costs of using common resources by subdivision k is represented by $-(\pi^1)^T L_k x^k$, and so the more it uses the common resources the worse the overall objective will be.

The proposal that gives the greatest promise in reducing costs for the company (i.e., that proposal that achieves $r_{\min} < 0$) is selected by the supervisor. If all proposals are such that $r_{\min} \geq 0$, then there are no further cost savings possible and the current production plan is optimal. However, if a proposal is selected and a new production plan is generated by the supervisor, since the weights of the current proposals must be adjusted (this is accomplished by the updating of the basic variables) so that the capacities of the common resources are not violated with the introduction of a new proposal. A new set of prices are generated and the process iterates.

Thus, the Dantzig-Wolfe decomposition represents a decentralized mechanism for resource allocation as the decision making for actual production lies within the subdivisions and the coordination is accomplished through the prices set by the supervisor which reflect supply and demand. The prices serve to guide the subdivisions to a production plan that is systemwide (companywide) cost optimal, which represents an equilibrium where supply and demand are balanced.

5.5.2 Initialization

To start the Dantzig-Wolfe decomposition, the master problem requires an initial basic feasible solution. The strategy is to develop an auxiliary problem similar to a Phase 1 approach for the revised simplex method in Chapter 3. First, an extreme point $x^k = v_1^k$ is generated for each subproblem SP_k using the Phase 1 approach for the revised simplex method. If any of the subproblems do not have a feasible extreme point, then the original problem is infeasible. Even if all subproblems admit an extreme point, it might be the case that the linking constraints might be violated. Like the Phase I procedure for the simplex method, artificial variables x_a can be added to the linking constraints and these artificial variables are then minimized. So the auxiliary problem for the master problem is

$$\begin{array}{lll} \textit{minimize} & e^T x_a & \\ \textit{subject to} & \sum\limits_{k=1}^{K}\sum\limits_{i=1}^{N_k} \lambda_i^k (L_k v_i^k) + \sum\limits_{k=1}^{K}\sum\limits_{j=1}^{l_k} \mu_j^k (L_k d_j^k) + x_a = b^0 & \\ & \sum_{i=1}^{N_k} \lambda_i^k = 1 & k = 1, ..., K \\ & \lambda_i^k \geq 0 \ , \ i = 1, ..., N_k & k = 1, ..., K \\ & \mu_j^k \geq 0 \ , \ j = 1, ..., l_k & k = 1, ..., K \\ & x_a \geq 0. & \end{array}$$

An initial basic feasible solution for the auxiliary problem is to let $\lambda_1^k = 1$ for all $k = 1, ..., K$, $\lambda_i^k = 0$ for $i \neq 1$, $\mu_j^k = 0$ for all k and j, and $x_a = b^0 - \sum\limits_{k=1}^{K} \lambda_1^k (L_k v_1^k)$. Recall that the points v_1^k are generated earlier by the Phase I method applied to SP_k. Therefore, if $e^T x_a > 0$, then the master problem is infeasible, else $e^T x_a = 0$ and the optimal solution to the auxiliary problem will provide an initial basic feasible solution for the master problem.

5.5.3 Bounds on Optimal Cost

It is found through computational experiments that the Dantzig-Wolfe decomposition can often take too much time before termination in solving very large problem instances. However, the method can be stopped before optimality has been achieved and one can evaluate to some extent how close the current basic feasible solution is from optimal. An objective function value based on a feasible solution obtained before optimality represents an upper bound (assuming a minimization problem) on the optimal objective function value. The idea is to now generate a feasible solution of the dual of the restricted master problem; the associated objective function value will give a lower bound. Thus, both upper and lower bounds can be obtained on the optimal cost.

Theorem 5.4

Suppose the master problem is consistent and bounded with optimal objective function value z^*. Let x_B be the basic variables from a feasible solution obtained from the Dantzig-Wolfe decomposition before termination and denote z as its corresponding objective function value. Further assume that the subproblems SP_k are bounded with optimal objective function value σ_k^*, then

$$z + \sum_{k=1}^{K} (\sigma_k^* - \pi_k^2) \leq z^* \leq z$$

where π_k^2 is the dual variable associated with the convexity constraint of subproblem k in the master problem.

Before the proof of Theorem 5.4 is presented, we give the following lemma.

Lemma 5.5

Let x be a non-optimal feasible solution for the master problem that is generated during an intermediate step of the Dantzig-Wolfe decomposition and suppose that all subproblems are bounded. Then, a feasible solution for the dual of the master problem exists whose objective function value is equal to the objective function value of master problem at x.

Proof:

Let z denote the objective function of the master problem at x. Without loss of generality, we consider the case that there are only two subproblems and all complicating constraints are equality constraints, and so the master problem is

$$\begin{array}{llll}
\textit{minimize} & \sum\limits_{i=1}^{N_1} \lambda_i^1 \mathrm{f}_i^1 + \sum\limits_{i=1}^{N_2} \lambda_i^2 \mathrm{f}_i^2 + \sum\limits_{j=1}^{l_1} \mu_i^1 \bar{\mathrm{f}}_j^1 + \sum\limits_{j=1}^{l_2} \mu_i^2 \bar{\mathrm{f}}_j^2 & & \\
\textit{subject to} & \sum\limits_{i=1}^{N_1} \lambda_i^1 L_1 v_i^1 & + \sum\limits_{i=1}^{N_2} \lambda_i^2 L_2 v_i^2 & = b^0 \\
& \sum_{i=1}^{N_1} \lambda_i^1 & & = 1 \\
& & \sum_{i=1}^{N_2} \lambda_i^2 & = 1 \\
& \lambda_i^1 \geq 0 & & i = 1, ..., N_1 \\
& & \lambda_i^2 \geq 0 & i = 1, ..., N_2.
\end{array}$$

The dual of the master problem is then

$$\begin{array}{llllll} \textit{minimize} & (\pi^1)^T b_0 & + \pi_1^2 & + \pi_2^2 & & \\ \textit{subject to} & (\pi^1)^T L_1 v_i^1 & + \pi_1^2 & & \leq f_i^1 & \forall v_i^1 \in P_1 \\ & (\pi^1)^T L_1 d_j^1 & & & \leq \bar{f}_j^1 & \forall d_j^1 \in P_1 \\ & (\pi^1)^T L_2 v_i^2 & & + \pi_2^2 & \leq f_i^2 & \forall v_i^2 \in P_2 \\ & (\pi^1)^T L_2 d_j^2 & & & \leq \bar{f}_j^2 & \forall d_j^2 \in P_2. \end{array}$$

Assuming that the master problem is solved using the revised simplex method, there is a dual solution (simplex multipliers) $\pi = \begin{bmatrix} \pi^1 & \pi_1^2 & \pi_2^2 \end{bmatrix}$ such that the objective function value of the dual $(\pi^1)^T b_0 + \pi_1^2 + \pi_2^2 = z$. However, π is infeasible except at optimality of the revised simplex method. To obtain a feasible dual solution, the boundedness of the subproblems are exploited.

The first subproblem SP_1 is

$$\begin{array}{lll} \textit{minimize} & \sigma_1 = ((c^1)^T - (\pi^1)^T L_1)x^1 & \\ \textit{subject to} & A_1 x^1 & \leq b^1 \\ & x^1 \geq 0. & \end{array}$$

The optimal solution is some extreme point $v_{i*}^1 \in P_1$ with corresponding finite objective function value

$$\sigma_1^* = ((c^1)^T - (\pi^1)^T L_1)v_{i*}^1 = (c^1)^T v_{i*}^1 - (\pi^1)^T L_1 v_{i*}^1 \ .$$

Furthermore, there is no direction in P_1 for which SP_1 is unbounded, and so we have for all directions $d_j^1 \in P_1$

$$\bar{f}_j^1 - (\pi^1)^T L_1 d_j^1 = (c^1)^T d_j^1 - (\pi^1)^T L_1 d_j^1 \geq 0.$$

This suggests that we can use the quantity σ_1^* instead of π_1^2 since π^1 and σ_1^* are feasible for the first two constraints of the dual problem. In a similar fashion, since the second subproblem SP_2 is bounded, we can use

$$\sigma_2^* = ((c^2)^T - (\pi^1)^T L_2)v_{i*}^2 = (c^2)^T v_{i*}^2 - (\pi^1)^T L_1 v_{i*}^2$$

in place of π_2^2. Therefore, the solution $\pi^f = \begin{bmatrix} \pi^1 & \sigma_1^* & \sigma_2^* \end{bmatrix}$ is feasible for the dual problem.■

Proof of Theorem 5.4:

Now the objective function value of the master problem at x is $z = (\pi^1)^T b_0 + \pi_1^2 + \pi_2^2$ where $\pi = \begin{bmatrix} \pi^1 & \pi_1^2 & \pi_2^2 \end{bmatrix}$ is the vector of simplex multipliers associated with x. By Lemma 5.5, the solution

$$\pi^f = \begin{bmatrix} \pi^1 & \sigma_1^* & \sigma_2^* \end{bmatrix} \text{ is dual feasible.}$$

Thus, by weak duality of linear programming we have

$$\begin{aligned} z^* &\geq (\pi^1)^T b_0 + \sigma_1^* + \sigma_2^* \\ &= [(\pi^1)^T b_0 + \pi_1^2 + \pi_2^2] + \sigma_1^* - \pi_1^2 + \sigma_2^* - \pi_2^2 \\ &= z + \sum_{i=1}^{2} (\sigma_i^* - \pi_i^2). \end{aligned}$$

■

Example 5.6

Consider the basic variables at the end of the second iteration in Example 5.2

$$x_B = \begin{bmatrix} s_1 \\ \lambda_2^1 \\ \lambda_1^1 \\ \lambda_2^2 \end{bmatrix} = \begin{bmatrix} 1 \\ 0.5 \\ 0.5 \\ 1 \end{bmatrix}, \text{ then } z = f_B^T x_B = \begin{bmatrix} 0 & -6 & 0 & -9 \end{bmatrix} \begin{bmatrix} 1 \\ 0.5 \\ 0.5 \\ 1 \end{bmatrix} = -12$$

Also, $\sum_{k=1}^{2} (\sigma_k^* - \pi_k^2) = (\sigma_1^* - \pi_1^2) + (\sigma_2^* - \pi_2^2) = (-4-0) + (-\frac{10}{3} - (-3)) = -4\frac{1}{3}$, so we have $-16\frac{1}{3} \leq z^* \leq -12$. Recall that for this problem $z^* = -14$.

5.6 Dantzig-Wolfe MATLAB® Code

A MATLAB function DantzigWolfeDecomp

function [**x, fval, bound**, **exit_flag**] = DantzigWolfeDecomp(**mast**, **sub**, **K**)

is given below that implements the Dantzig-Wolfe decomposition. The argument **mast** is a struct that contains the coefficients of the linking constraints L_i and **sub** is a struct that contains the coefficients of the subproblems, i.e., c^i, A^i, and b^i as well as the initial extreme points for each subproblem. **K** is the number of subproblems.

Consider the linear program in Example 5.1:

$$\begin{array}{llllll} \textit{minimize} & -2x_1 & -3x_2 & -5x_3 & -4x_4 & \\ \textit{subject to} & x_1 + & x_2 + & 2x_3 & & \leq 4 \\ & & x_2 + & x_3 + & x_4 & \leq 3 \\ & 2x_1 + & x_2 & & & \leq 4 \\ & x_1 + & x_2 & & & \leq 2 \\ & & & x_3 + & x_4 & \leq 2 \\ & & & 3x_3 + & 2x_4 & \leq 5 \\ & x_1 \geq 0 & x_2 \geq 0 & x_3 \geq 0 & x_4 \geq 0. & \end{array}$$

The LP is in the form

$$\begin{array}{ll} \textit{minimize} & c^T x \\ \textit{subject to} & Ax = b \\ & x \geq 0 \end{array}$$

where that data c, A, and b are entered in MATLAB as

```
c=[-2; -3; -5; -4];
A= [1 1 2 0;
    0 1 1 1;
    2 1 0 0;
     1 1 0 0;
    0 0 1 1;
    0 0 3 2];
b=[4; 3; 4; 2; 2; 5];
```

There are two subproblems, and so K is set equal to 2, i.e., in MATLAB we have

K=2; %number of subproblems

Now, the first two constraints of the LP are the linking constraints and the submatrices and corresponding right-hand side values are

$$L_1 = \begin{bmatrix} 1 & 1 \\ 0 & 1 \end{bmatrix}, L_2 = \begin{bmatrix} 2 & 0 \\ 1 & 1 \end{bmatrix}, b^0 = \begin{bmatrix} 4 \\ 3 \end{bmatrix},$$

and are written in MATLAB as

```
mast.L{1}=A(1:2, 1:2);
mast.L{2}=A(1:2, 3:4);
mast.b=b(1:2);
```

The cost coefficient vector, constraint matrix, and right-hand coefficient vector of the first subproblem are given as

$$c^1 = \begin{bmatrix} -2 \\ -3 \end{bmatrix}, A^1 = \begin{bmatrix} 2 & 1 \\ 1 & 1 \end{bmatrix}, \text{ and } b^1 = \begin{bmatrix} 4 \\ 2 \end{bmatrix}$$

and the corresponding MATLAB statements to represent these matrices are

```
sub.c{1}=c(1:2);
sub.A{1}=A(3:4, 1:2);
sub.b{1}=b(3:4);
```

The cost coefficent vector, constraint matrix and right hand coefficient vector of the second subproblem are given as

$$c^2 = \begin{bmatrix} -5 \\ -4 \end{bmatrix}, A^2 = \begin{bmatrix} 1 & 1 \\ 3 & 2 \end{bmatrix}, \text{ and } b^2 = \begin{bmatrix} 2 \\ 5 \end{bmatrix}$$

and the corresponding MATLAB statements to represent these matrices are

sub.c{2}=c(3:4);
sub.A{2}=A(5:6, 3:4);
sub.b{2}=b(5:6);

The initial extreme points for each subproblem are set to the origin in the feasible set for each subproblem and in MATLAB are written as

sub.v{1}=zeros(length(sub.c{1}),1); sub.v{2}=zeros(length(sub.c{2}),1);.

Finally, once the data is entered as above, the function DantzigWolfeDecomp can be called by the following MATLAB statement

[x_DanWof, fval_DanWof, iter, exitflag_DanWof] = DantzigWolfeDecomp(mast, sub, K)

The optimal solution can be accessed in MATLAB by entering x_DanWof at the prompt, i.e.,

>> x_DanWof

x_DanWof =

1.0000
1.0000
1.0000
1.0000

The optimal objective function can be accessed by typing fval_DanWof at the prompt, i.e.,

>> fval_DanWof

fval_DanWof =

-14.0000

5.6.1 DantzigWolfeDecomp MATLAB Code

```
function [x, fval, bound, exit_flag] = DantzigWolfeDecomp(mast, sub, K)
%DantzigWolfeDecomp solves a linear programming with a special structure:
%                         min  c'*x
%                        s.t.  A*x <= b
%                                x >= 0
%where A is an m*n matrix which can be write as block angular form:
%          --              --        --    --          --    --          --    --
%       | L1 L2 ... Lk    |        |  x1   |        |  c1   |        |  b0    |
```

```
%       | A1                 |      |  x2  |     |  c2  |      |  b1   |
%   A = |     A2             |, x = |   :  |, c =|   :  |, b = |  b2   |
%       |         ...        |      |   :  |     |   :  |      |   :   |
%       |__           Ak __|      |__ xk __|     |__ ck __|      |__ bk __|
%so the LP can be decomposed into a master problem(MP) and k subproblems(SPk),
%we can rewrite the MP as a restricted master problem by Resolution Theorem
%
% Inputs:
%  mast is a struct includes MP's coefficients, i.e. L1,...,Lk, b0
%  sub is a struct includes coefficients of SPks, i.e. c1,...,ck, A1,...,Ak,
%    b1,...,bk and the initial extreme points v1, ..., vk.
%  K is the number of subproblems
%
% Outputs:
%  x = n*1 vector, final solution
%  fval is a scalar, final objective value
%  bound is a matrix includes all LBs and UBs for each iteration
%  exit_flag describes the exit condition of the problem as follows:
%           0  - optimal solution
%           1  - LP is unbounded

x=[]; fval=[]; bound=[]; exit_flag=[];
%% Step 0: Initialization
%%%%%%%%%%%%%%%%%%%%%%%%%%%%%%
% Generate an initial basis B for the master problem.
% Let x_B be the basic variables and B_bar the index set of the basic variables
% Set all other variables to zero to get the restricted master problem.
% Go to STEP 1.
s=mast.b;                 %slack variables for inequality constraints
x_B=[s; ones(K,1)];       %initial basic variables
%x_Bflag is an index of the basic variables in the restricted master problem
%associated with linking constraints and subproblem k, i.e. slack variables of
%linking constraints are initially basic and other basic variables
%associated with subproblems are set to 1.
x_Bflag=[zeros(length(s),1); [1:K]'];
f_sub=[];
for k=1:K
    %obtain initial extreme points from struct sub for subproblems these are
    %zero vectors, so initial objective function values will be zero, v(k)
    %is initial extreme point of subproblem k
    v_sub{k}=sub.v{k};
    f_sub=cat(1, f_sub, sub.c{k}'*v_sub{k});
    for a=1:length(s)
        %generating initial extreme point for linking constraint
        %v_L{a,k}= initial extreme point of linking constraint of Lk
```

```
        %b(k) is the RHS vector in A_k*x^k=b^k
        v_L{a,k}=zeros(length(sub.b{k}),1);
    end
end
f_s=zeros(length(s),1);
f_B=[f_s; f_sub];       %initial f_B i.e. the objective coefficient of
                        %the restricted MP
B=eye(length(x_B));     %initial basis for master problem
iter_num=0;             %counter
options=optimset('LargeScale', 'on', 'Simplex', 'off');%choose largescale LP
                                                       %solver in linprog
while iter_num>=0
    %% Step 1: Simplex Multiplier Generation
    %%%%%%%%%%%%%%%%%%%%%%%%%%%%%%%%%%%%%%%%%%%%%%%%
    % Solve for duals by solving the system B^T*pie=f_B, then Go to STEP2.
    pie=B'\f_B;                 %solve B^T*pie=f_B
    pie_sub=pie(end-K+1:end);%duals of kth convexity constraints, pie_k_2
    pie(end-K+1:end)=[];        %duals of linking constraints, pie_1
    %% Step 2: Optimality Check
    %%%%%%%%%%%%%%%%%%%%%%%%%%%%%%%%%%
    % For each k=1,...,K, using the revised simplex method to solve SP_k i.e.
    %           min sig_k=[(c^k)^T-(pie^1)^T*L_k]*x^k
    %           s.t.A_k*x^k <= b^k
    %                   x^k >= 0
    % If SP_k is unbounded, then Go to STEP3, else let x^k=(v_i*)^k denote the
    % optimal basic feasible solution, compute(r^k)_* = (sig_k)^* - pie_k^2.
    % If r_min={(r^k)_*}>=0, then the current basis B is optimal,
    % else Go to STEP3.
    for k=1:K
        c_sub=[sub.c{k}'-pie'*mast.L{k}]';  %update the objective coefficient
        [x_sub{iter_num+1, k}, sig(k) exitflag(k)] = ... %call linprog solver
            linprog(c_sub, sub.A{k}, sub.b{k},[],[],...
                    zeros(length(c_sub),1),[],[],options);
        sig(k)=sig(k)-pie_sub(k);                       %computer (r^k)_*
    end
    r_min=min(sig);                         %minimum ratio test to obtain r_min
    r_minflag=find(sig==r_min);
    if isempty(find(r_min < 0)) || abs(r_min) <= 1e-8 %reduced cost>=0, optimal
        disp('problem solved')
        fval=0;
        x_Bflag_s=x_Bflag(1:length(s));
        for k=1:K                   %convert to optimal solution for original LP
            x{k,1}=x_B(length(s)+k)*v_sub{k};
            for a=1:length(x_Bflag_s)
                if x_Bflag_s(a)==k
```

```
                x{k}=x{k}+x_B(a)*v_L{a, k};
            end
        end
       %convert to optimal obj val for original LP
        fval=fval+sub.c{k}'*x{k};
    end
    x=cell2mat(x);
    exit_flag=0;
    break
else
    %% Step 3: Column Generation
    %%%%%%%%%%%%%%%%%%%%%%%%%%%%%%%%%%
    % If all subproblems are bounded and r_min={(r^k)_*}<0, then let t be
    % the index of k in SP_k such that r_min=(r^t)_*.
    % Let alpha_bar=[q_i*^t e_t]^T =[L_t*v_i*^t e_t]^T where v_i*^t is the
    % optimal extreme points of SP_t and Go to STEP4. Else there is a
    % subproblem SP_s that is unbounded and so an extreme direction d_j*^s
    % will be generated such that [(c^s)^T-(pie^1)^T*L_s]*d_j*^s < 0 and
    % so let alpha_bar = [(q_bar)_j*^s 0]^T = [L_s*d_j*^s 0]^T and
    % go to STEP4.
    if length(find(exitflag==1)) == K %if subproblems bounded and r_min<0
        t=r_minflag(1);               %subproblem t such that r_min=r_*_t
        q_t=mast.L{t}*x_sub{iter_num+1,t}; %generate q_i*_t
        e=zeros(length(x_B)-length(q_t),1);
        e(t)=1;                            %generate e_t
        alpha_bar=[q_t; e];                %generate alpha_bar
    else                              %if any subproblems is unbounded
        disp('unbouded subproblem exist')
        unboundflag=find(exitflag==-3);
        t=unboundflag(1);     %subproblem s with extreme direction d_j*^s
        q_t=mast.L{t}*x_sub{iter_num+1,t}; %generate (q_bar)_j*^s
        %generate alpha_bar
        alpha_bar=[q_t; zeros(length(x_B)-length(q_t),1)];
    end
end
%% Step 4: Descent Direction Generation
%%%%%%%%%%%%%%%%%%%%%%%%%%%%%%%%%%%%%%%%%%%%%
% Solve for d in the linear system Bd = -alpha_bar.
% If d>=0, then the LP is unbounded STOP, else go to STEP5.
d=-B\alpha_bar;     %solve Bd=-alpha_bar
d_flag=find(d<0);
if isempty(d_flag) %if d>=0, unbounded
    disp('unbounded LP')
    exit_flag=1;
    return
```

```
    else                  %else Go to STEP 5
        %% Step 5: Step Length Generation
        %%%%%%%%%%%%%%%%%%%%%%%%%%%%%%%%%%%%%%%
        % Computer the step length alpha by minimum ratio test. Let l* be
        % the index of the basic variables then attains the minimum ratio
        % alpha. Go to STEP 6.
        alpha=min(x_B(d_flag)./abs(d(d_flag))); %minimum ratio test
        %% Step 6: Update Basic Variables and Basis
        %%%%%%%%%%%%%%%%%%%%%%%%%%%%%%%%%%%%%%%%%%%%%%%%%%%
        % Let x_B = x_B + alpha*d. Go to STEP 7.
        x_B=x_B+alpha*d;                    %get new basis variables
        delta=1e-30;                        %computation error tolerance
        leave=find(abs(x_B)<=delta);        %index of leave variable
        while isempty(leave)
            delta=10*delta;
            leave=find(abs(x_B)<=delta);
        end
        x_B(leave(1))=alpha;
        x_Bflag(leave(1))=t;
        if leave(1) <= length(s)            %update f_s and extreme point
            f_s(leave(1))=sub.c{t}'*x_sub{iter_num+1,t};
            v_L{leave(1),t}=x_sub{iter_num+1,t};
        else
            f_sub(leave(1)-length(s))=sub.c{t}'*x_sub{iter_num+1,t};
            v_sub{leave(1)-length(s)}=x_sub{iter_num+1,t};
        end
        %% Step 7: Basis Update
        %%%%%%%%%%%%%%%%%%%%%%%%%%%%%%
        % Let B_l* be the column associated with the leaving variable x_l*.
        % Update the basis matrix B by removing B_l* and adding the column
        % alpha_bar, and update B_set.
        B(:,leave(1))=alpha_bar;                 %update the basis B
    end
    iter_num=iter_num+1;
    f_B=[f_s; f_sub];                           %update f_B for next iteration
    bound(:,iter_num)=[f_B'*x_B + sum(sig); f_B'*x_B];%new lower/upper bound
end %Go to STEP 1
```

5.7 Exercises

Exercise 5.1

Consider the following linear program:

$$
\begin{array}{lrrrrl}
\textit{minimize} & -2x_1 & -3x_2 & -5x_3 & -4x_4 & \\
\textit{subject to} & x_1 + & 2x_2 + & 3x_3 & & \leq 8 \\
 & & x_2 + & x_3 + & x_4 & \leq 7 \\
 & 3x_1 + & 2x_2 & & & \leq 6 \\
 & x_1 + & x_2 & & & \leq 3 \\
 & & & x_3 + & x_4 & \leq 3 \\
 & & & 4x_3 + & 3x_4 & \leq 6 \\
 & x_1 \geq 0 & x_2 \geq 0 & x_3 \geq 0 & x_4 \geq 0. &
\end{array}
$$

Exploit the structure of the linear program to solve for the optimal solution without using Dantzig-Wolfe decomposition.

Exercise 5.2
Consider the following linear programming problem:

$$
\begin{array}{llr}
\textit{minimize} & -5x_1 - 3x_2 - x_3 & \\
\textit{subject to} & x_1 + 2x_2 + x_3 & \leq 10 \\
 & x_1 & \leq 3 \\
 & 2x_2 + x_3 & \leq 8 \\
 & x_2 + x_3 & \leq 5 \\
 & x_1 \geq 0, x_2 \geq 0, x_3 \geq 0. &
\end{array}
$$

Solve by using the Dantzig-Wolfe decomposition.

Exercise 5.3
Consider the following linear programming problem:

$$
\begin{array}{llr}
\textit{minimize} & -8x_1 - 7x_2 - 6x_3 - 5x_4 & \\
\textit{subject to} & 8x_1 + 6x_2 + 7x_3 + 5x_4 & \leq 80 \\
 & 4x_1 + x_2 & \leq 12 \\
 & 5x_1 + x_2 & \leq 15 \\
 & 7x_3 + 2x_4 & \leq 10 \\
 & x_3 + x_4 & \leq 4 \\
 & x_1 \geq 0, x_2 \geq 0, x_3 \geq 0, x_4 \geq 0. &
\end{array}
$$

(a) Solve by Dantzig-Wolfe decomposition.

(b) Show the progress of the primal objective function (which is an upper bound) along with the lower bound obtained from Theorem 5.4 after each iteration.

Exercise 5.4
Consider the following linear programming problem:

$$
\begin{array}{lllr}
\textit{maximize} & 2x_1 + 2x_2 + 3x_3 + 2x_4 & & \\
\textit{subject to} & x_1 + 2x_2 + 2x_3 + x_4 & \leq & 40 \\
& -x_1 + x_2 + x_3 + x_4 & \leq & 10 \\
& 2x_1 + 3x_2 & \leq & 29 \\
& 5x_1 + x_2 & \leq & 25 \\
& x_3 & \leq & 10 \\
& x_4 & \leq & 8 \\
& x_3 + x_4 & \leq & 15 \\
& x_1 \geq 0, x_2 \geq 0, x_3 \geq 0, x_4 \geq 0. & &
\end{array}
$$

Solve by Dantzig-Wolfe decomposition.

Exercise 5.5

Consider the following linear program:

$$
\begin{array}{lllr}
\textit{minimize} & x_1 - x_2 - 3x_3 - x_4 & & \\
\textit{subject to} & 2x_1 + x_2 + x_3 + x_4 & \leq & 12 \\
& -x_1 + x_2 & \leq & 2 \\
& 3x_1 - 4x_3 & \leq & 3 \\
& x_3 + x_4 & \leq & 4 \\
& -x_3 + x_4 & \leq & 2 \\
& x_1 \geq 0, x_2 \geq 0, x_3 \geq 0, x_4 \geq 0. & &
\end{array}
$$

Solve using Dantzig-Wolfe decomposition.

Exercise 5.6

Consider the following transportation problem where there are two warehouses that store a product and there are three retailers that need amounts of the product from the warehouses. The shipping cost per unit c_{ij} from a warehouse i to a retailer j is given as follows.

	Retailer 1	Retailer 2	Retailer 3
Warehouse 1	2	5	3
Warehouse 2	4	2	2

Warehouse 1 has a supply of 1000 units of the product and Warehouse 2 has a supply of 400 units of the product. The demand d_i of Retailer i is given as follows.

	Demand
Retailer 1	300
Retailer 2	750
Retailer 3	350

Solve this problem by using Dantzig-Wolfe decomposition. (Hint: the master problem involves the warehouse capacity constraints and the sub-problems involve retailers' demand constraints). Give an economic interpretation of each iteration of the decomposition.

Exercise 5.7
Dantzig-Wolfe decomposition is just a special case of the revised simplex method and so in principle, without an anti-cycling method, Dantzig-Wolfe decomposition can cycle. Comment on the whether Bland's method (see Chapter 3) can be used in the context of Dantzig-Wolfe decomposition.

Exercise 5.8
Consider linear programs of the following form

$$\begin{array}{lllllllll} \textit{minimize} & c_0^T x_0 & + & c_1^T x_1 & + & c_2^T x_2 & + & \cdots & + & c_N^T x_N & \\ \textit{subject to} & T_0 x_0 & + & A_1 x_1 & + & A_2 x_2 & + & & + & A_N x_N & = b_0 \\ & T_1 x_0 & + & W_1 x_1 & & & & & & & = b_1 \\ & T_2 x_0 & & & + & W_2 x_2 & & & & & = b_2 \\ & \vdots & & & & & & \ddots & & & \vdots \\ & T_N x_0 & & & & & & & + & W_T x_T & = b_T \\ & x_0 \geq 0 & & x_1 \geq 0 & & x_2 \geq 0 & , & \cdots & , & x_T \geq 0. & \end{array}$$

Discuss how Dantzig-Wolfe decomposition can be applied to linear programming problems with the structure above.

Exercise 5.9
Solve the linear program in Exercise 5.2 by using the Dantzig-Wolfe MATLAB code.

Exercise 5.10
Solve the linear program in Exercise 5.6 by using the Dantzig-Wolfe MATLAB code.

Exercise 5.11
The multi-commodity flow problem is a generalization of the minimum cost network flow problem (see Chapter 1 and Exercise 5.6) where more than one commodity (item) can flow on a single directed network $G = (N, E)$. Each commodity will have its own set of demand and supply nodes and the different commodities share the capacities on the edges. Suppose that there are K different commodities. Then, flow variables x_{ij} and cost (or benefit) coefficients c_{ij} will be indexed by k indicating the flow and costs (benefits) incurred of commodity k along the network. Each b_i (recall that $b_i > 0$ if i is a supply node, $b_i < 0$ if i is a demand node, $b_i = 0$ if i is a transshipment node) will also be indexed by k indicating whether the node $i \in N$ is a supply, demand, or transshipment node for commodity k. Then, the formulation of the multi-commodity flow problem is

$$\text{minimize} \quad \sum_{k \in K} \sum_{(i,j) \in E} c_{ij}^k x_{ij}^k$$

(flow balance constraints)

$$\text{subject to} \quad \sum_{\{j:(i,j) \in E\}} x_{ij}^k - \sum_{\{j:(j,i) \in E\}} x_{ji}^k = b_i^k \quad \text{for all } i \in N \text{ and } k \in K$$

(edge capacity constraints)

$$\sum_{k \in K} x_{ij}^k \leq u_{ij} \quad \text{for all } (i,j) \in E$$

(non-negativity constraints)

$$x_{ij}^k \geq 0 \quad \text{for all } (i,j) \in E \text{ and } k \in K.$$

The model is amenable to decomposition by observing that the edge capacity constraints are the linking constraints without which the model would decompose into K individual minimum cost flow problems, one for each commodity $k \in K$. Thus, in a Dantzig-Wolfe decomposition, a subproblem will be formed for each commodity. The master problem will contain the capacity constraints.

Consider the following multi-commodity transportation problem where there are $K = 2$ types of commodities to be shipped from 2 warehouses to 3 retailers. The demand of each commodity from a retailer is given as

Demand	Commodity 1	Commodity 2
Retailer 1	10	9
Retailer 2	19	15
Retailer 3	8	10

The supply of each commodity at each warehouse is given as

Supply	Commodity 1	Commodity 2
Warehouse 1	12	16
Warehouse 2	25	18

The benefit per unit of commodity 1 shipped from a warehouse to a retailer is given as

Benefit	Retailer 1	Retailer 2	Retailer 3
Warehouse 1	35	54	66
Warehouse 2	62	43	39

The benefit per unit of commodity 2 shipped from a warehouse to a retailer is given as

Benefit	Retailer 1	Retailer 2	Retailer 3
Warehouse 1	67	22	41
Warehouse 2	58	37	72

The maximum that can be shipped from a warehouse to a retailer is given as

	Retailer 1	Retailer 2	Retailer 3
Warehouse 1	16	27	14
Warehouse 2	9	18	22

(a) Formulate the problem of finding the shipments of each commodity from the warehouses to retailers that maximizes total benefit.

(b) Solve the formulation in (a) by using Dantzig-Wolfe decomposition showing for each iteration a lower bound and upper bound on the optimal objective function value.

Notes and References

Dantzig-Wolfe decomposition was developed by Dantzig and Wolfe (1960) and was motivated by the work in multi-commodity flows by Ford and Fulkerson (1958). The importance of the decomposition waned in the following years with the advent of more powerful simplex algorithms. Despite this, the decomposition strategy has considerable value and is an important technique in many combinatorial optimization problems, such as vehicle routing and crew scheduling, where such a problem can be formulated as a linear integer program (i.e., a linear program, but with restrictions that some variables take on discrete values) and then the linear programming relaxation is considered and is solved by a column generation technique such as Dantzig-Wolfe decomposition; see Desrochers, Desrosiers, and Solomon (1992), Barnhart et al. (1998), and Simchi-Levi, Chen, and Bramel (2005). The linear programming relaxations for these routing and scheduling problems otherwise would not be able to be stored on a computer due to an exponential number of columns.

The cutting stock problem (see Gilmore and Gomory 1961), involved column-generation techniques similar to that in the Dantzig-Wolfe decomposition. Recent computational studies of the Dantzig-Wolfe decomposition method can be found in Tebboth (2001). The presentation of Dantzig-Wolfe in this chapter is motivated by exploiting the structure of linear programs, e.g., block angular. Dantzig-Wolfe decomposition can be developed in a more general context where the only requirements are that it is possible to separate constraints into a hard and easy classification, see Bazaraa, Jaris, and Sherali (1977) and Chvatal (1983). Classical references for Dantzig-Wolfe include those by Dantzig (1963) and Lasdon (1970). Running the Dantzig-Wolfe decomposition on the dual of a block angular linear program results in the method called Benders decomposition; see Benders (1962), which is a constraint generation method and is an important framework for solving stochastic linear programs, as will be seen in Chapter 8. A more recent reference highlighting the importance of Dantzig-Wolfe decomposition in many engineering applications in energy and other areas can be found in Conejo et al. (2006).

6

Interior Point Methods

6.1 Introduction

In this chapter, we consider a class of methods for linear programming that finds an optimal solution by going through the interior of the feasible set as opposed to traversing extreme points as in the simplex method. This class of methods goes by the name of interior point methods referring to the strategy of going through the interior of the feasible set. Interior point methods have become an important part of the optimization landscape over the last twenty years, and for linear programming these methods have become very competitive with the simplex method.

The general appeal of interior point methods is that they can exhibit both good practical and theoretical properties. For example, although the simplex method performs very well in practice, it can, (as seen in Chapter 3) in the worst case, explore most or all extreme points thereby exhibiting exponential worst-case complexity. For this chapter, we develop a variant of interior point methods called primal-dual path-following methods, which is found to be very effective in solving linear programs and exhibits polynomial worst-case complexity, i.e., the number of basic computational steps needed to find an optimal solution is bounded in the worst case by a polynomial whose value is a function of the size of the problem and desired level of accuracy.

6.2 Linear Programming Optimality Conditions

Consider a linear program in standard form and call this the primal problem (P)

$$\begin{array}{ll} \textit{minimize} & c^T x \\ \textit{subject to} & Ax = b \\ & x \geq 0 \end{array}$$

where $x \in R^n$ and $b \in R^m$. Then, the dual problem (D) is

$$\begin{array}{ll} \textit{minimize} & b^T\pi \\ \textit{subject to} & A^T\pi \leq c \end{array}$$

by adding slack variables we get

$$\begin{array}{ll} \textit{minimize} & b^T\pi \\ \textit{subject to} & A^T\pi + z = c \\ & z \geq 0. \end{array}$$

From Chapter 4, we know that x^* is an optimal solution to the primal problem (P) if and only if

x^* is primal feasible i.e. $Ax^* = b$. (6.1)

There are vectors π^* and z^* such that $A^T\pi^* + z^* = c$. (6.2)

Complementary slackness holds i.e. $x_i^* z_i^* = 0$ for $i = 1, ..., n$. (6.3)

Non-negativity i.e. $x^* \geq 0$ and $z^* \geq 0$. (6.4)

Conditions (6.1), (6.2), (6.3), and (6.4) can in general be written as

$$\begin{array}{c} Ax = b \\ A^T\pi + z = c \\ XZe = 0 \\ x \geq 0, z \geq 0 \end{array}$$

where $X = diag(x)$, i.e.,

$$X = \begin{bmatrix} x_1 & 0 & 0 & 0 \\ 0 & x_2 & 0 & 0 \\ 0 & 0 & \ddots & 0 \\ 0 & 0 & 0 & x_n \end{bmatrix},$$

and $Z = diag(z)$, i.e.,

$$Z = \begin{bmatrix} z_1 & 0 & 0 & 0 \\ 0 & z_2 & 0 & 0 \\ 0 & 0 & \ddots & 0 \\ 0 & 0 & 0 & z_n \end{bmatrix}.$$

e is a vector in R^n, all of whose entries are 1. Conditions (6.1)–(6.4) are also known as the Karush Kuhn Tucker (KKT) conditions for linear programming. Thus, the KKT conditions are both necessary and sufficient for optimality of linear programs.

Except for the non-negativity condition (6.4), the KKT conditions constitute a system of equations where conditions (6.1) and (6.2) represent linear equations and condition (6.3) represents a system of non-linear equations. Then, collectively, conditions (6.1), (6.2), and (6.3) represent a system of non-linear equations. Any strategy for obtaining an optimal solution needs to generate a solution that will satisfy this system of non-linear equations along with the non-negativity requirements. The Newton-Raphson method is an important iterative technique for solving systems of non-linear equations and will be used as a vital component of interior point methods. We now give a brief overview of this important method.

6.2.1 Newton-Raphson Method

Suppose that we wish to solve

$$f(x) = 0$$

where $f(x)$ is a function from R^n to R^n, i.e., it is a vector valued function. In particular,

$$f(x) = \begin{bmatrix} f_1(x) \\ f_2(x) \\ \vdots \\ f_n(x) \end{bmatrix} = \begin{bmatrix} 0 \\ 0 \\ \vdots \\ 0 \end{bmatrix}$$

where $f_i(x)$ is a function from R^n to R, i.e., a real-valued function. If x^* is such that $f(x^*) = 0$, then x^* is called a root of $f(x)$.

Example 6.1
Consider the following non-linear system of equations:

$$\begin{aligned} x_1^2 + x_2^2 + x_3^2 &= 3 \\ x_1^2 + x_2^2 - x_3 &= 1 \\ x_1 + x_2 + x_3 &= 3, \end{aligned}$$

then, the non-linear equations can be represented as

$$f(x) = \begin{bmatrix} f_1(x) \\ f_2(x) \\ f_3(x) \end{bmatrix} = \begin{bmatrix} x_1^2 + x_2^2 + x_3^2 - 3 \\ x_1^2 + x_2^2 - x_3 - 1 \\ x_1 + x_2 + x_3 - 3 \end{bmatrix} = \begin{bmatrix} 0 \\ 0 \\ 0 \end{bmatrix}.$$

The Newton-Raphson method starts with an initial guess of the root of $f(x)$ and then produces iterates that each represent a successive approximation of the root of $f(x)$. The idea in Newton-Raphson is to generate the next iterate by linearizing the function $f(x)$ at the current iterate $x^{(k)}$ and then solving for the root of the linearized function. Then, under suitable conditions the sequence $\{x^{(k)}\}$ will converge to x^* where x^* is a root of $f(x)$.

The linearization of $f(x)$ is accomplished by taking the Taylor series approximation at $x^{(k)}$. Let $d \in R^n$ be a direction vector, then

$$f(x^{(k)} + d) \approx f(x^{(k)}) + \nabla f(x^{(k)})d$$

where $\nabla f(x^{(k)})$ is the $n \times n$ Jacobian matrix of $f(x)$ at $x^{(k)}$, i.e., the ith row is the transpose of the gradient of $f_i(x)$ at $x^{(k)}$

$$\nabla f(x^{(k)}) = \begin{bmatrix} \nabla f_1(x^{(k)})^T \\ \nabla f_2(x^{(k)})^T \\ \vdots \\ \nabla f_n(x^{(k)})^T \end{bmatrix} = \begin{bmatrix} \frac{df_1}{dx_1} & \frac{df_1}{dx_2} & \cdots & \frac{df_1}{dx_n} \\ \frac{df_2}{dx_1} & \frac{df_2}{dx_2} & \cdots & \frac{df_2}{dx_n} \\ & & \ddots & \\ \frac{df_n}{dx_1} & \frac{df_n}{dx_2} & \cdots & \frac{df_n}{dx_n} \end{bmatrix}.$$

Now we solve for the direction d that would give the next approximate root $x^{(k+1)}$ by finding d such that

$$f(x^{(k+1)}) \approx f(x^{(k)}) + \nabla f(x^{(k)})d = 0$$

to get

$$d = -[\nabla f(x^{(k)})]^{-1} f(x^{(k)}).$$

The vector d is called the Newton direction and the following formula

$$x^{(k+1)} = x^{(k)} + d = x^{(k)} - [\nabla f(x^{(k)})^T]^{-1} f(x^{(k)})$$

is known as the $(k+1)st$ Newton-Raphson iterate. Now, the Newton-Raphson method can be summarized as follows.

Newton-Raphson Method

Step 0: (Initialization)
Let $x^{(0)}$ be an initial point and $k = 0$.

Step 1: (Generating Newton direction)
Compute $d^{(k)} = -[\nabla f(x^{(k)})]^{-1} f(x^{(k)})$.
If $d^{(k)} \approx 0$, then stop, else go to Step 2.

Step 2: (Step Length)
Set $\alpha^{(k)} = 1$ and go to Step 3.

Step 3: (Generating next iterate)
$x^{(k+1)} = x^{(k)} + \alpha^{(k)} d^{(k)}, k = k + 1$
Go to Step 1.

Example 6.2
Consider the system of non-linear equations in Example 6.1, i.e.,

$$f(x) = \begin{bmatrix} f_1(x) \\ f_2(x) \\ f_3(x) \end{bmatrix} = \begin{bmatrix} x_1^2 + x_2^2 + x_3^2 - 3 \\ x_1^2 + x_2^2 - x_3 - 1 \\ x_1 + x_2 + x_3 - 3 \end{bmatrix}.$$

The Jacobian of $f(x)$ is

$$\nabla f(x) = \begin{bmatrix} 2x_1 & 2x_2 & 2x_3 \\ 2x_1 & 2x_2 & -1 \\ 1 & 1 & 1 \end{bmatrix}.$$

Step 0: Let $x^{(0)} = (1, 0, 1)^T$.

Iteration 1

Step 1: Compute $d^{(0)} = -[\nabla f(x^{(0)})]^{-1} f(x^{(0)})$, or equivalently solve for $d^{(0)}$ in

$$[\nabla f(x^{(0)})] d^{(0)} = -f(x^{(0)}), \text{ i.e.,}$$

$$\begin{bmatrix} 2 & 0 & 2 \\ 2 & 0 & -1 \\ 1 & 1 & 1 \end{bmatrix} \begin{bmatrix} d_1^{(0)} \\ d_2^{(0)} \\ d_3^{(0)} \end{bmatrix} = - \begin{bmatrix} -1 \\ -1 \\ -1 \end{bmatrix},$$

so

$$d^{(0)} = \begin{bmatrix} 1/2 \\ 1/2 \\ 0 \end{bmatrix}.$$

Go to Step 2.

Step 2: $x^{(1)} = x^{(0)} + (1)d^{(0)} = \begin{bmatrix} 1 \\ 0 \\ 1 \end{bmatrix} + (1) \begin{bmatrix} 1/2 \\ 1/2 \\ 0 \end{bmatrix} = \begin{bmatrix} 3/2 \\ 1/2 \\ 1 \end{bmatrix}.$

Go to Step 1.

Iteration 2

Step 1: Solve $[\nabla f(x^{(1)})] d^{(1)} = -f(x^{(1)})$, i.e.,

$$\begin{bmatrix} 3 & 1 & 2 \\ 3 & 1 & -1 \\ 1 & 1 & 1 \end{bmatrix} \begin{bmatrix} d_1^{(1)} \\ d_2^{(1)} \\ d_3^{(1)} \end{bmatrix} = - \begin{bmatrix} 1/2 \\ 1/2 \\ 0 \end{bmatrix},$$

so

$$d^{(0)} = \begin{bmatrix} -1/4 \\ 1/4 \\ 0 \end{bmatrix}.$$

Go to Step 2.

Step 2: $x^{(2)} = x^{(1)} + (1)d^{(1)} = \begin{bmatrix} 3/2 \\ 1/2 \\ 1 \end{bmatrix} + (1)\begin{bmatrix} -1/4 \\ 1/4 \\ 0 \end{bmatrix} = \begin{bmatrix} 5/4 \\ 3/4 \\ 1 \end{bmatrix}$.

Table 6.1 shows the first 6 of the first 20 iterates of the Newton-Raphson method. It should be noted that the error between successive iterates, i.e., $\|x^{(k+1)} - x^{(k)}\|$ is reduced by half, each iteration, which results in rapid convergence toward the root $x^* = \begin{bmatrix} 1 & 1 & 1 \end{bmatrix}$.

Table 6.1 First 20 Newton-Raphson iterates for Example 6.1

k	$x_1^{(k)}$	$x_2^{(k)}$	$x_3^{(k)}$	$d_1^{(k)}$	$d_2^{(k)}$	$d_3^{(k)}$	$\|x^{(k+1)} - x^{(k)}\|$
0	1	0	1	0.5	0.5	0	0.707107
1	1.5	0.5	1	−0.25	0.25	0	0.353553
2	1.25	0.75	1	−0.125	0.125	0	0.176777
3	1.125	0.875	1	−0.0625	0.0625	0	0.088388
4	1.0625	0.9375	1	−0.03125	0.03125	0	0.044194
5	1.03125	0.96875	1	−0.015625	0.015625	0	0.022097
6	1.015625	0.984375	1	−0.007813	0.007813	0	0.011049
⋮	⋮	⋮	⋮	⋮	⋮	⋮	⋮
20	1.000001	0.999999	1	−4.77E-07	4.77E-07	0	6.74E-07

Comments

(1) The Newton-Raphson method assumes that $\nabla f(x^{(k)})^T$ is invertible at each iteration.

(2) It is not guaranteed that $x^{(k+1)}$ is a better approximate root than $x^{(k)}$, and the success of the method will depend on the initial point $x^{(0)}$.

(3) In computing the direction vector, it is better to solve $[\nabla f(x^{(k)})]d^{(k)} = -f(x^{(k)})$ instead of calculating the inverse of the Jacobian directly.

(4) A step length of $\alpha^{(k)} = 1$ is not necessary, e.g., line search methods can be used to find the step lengths at each iteration that do not have to be 1; see Nocedal and Wright (1999).

Convergence

The Newton-Raphson method produces a sequence of iterates. One important issue is whether the sequence will converge toward a root x^* of the function $f(x)$. For the Newton-Raphson method to work well, it is critical that the initial point $x^{(0)}$ is sufficiently close to the root or the method might breakdown. If an initial point is close enough to a root, then the method will converge quadratically toward the root; see Dennis and Schnabel (1983) or Ortega and Rheinboldt (1970).

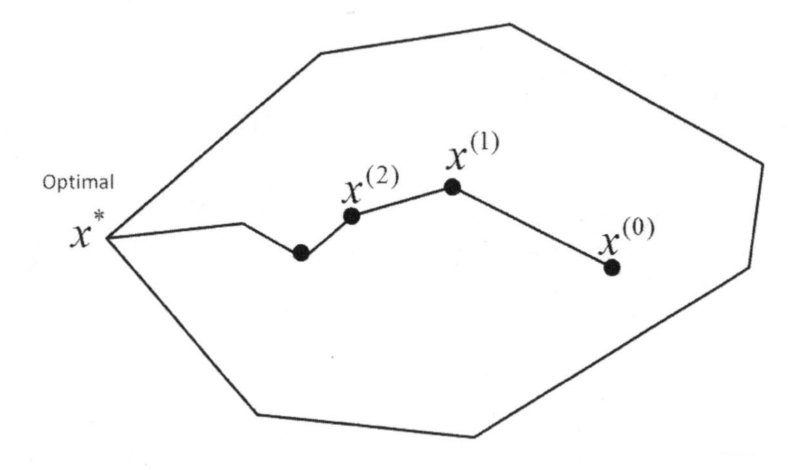

FIGURE 6.1
Interior point trajectory for solving a linear program.

6.3 Primal-Dual Interior Point Strategy

Recall that the simplex method is a strategy that starts at an extreme point. moves to an adjacent extreme point, and iterates until at optimality finds vectors x, π, and z that satisfies the KKT conditions. In particular, at each iteration it was seen that the simplex method generates an extreme point that is primal feasible and satisfies complementary slackness and at termination when optimality is achieved a dual feasible solution is generated, thereby satisfying all of the KKT conditions. Simplex methods are called boundary point methods since only extreme points are considered and are points on the edge or boundary of the feasible set. The foundation for the strategy for interior point methods is the same as for simplex methods. That is, both classes of methods seek to satisfy the KKT conditions for linear programming.

By contrast, the interior point method strategy that we consider generates points that are interior to the feasible set and ultimately converge to an optimal extreme point of the original linear programming problem; see Figure 6.1. The interior point strategy will start with an initial interior solution $(x^{(0)}, \pi^{(0)}, z^{(0)})$, i.e., $x^{(0)} > 0, z^{(0)} > 0$, with $Ax^{(0)} = b$, i.e., satisfies primal feasibility, and dual feasibility is satisfied, i.e., $A^T\pi^{(0)} + z^{(0)} = c$ and seeks to iteratively solve the KKT conditions of the original primal problem by maintaining at each iteration primal and dual feasibility while reducing the duality gap (i.e., getting closer to maintaining complementary slackness).

An important component of the interior point strategy is the use of the Newton-Raphson method to solve the system of non-linear equations that represent the KKT conditions of the linear programming problem. However, the Newton-Raphson method will not be directly applied to the KKT conditions of the original linear programming problem. Applying the method as described earlier to conditions (6.1),(6.2), and (6.3) will often produce vectors that have zero or negative components since the step length, which is always set at $\alpha = 1$ will be too long in many instances.

The idea to overcome the limitations of using the Newton-Raphson method directly on the original linear program is to modify the original linear program in a way that will facilitate the generation of interior points while allowing good progress toward reaching optimality. This will require a modification of the Newton-Raphson method to allow variable step lengths along directions that will offer good improvement toward reaching optimality. We first discuss the modification of the original linear program.

6.3.1 Barrier Reformulation

The key idea behind the primal-dual interior point strategy is to modify the primal problem by adding a logarithmic barrier term in the objective function in place of the non-negativity requirement of the primal variables. The modified primal problem is

$$\begin{array}{ll} \textit{maximize} & c^T x - \mu \sum_{i=1}^{n} \ln(x_i) \\ \textit{subject to} & Ax = b \end{array} \tag{6.5}$$

where $\mu > 0$ is a positive constant and this problem will be denoted by P_μ.

Observe that as x_i gets close to 0, then $\ln(x_i) \to -\infty$ and so the objective function prevents the variables x from becoming zero. P_μ is no longer a linear programming problem since the logarithmic terms are non-linear. The strategy is now to take the Lagrangian of (6.5) and set up the equations that define the critical points of the Lagrangian.

The Lagrangian of the (6.5) is

$$L(x, \pi) = c^T x - \mu \sum_{i=1}^{n} \ln(x_i) - \pi^T (b - Ax),$$

then we have the equations that define critical points as follows

$$\frac{dL}{dx} = c - \mu X^{-1} e - A^T \pi = 0 \tag{6.6}$$

$$\frac{dL}{d\pi} = b - Ax = 0. \tag{6.7}$$

Now let $z = \mu X^{-1} e$, then $Xz = \mu e$ or $XZe = \mu e$ where $Z = diag(z)$. Then, the conditions (6.6) and (6.7) become

$$A^T \pi + z = c \tag{6.8}$$

$$Ax = b \tag{6.9}$$

$$XZe = \mu e. \tag{6.10}$$

The conditions (6.8), (6.9), and (6.10) represent the KKT conditions for the barrier problem for a given penalty parameter μ. Condition (6.8) is the dual feasibility requirement and condition (6.9) is the primal feasibility requirement, and these two conditions are the same as in the KKT conditions. The major difference is that $\mu > 0$ implies that condition (6.10) will only approximately enforce complementary slackness, however it will ensure that x and z will be positive and therefore in the interior of the feasible set and so the non-negativity requirements on x and z are satisfied if condition (6.10) is satisfied. We now define an important subset of interior feasible solutions.

Definition 6.3

The central path Γ is the set of all vectors satisfying conditions (6.8), (6.9), and (6.10) for some $\mu > 0$, i.e.,

$$\Gamma = \{(x, \pi, z)|\ A^T\pi + z = c, Ax = b, XZ = \mu e, \mu > 0\}.$$

Γ is a curve in the interior of the feasible set of the original primal linear program. A point on the curve of Γ is an optimal solution for P_μ for some $\mu > 0$. Intuitively, Γ is a path that is centered in the interior of the feasible set of the primal problem (and dual problem) since both the primal x and dual solutions π are kept away from the boundary of the "walls" of their respective feasible sets; see Figure 6.2. As $\mu \to 0$, the points on the central path converge toward the optimal solution of the original primal linear programming problem and its dual. This suggests that the barrier problem P_μ should be solved repeatedly, each time using a smaller positive constant μ. That is, one should repeatedly solve conditions (6.8), (6.9), and (6.10) for successively smaller values of μ. These solutions will be iterates that should ultimately converge toward satisfying the KKT conditions of the original primal problem.

6.3.1.1 Newton-Raphson Method for KKT Conditions of the Barrier Problem

The conditions (6.8), (6.9), and (6.10) for a given value of μ can be represented in functional form as

$$F(x, \pi, z) = \begin{bmatrix} A^T\pi + z - c \\ Ax - b \\ XZe \end{bmatrix} = \begin{bmatrix} 0 \\ 0 \\ \mu e \end{bmatrix} \quad (6.11)$$

$$x \geq 0 \text{ and } z \geq 0.$$

F is a mapping from R^{m+2n} to R^{m+2n}. This functional form represents a non-linear system of equations due to the XZe term, and so in general cannot be solved exactly, so the Newton-Raphson method will be used to find an approximate solution to (6.11). Then, the Newton direction d is obtained by solving the system

$$[\nabla F(x, \pi, z)]d = -F(x, \pi, z),$$

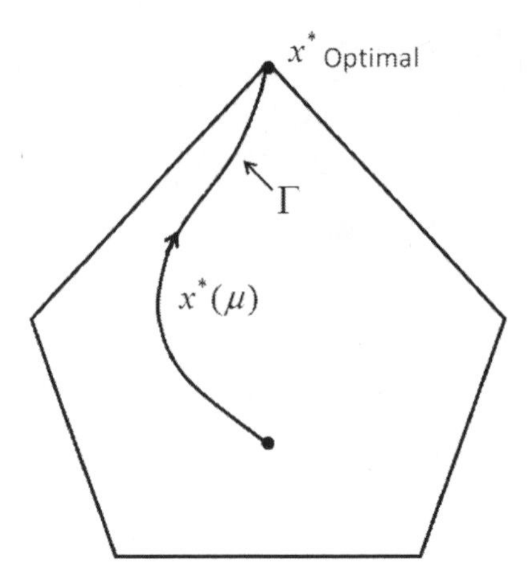

FIGURE 6.2
Central path for a linear program.

which is the equivalent to

$$\begin{bmatrix} 0 & A^T & I \\ A & 0 & 0 \\ Z & 0 & X \end{bmatrix} \begin{bmatrix} d_x \\ d_\pi \\ d_z \end{bmatrix} = \begin{bmatrix} 0 \\ 0 \\ -XZe + \mu e \end{bmatrix}. \tag{6.12}$$

Biased Newton-Raphson Directions

As mentioned, the Newton-Raphson method will be applied to the non-linear system (6.11) for successively smaller positive values of μ, which will give rise to successive systems of the form of (6.12). An important consideration is how to decrease μ from one iteration to the next since this value will impact the generation of a search direction d. There will be two factors that will influence the decrement: (1) the need to have iterates in the interior of the feasible set and (2) to have iterates move in directions that enable reduction in the duality gap, i.e., make progress toward the optimal solution. In regard to (1), the iterates should be near the central path Γ, since this is curve that is in the center of the interior of the feasible set, and by being near the central path iterates will be positioned well for long moves toward the optimal solution, which will help with achieving (2).

Thus, μ will be expressed as product of a centering parameter $\tau \in (0, 1)$ that will influence iterations to be nearer the central path the closer τ is to 1, and a duality measure y that measures the progress of the primal-dual iterates toward achieving optimality. In particular,

$$y = \tfrac{1}{n} \sum_{i=1}^{n} x_i z_i = \tfrac{x^T z}{n}$$

That is, y is the average value of the terms $x_i z_i$. A smaller average indicates more proximity to optimality. Then

$$\mu = \tau y,$$

and thus (6.12) becomes

$$\begin{bmatrix} 0 & A^T & I \\ A & 0 & 0 \\ Z & 0 & X \end{bmatrix} \begin{bmatrix} d_x \\ d_\pi \\ d_z \end{bmatrix} = \begin{bmatrix} 0 \\ 0 \\ -XZe + \tau y e \end{bmatrix}. \tag{6.13}$$

The centering parameter τ can be selected to be less than or equal to 1 (but greater than or equal to 0) to allow a tradeoff between moving toward the central path and reducing y.

6.3.2 General Primal-Dual Interior Point Method

We now present the general primal-dual interior point framework. The general iterative strategy is a follows.

Step 0: Obtain an initial interior primal-dual solution $(x^{(0)}, \pi^{(0)}, z^{(0)})$ such that $x^{(0)} > 0, z^{(0)} > 0$, $Ax^{(0)} = b$, and $A^T \pi^{(0)} + z^{(0)} = c$. Let $k = 0$ and ε be some small positive number. Go to Step 1.

Step 1: Choose $\tau^{(k)} \in [0, 1]$ and let $y^{(k)} = \frac{(x^{(k)})^T z^{(k)}}{n}$ and $\mu^{(k)} = \tau^{(k)} y^{(k)}$.
Solve the following system for $d^{(k)}$

$$\begin{bmatrix} 0 & A^T & I \\ A & 0 & 0 \\ Z^{(k)} & 0 & X^{(k)} \end{bmatrix} \begin{bmatrix} d_x^{(k)} \\ d_\pi^{(k)} \\ d_z^{(k)} \end{bmatrix} = \begin{bmatrix} 0 \\ 0 \\ -X^{(k)} Z^{(k)} e + \tau^{(k)} y^{(k)} e \end{bmatrix}$$

where

$$X^{(k)} = diag(x^{(k)}) \text{ and } Z^{(k)} = diag(z^{(k)}).$$

Go to Step 2.

Step 2: Let $\begin{bmatrix} x^{(k+1)} \\ \pi^{(k+1)} \\ z^{(k+1)} \end{bmatrix} = \begin{bmatrix} x^{(k)} \\ \pi^{(k)} \\ z^{(k)} \end{bmatrix} + \alpha^{(k)} \begin{bmatrix} d_x^{(k)} \\ d_\pi^{(k)} \\ d_z^{(k)} \end{bmatrix}$,

where $\alpha^{(k)}$ is selected so that $x^{(k+1)} > 0$ and $z^{(k+1)} > 0$.

If the stopping criteria is met, i.e.,

$$\left\| Ax^{(k+1)} - b \right\| \leq \varepsilon$$
$$\left\| A^T \pi^{(k+1)} + z^{(k+1)} - c \right\| \leq \varepsilon$$
$$(x^{(k+1)})^T z^{(k+1)} \leq \varepsilon$$

for some small tolerance $\varepsilon > 0$, then STOP
Else let $k = k + 1$ and go to Step 1.

It is clear that as $k \to \infty$ and $\mu^{(k)} \to 0$, then

$$Ax^{(k)} \to b$$
$$A^T \pi^{(k)} + z^{(k)} \to c$$
$$X^{(k)} Z^{(k)} e \to 0$$
$$\text{and } x^{(k)} > 0, z^{(k)} > 0.$$

That is, the iterates will converge toward satisfying the KKT conditions for the original linear program.

The step length $\alpha^{(k)}$ can be applied to all components of the Newton-Raphson direction $d^{(k)}$ or the step lengths can be individualized, i.e., one can define a step length parameter $\alpha_x^{(k)}$ for $x^{(k)}$, $\alpha_\pi^{(k)}$ for $\pi^{(k)}$, and $\alpha_z^{(k)}$ for $z^{(k)}$ to get

$$\begin{aligned} x^{(k+1)} &= x^{(k)} + \alpha_x^{(k)} d_x^{(k)} \\ \pi^{(k+1)} &= \pi^{(k)} + \alpha_\pi^{(k)} d_\pi^{(k)} \\ z^{(k+1)} &= z^{(k)} + \alpha_z^{(k)} d_z^{(k)}. \end{aligned}$$

6.3.2.1 Starting with an Infeasible Interior Point

The primal-dual framework can be modified to start with an infeasible interior solution (i.e., with $x^{(0)} > 0$ and $z^{(0)} > 0$, but $Ax^{(0)} \neq b$ and $A^T \pi^{(0)} + z^{(0)} \neq c$) in case it is not straightforward to obtain an feasible interior solution. The construction of the search direction can be made to improve feasibility as well as improve the centering of iterates to be near the central path. To this end, we define the residuals

$$r_p^{(k)} = Ax^{(k)} - b \text{ (primal residuals)}$$
$$r_d^{(k)} = A^T \pi^{(k)} + z^{(k)} - c \text{ (dual residuals)}$$

and modify the system of equations (6.12) to get

$$\begin{bmatrix} 0 & A^T & I \\ A & 0 & 0 \\ Z^{(k)} & 0 & X^{(k)} \end{bmatrix} \begin{bmatrix} d_x^{(k)} \\ d_\pi^{(k)} \\ d_z^{(k)} \end{bmatrix} = \begin{bmatrix} -r_p^{(k)} \\ -r_d^{(k)} \\ -X^{(k)} Z^{(k)} e + \tau^{(k)} y^{(k)} e \end{bmatrix}. \tag{6.13}$$

The direction $d^{(k)}$ that is generated at the *kth* iteration remains a Newton-Raphson direction toward the point $(x, \pi, z) \in \Gamma$, which is associated with the value $\mu^{(k)} = \tau^{(k)}y^{(k)}$ and will also help drive the residuals to zero.

6.3.3 Complexity of General Primal-Dual Interior Path Following Methods

One of the major advantages of primal-dual path-following methods is that they offer not only excellent practical performance, but a theoretical guarantee of performance in terms of the amount of computing time or resources required to solve a linear program. In particular, primal-dual path-following methods can exhibit polynomial (worst-case) complexity, meaning that the amount of time required to solve a linear program is in the worst case a polynomial function of some aspect of the size of a linear program.

The essence of achieving polynomial complexity is to ensure that an initial point is selected so that the initial duality measure $y^{(0)}$ is not too large and that this measure gets suitably reduced after each iteration and depends on the size (dimension) of the problem. We have the following important technical lemma that relates these requirements to polynomial complexity for path-following type methods.

Lemma 6.4

Suppose that $\varepsilon \in (0, 1)$ *and that an initial point* $(x^{(0)}, \pi^{(0)}, z^{(0)})$ *is such that* $y^{(0)} \leq \frac{1}{\varepsilon^{\gamma}}$ *for some constant* $\gamma > 0$*. Furthermore, suppose that the general primal-dual interior point method generates a sequence of iterates that satisfy*

$$y^{(k+1)} \leq (1 - \tfrac{\phi}{n^{\chi}})y^{(k)} \textit{ for } k = 0, 1, 2, ...,$$

for some constants $\phi > 0$ *and* $\chi > 0$.

Then, there exists an index K *with*

$$K \leq Cn^{\chi}|\log \varepsilon| \textit{ for some constant } C > 0$$

such that

$$y^{(k)} \leq \varepsilon \text{ for all } k \geq K.$$

The proof of Lemma 6.4 requires an inequality result concerning an approximation to the natural log function. In particular, we have the following.

Lemma 6.5

$\log(1 + \Phi) \leq \Phi$ *for all* $\Phi > -1$.

Proof: Left for the reader. ■

Now we prove Lemma 6.4.

Proof: Take the logarithms on both sides of

$$y^{(k+1)} \leq (1 - \tfrac{\phi}{n\chi})y^{(k)}$$

to get

$$\log(y^{(k+1)}) \leq \log(1 - \tfrac{\phi}{n\chi})y^{(k)} = \log(1 - \tfrac{\phi}{n\chi}) + \log y^{(k)}.$$

Now using the inequality repeatedly along with the assumption that $y^{(0)} \leq \frac{1}{\varepsilon^{\gamma}}$ gives the following

$$\begin{aligned} \log(y^{(k)}) &\leq k\log(1 - \tfrac{\phi}{n\chi}) + \log y^{(0)} \\ &\leq k\log(1 - \tfrac{\phi}{n\chi}) + \log \tfrac{1}{\varepsilon^{\gamma}} \\ &= k\log(1 - \tfrac{\phi}{n\chi}) + \gamma \log \tfrac{1}{\varepsilon}. \end{aligned}$$

Therefore, by Lemma 6.5 we have

$$\log(y^{(k)}) \leq k(-\tfrac{\phi}{n\chi}) + \gamma \log \tfrac{1}{\varepsilon},$$

so $y^{(k)} \leq \varepsilon$ will hold if

$$\log(y^{(k)}) \leq k(-\tfrac{\phi}{n\chi}) + \gamma \log \tfrac{1}{\varepsilon} \leq \log \varepsilon$$

and will be satisfied when

$$k \geq (\gamma \log \tfrac{1}{\varepsilon} - \log \varepsilon)/\tfrac{\phi}{n\chi}$$

or equivalently when

$$k \geq (1+\gamma)\tfrac{n\chi}{\phi} \log \tfrac{1}{\varepsilon},$$

and thus the result follows. ■

We now give another result that pertains to primal-dual interior point methods that solves the system (6.13) to generate a search direction. This result gives an expression of the particular form of the reduction of the duality measure after each iteration that is a result of solving system (6.13). Then, with certain values of the parameters, the form of the reduction can be shown to satisfy the conditions of Lemma 6.4, namely the condition $y^{(k+1)} \leq (1 - \frac{\phi}{n\chi})y^{(k)}$ for $k = 0, 1, ...$, which then implies polynomical complexity for a primal-dual path-following interior point method with those parameters.

Lemma 6.6

Suppose the vector $[x^{(k+1)}, \pi^{(k+1)}, z^{(k+1)}]^T$ *is obtained after Step 2 of the general primal-dual interior point method, i.e.,*

$$\begin{bmatrix} x^{(k+1)} \\ \pi^{(k+1)} \\ z^{(k+1)} \end{bmatrix} = \begin{bmatrix} x^{(k)} \\ \pi^{(k)} \\ z^{(k)} \end{bmatrix} + \alpha^{(k)} \begin{bmatrix} d_x^{(k)} \\ d_\pi^{(k)} \\ d_z^{(k)} \end{bmatrix},$$

then

(1) $(d_x^{(k)})^T d_z^{(k)} = 0$
(2) $y^{(k+1)} = (1 - \alpha^{(k)}(1 - \tau^{(k)}))y^{(k)}$
Proof:
Part (1)
By Exercise 6.3 we have

$$d_x^{(k)} = (I - X^{(k)}(Z^{(k)})^{-1}(A^{(k)})^T(A^{(k)}X^{(k)}(Z^{(k)})^{-1}(A^{(k)})^T)^{-1}A^{(k)})(-x^{(k)} + \mu^{(k)}(Z^{(k)})^{-1}e)$$

and

$$d_z^{(k)} = (A^{(k)})^T(A^{(k)}X^{(k)}(Z^{(k)})^{-1}(A^{(k)})^T)^{-1}A^{(k)}(-x^{(k)} + \mu^{(k)}(Z^{(k)})^{-1}e)$$

and it is then straightforward to verify that

$$\begin{gathered}(d_x^{(k)})^T d_z^{(k)} \\ = [(I - X^{(k)}(Z^{(k)})^{-1}(A^{(k)})^T(A^{(k)}X^{(k)}(Z^{(k)})^{-1}(A^{(k)})^T)^{-1}A^{(k)})(-x^{(k)} + \\ \mu^{(k)}(Z^{(k)})^{-1}e)]^T \\ [(A^{(k)})^T(A^{(k)}X^{(k)}(Z^{(k)})^{-1}(A^{(k)})^T)^{-1}A^{(k)}(-x^{(k)} + \mu^{(k)}(Z^{(k)})^{-1}e)\ ] \\ = 0.\end{gathered}$$

Part (2)
By the third equation in (6.13) we have

$$Z^{(k)}d_x^{(k)} + X^{(k)}d_z^{(k)} = -X^{(k)}Z^{(k)}e + \tau^{(k)}y^{(k)}e$$

or equivalently

$$z^{(k)T}d_x^{(k)} + x^{(k)T}d_z^{(k)} = -(1 - \tau^{(k)})x^{(k)T}z^{(k)}.$$

Now

$$\begin{gathered}(x^{(k+1)})^T z^{(k+1)} = (x^{(k)} + \alpha^{(k)}d_x^{(k)})^T(z^{(k)} + \alpha^{(k)}d_z^{(k)}) \\ = (x^{(k)})^T z^{(k)} + \alpha^{(k)}(z^{(k)T}d_x^{(k)} + x^{(k)T}d_z^{(k)}) + (\alpha^{(k)})^2(d_x^{(k)})^T d_z^{(k)} \\ = (x^{(k)})^T z^{(k)} + \alpha^{(k)}(-(1 - \tau^{(k)})x^{(k)T}z^{(k)}) \\ = (x^{(k)})^T z^{(k)}(1 - \alpha^{(k)}(1 - \tau^{(k)})).\end{gathered}$$

Therefore

$$(x^{(k+1)})^T z^{(k+1)}/n = (x^{(k)})^T z^{(k)}(1 - \alpha^{(k)}(1 - \tau^{(k)}))/n$$

or equivalently

$$y^{(k+1)} = (1 - \alpha^{(k)}(1 - \tau^{(k)}))y^{(k)}. \ \blacksquare$$

6.3.3.1 Polynomial Complexity of Short-Step Path-Following Methods

Now we develop the results leading to the polynomial complexity of a particular variant of a path-following method called the short-step primal-dual path-following method as an illustration of use of the lemmas above to show polynomial complexity for primal-dual interior point methods. We will refer to this method in shorthand as SSPF. Recall that general primal-dual interior point strategies follow the central path Γ in the direction of decreasing μ, and in particular at an iteration k will generate a search direction that is a Newton-Raphson direction toward a point $(x^{(k+1)}, \pi^{(k+1)}, z^{(k+1)})$ on the central path Γ. By construction, the point will have a duality measure that is less than or equal to the current measure $y^{(k)}$. However, since the Newton-Raphson iterate is only an approximation to the equation (6.10), i.e.,

$$XZe = \mu e,$$

(note that (6.8) and (6.9) will be satisfied at every iteration along with non-negativity of x and z), the primal-dual interior point strategy will not necessarily generate iterates that are on the central path Γ. In particular, the products $x_i z_i$ will not in general be equal, which results in a violation of equation (6.10).

The amount of deviation from equation (6.10) can be measured by using the difference between $y^{(k)}$, the average value of the products, and the actual products XZe. The following scaled norm

$$\tfrac{1}{\mu}\,\|XZe - \mu e\|$$

will be used to measure the deviation where the 2-norm (Euclidean norm) will be assumed. Now we add a further restriction that the deviation be less than a constant θ. Let

$$F_0 = \{(x,\, \pi,\, z)|Ax = b, A^T\,\pi + z = c, x > 0, z > 0\},$$

which is the set of primal and dual feasible solutions that are in the interior of the feasible set of the linear program. Then, let

$$N_2(\theta) = \{(x,\, \pi,\, z) \in F_0|\tfrac{1}{\mu}\,\|XZe - \mu e\| \leq \theta\}.$$

Short-Step Path-Following Method

Now we specify the short-step path-following (SSPF) method, which is an instance of the general primal-dual interior point method of section 3.2 where

(1) The initial point $(x^{(0)}, \pi^{(0)}, z^{(0)})$ is in $N_2(\theta)$ for $\theta = 0.4$.
(2) $\tau^{(k)} = 1 - \frac{0.4}{\sqrt{n}}$for all k.
(3) The step length $\alpha^{(k)} = 1$ for all k.

Then, we have the following polynomial complexity result of SSPF.

Theorem 6.7

Let $\varepsilon > 0$ and suppose that the initial point $(x^{(0)}, \pi^{(0)}, z^{(0)}) \in N_2(\theta)$ in the SSPF method is such that

$$y^{(0)} \leq \frac{1}{\varepsilon^{\gamma}} \text{ for some constant } \gamma > 0.$$

Then, there is an index K such that

$$y^{(k)} \leq \varepsilon \text{ for all } k \geq K,$$

where $K = O(\sqrt{n} \log \frac{1}{\varepsilon})$.

Proof:

For the SSPF method we have, from Lemma 6.6, that

$$\begin{aligned} y^{(k+1)} &= (1 - \alpha^{(k)}(1 - \tau^{(k)}))y^{(k)} \\ &= (1 - \tfrac{0.4}{\sqrt{n}})y^{(k)} \text{ for } k = 0, 1, 2, ..., \end{aligned}$$

since $\alpha^{(k)} = 1$ and $\tau^{(k)} = 1 - \frac{0.4}{\sqrt{n}}$. So now let $\phi = 0.4$ and $\chi = 0.5$ in Lemma 6.4 and then the result follows. ■

6.4 The Predictor-Corrector Variant of the Primal-Dual Interior Point Method

We now present a variant of the primal-dual interior point strategy due to Mehrotra (1992), which enjoys excellent practical performance. The key modifications involve (1) a predictor step that computes a search direction without centering, (2) a corrector step that uses a second-order approximation to the central path that adjusts the search direction to more closely follow a trajectory to the optimal solution, and (3) incorporates an adaptive choice for the centering parameter τ at each iteration. For the development below, the superscripts k will be omitted until the algorithm is presented later in full detail.

6.4.1 Predictor Step

A search direction is first computed where the centering parameter τ is set to 0. The idea is that if the duality measure can be reduced sufficiently well along this direction, then a large centering value is not necessary. Let this direction (also known as the affine scaling direction) be denoted by d^{aff} where it is computed by solving the system (6.13) but with the term involving τ set to 0

$$\begin{bmatrix} 0 & A^T & I \\ A & 0 & 0 \\ Z & 0 & X \end{bmatrix} \begin{bmatrix} d_x^{aff} \\ d_\pi^{aff} \\ d_z^{aff} \end{bmatrix} = \begin{bmatrix} -r_p \\ -r_d \\ -XZe \end{bmatrix}. \tag{6.14}$$

6.4.2 Setting the Centering Parameter

After the affine scaling direction is computed, the quality of this predictor direction is evaluated and the centering parameter is set based on how good this direction is. To evaluate the predictor direction the longest possible step lengths along this direction are computed before the non-negativity constraints are violated. The maximum step length allowed is 1 and the formulas for the step lengths are analogous to minimum ratio tests and are given as

$$\alpha_x^{aff} = \min\{1, \min_{i:(d_x^{aff})_i<0} -\frac{x_i}{(d_x^{aff})_i}\} \tag{6.15}$$

$$\alpha_z^{aff} = \min\{1, \min_{i:(d_z^{aff})_i<0} -\frac{z_i}{(d_z^{aff})_i}\}. \tag{6.16}$$

The step lengths are used to compute the iterate that would be at the boundary of the feasible set by moving in the predictor direction with the step lengths above. In particular, the duality measure y_{aff} is computed for this iterate where

$$y_{aff} = \frac{(x+\alpha_x^{aff} d_x^{aff})^T(z+\alpha_z^{aff} d_z^{aff})}{n}.$$

Now the centering parameter τ will be set to the value

$$\tau = \left(\frac{y_{aff}}{y}\right)^3.$$

Thus, when the predictor direction is good then y_{aff} is less than y and the centering parameter will be smaller; when the direction is not as good, the parameter will be larger.

6.4.3 Corrector and Centering Step

Recall that the approximate complementary slackness condition is $(x)^T z = \mu$ and can be expressed as $(x + d_x^{aff})^T(z + d_z^{aff}) = \mu$. Then, the condition can be written as

$$x^T z + x^T d_z^{aff} + (\triangle x)^T d_z^{aff} + (d_x^{aff})^T d_z^{aff} = \mu \tag{6.17}$$

or equivalently

$$x^T d_z^{aff} + (d_x^{aff})^T z = \mu - x^T z - (d_x^{aff})^T d_z^{aff}. \tag{6.18}$$

Observe that if the non-linear term $(d_x^{aff})^T d_z^{aff}$ is dropped, then (6.18) can be written as

$$\begin{bmatrix} Z & 0 & X \end{bmatrix} \begin{bmatrix} d_x^{aff} \\ d_\pi^{aff} \\ d_z^{aff} \end{bmatrix} = [-XZe + \mu e] = [-XZe + \tau y e],$$

which is the third (block) equation in system (6.12) (Note: In (6.12) the directions are affine scaling directions but do not have the aff superscript, whereas here we use these supercripts.) Observe that the presence of the term $\mu e = \tau y e$ implies that the centering step is included. Now the key idea in the corrector step is to add back the term $(d_x)^T d_z$ and represent it as $D_x D_z e$ where $D_x = diag(d_x^{aff})$ and $D_z = diag(d_z^{aff})$ to get a modified system

$$\begin{bmatrix} 0 & A^T & I \\ A & 0 & 0 \\ Z & 0 & X \end{bmatrix} \begin{bmatrix} d_x \\ d_\pi \\ d_z \end{bmatrix} = \begin{bmatrix} -r_p \\ -r_d \\ -XZe + D_x D_z e + \tau y e \end{bmatrix}. \tag{6.19}$$

The maximum step lengths along this direction are similar to (6.15) and (6.16) and are defined as

$$\alpha_x^{\max} = \min\{1, \min_{i:(d_x)_i<0} -\tfrac{x_i}{(d_x)_i}\} \tag{6.20}$$

$$\alpha_z^{\max} = \min\{1, \min_{i:(d_z)_i<0} -\tfrac{z_i}{(d_z)_i}\}. \tag{6.21}$$

Then the step lengths used in the predictor-corrector method are dampened versions of the maximum step lengths (6.20) and (6.21)

$$\alpha_x = \min\{1, \eta\alpha_x^{\max}\} \tag{6.22}$$

$$\alpha_z = \min\{1, \eta\alpha_z^{\max}\} \tag{6.23}$$

where η is a dampening parameter such that $\eta \in [0.9, 1)$; see Mehrotra (1992) for more details.

6.4.4 Computational Overhead

A crucial observation is that the system (6.19) incorporates the predictor, corrector, and centering steps in one system. The centering step is embodied in the term $\tau y e$. Observe that a factorization used to solve the affine scaling direction d^{aff} in (6.14) can be used to solve the direction in system (6.19). This is an important aspect of Mehrotra's predictor-corrector scheme since extra computational overhead is only moderately increased in the additional steps that are added to the general primal-dual framework.

6.4.5 Predictor-Corrector Algorithm

We now present the details of the predictor-corrector method.

Step 0: Obtain an initial interior solution $(x^{(0)}, \pi^{(0)}, z^{(0)})$ such that $x^{(0)} > 0$ and $z^{(0)} > 0$ (see below). Let $k = 0$ and ε be some small positive number (tolerance). Go to Step 1.

Step 1: Solve

$$\begin{bmatrix} 0 & A^T & I \\ A & 0 & 0 \\ Z^{(k)} & 0 & X^{(k)} \end{bmatrix} \begin{bmatrix} d_x^{aff} \\ d_\pi^{aff} \\ d_z^{aff} \end{bmatrix} = \begin{bmatrix} -r_p^{(k)} \\ -r_d^{(k)} \\ -X^{(k)}Z^{(k)}e \end{bmatrix}$$

for $\begin{bmatrix} d_x^{aff} \\ d_\pi^{aff} \\ d_z^{aff} \end{bmatrix}$. Go to Step 2.

Step 2: Compute α_x^{aff}, α_z^{aff},$y^{(k)}$, and $y_{aff}^{(k)}$ and let $\tau^{(k)} = \left(\frac{y_{aff}^{(k)}}{y^{(k)}}\right)^3$ and solve for $\begin{bmatrix} d_x \\ d_\pi \\ d_z \end{bmatrix}$ in

$$\begin{bmatrix} 0 & A^T & I \\ A & 0 & 0 \\ Z^{(k)} & 0 & X^{(k)} \end{bmatrix} \begin{bmatrix} d_x \\ d_\pi \\ d_z \end{bmatrix} = \begin{bmatrix} -r_p^{(k)} \\ -r_d^{(k)} \\ -X^{(k)}Z^{(k)}e + D_{x^{(k)}}D_{z^{(k)}}e + \tau^{(k)}y^{(k)}e \end{bmatrix},$$

where $D_{x^{(k)}} = diag(d_x^{aff})$ and $D_{z^{(k)}} = diag(d_z^{aff})$.

Go to Step 3.

Step 3: Compute α_x and α_z and let

$$\begin{aligned} x^{(k+1)} &= x^{(k)} + \alpha_x d_x \\ \pi^{(k+1)} &= \pi^{(k)} + \alpha_z d_\pi \\ z^{(k+1)} &= z^{(k)} + \alpha_z d_z. \end{aligned}$$

If the stopping criteria is met, i.e.,

$$\left\|Ax^{(k+1)} - b\right\| \leq \varepsilon$$

$$\left\|A^T\pi^{(k+1)} + z^{(k+1)} - c\right\| \leq \varepsilon$$

$$(x^{(k+1)})^T z^{(k+1)} \leq \varepsilon,$$

then STOP. Else $k = k + 1$ and go to Step 1.

6.4.5.1 Initial Point Generation

The following strategy attempts to generate an initial solution that will be close to minimizing the norms in the stopping condition while ensuring that $x^{(0)} > 0$ and $z^{(0)} > 0$. The strategy is as follows. First compute the following quantities

$$\bar{\pi} = (AA^T)^{-1}Ac$$

$$\bar{z} = c - A^T\bar{\pi}$$

$$\bar{x} = A^T(AA^T)^{-1}b$$

$$\delta_x = \max(-1.5 * \min\{\bar{x_i}\}, 0)$$

and

$$\delta_z = \max(-1.5 * \min\{\bar{z_i}\}, 0),$$

and then form the quantities

$$\bar{\delta_x} = \delta_x + [(\bar{x} + \delta_x e)^T(\bar{z} + \delta_z e)]/[2\sum_{i=1}^{n}(\bar{z_i} + \delta_z)]$$

and

$$\bar{\delta_z} = \delta_x + [(\bar{x} + \delta_x e)^T(\bar{z} + \delta_z e)]/[2\sum_{i=1}^{n}(\bar{x_i} + \delta_x)].$$

Then, the initial point is $x_i^{(0)} = \bar{x_i} + \delta_x$ and $z_i^{(0)} = \bar{z_i} + \delta_z$ for $i = 1, ..., n$ and $\pi^{(0)} = \bar{\pi}$.

By construction of δ_x and δ_z the initial point generated in this manner will be such that $x^{(0)} \geq 0$ and $z^{(0)} \geq 0$. Furthermore, if $\bar{\delta_x}$ and $\bar{\delta_z}$ are positive, then $x^{(0)} > 0$ and $z^{(0)} > 0$; see Exercise 6.6.

Example 6.4
Consider the linear program

$$\begin{array}{lllll} \textit{minimize} & -3x_1 - 2x_2 & & & \\ \textit{subject to} & x_1 & +2x_2 + x_3 & & = 20 \\ & 2x_1 & +2x_2 & +x_4 & = 15 \\ & x_1 \geq 0, x_2 \geq 0, x_3 \geq 0, x_4 \geq 0. & & & \end{array}$$

We show an iteration of the predictor-corrector primal-dual interior point method applied to this linear program.

Step 0: Now $A = \begin{bmatrix} 1 & 2 & 1 & 0 \\ 2 & 1 & 0 & 1 \end{bmatrix}$, $b = \begin{bmatrix} 20 \\ 15 \end{bmatrix}$, and $c = \begin{bmatrix} -3 \\ -2 \\ 0 \\ 0 \end{bmatrix}$. We select as tolerance $\varepsilon = 10^{-6}$ and by using the initial point generation technique we get

$$x^{(0)} = \begin{bmatrix} 5.3125 \\ 7.8125 \\ 4.3125 \\ 1.8125 \end{bmatrix}, \; z^{(0)} = \begin{bmatrix} 0.6250 \\ 1.1250 \\ 1.6250 \\ 2.1250 \end{bmatrix}, \text{ and } \pi^{(0)} = \begin{bmatrix} -0.5000 \\ -1.0000 \end{bmatrix}.$$

Go to Step 1.

Iteration 1

Step 1: Solve for $\begin{bmatrix} d_x^{aff} \\ d_\pi^{aff} \\ d_z^{aff} \end{bmatrix}$ is the system

$$\begin{bmatrix} 0 & A^T & I \\ A & 0 & 0 \\ Z^{(0)} & 0 & X^{(0)} \end{bmatrix} \begin{bmatrix} d_x^{aff} \\ d_\pi^{aff} \\ d_z^{aff} \end{bmatrix} = \begin{bmatrix} -r_p^{(0)} \\ -r_d^{(0)} \\ -X^{(0)}Z^{(0)}e \end{bmatrix}$$

where

$$X^{(0)} = diag(x^{(0)}) \text{ and } Z^{(0)} = diag(z^{(0)}).$$

Now
$-r_p^{(0)} = Ax^{(0)} - b = \begin{bmatrix} -5.2500 \\ -5.2500 \end{bmatrix}$, $-r_d^{(0)} = A^T\pi^{(0)} + z^{(0)} - c = \begin{bmatrix} -1.1250 \\ -1.1250 \\ -1.1250 \\ -1.1250 \end{bmatrix}$, and

$-X^{(0)}Z^{(0)}e = \begin{bmatrix} -3.32031250 \\ -8.78906250 \\ -7.00781250 \\ -3.85156250 \end{bmatrix}$. Then, $d_x^{aff} = \begin{bmatrix} -1.33818929 \\ -1.40434057 \\ -1.10312958 \\ -1.16928086 \end{bmatrix}$,

$d_\pi^{aff} = \begin{bmatrix} 0.08432798 \\ -0.37088101 \end{bmatrix}$, and $d_z^{aff} = \begin{bmatrix} -0.46756597 \\ -0.92277496 \\ -1.20932798 \\ -0.75411899 \end{bmatrix}$. Go to Step 2.

Step 2: Now

$$\alpha_{x^{(0)}}^{aff} = \min\{1, \min_{i:(d_{x^{(0)}}^{aff})_i<0} - \frac{x_i^{(0)}}{(d_{x^{(0)}}^{aff})_i}\} = 1,$$

$$\alpha_{z^{(0)}}^{aff} = \min\{1, \min_{i:(d_{z^{(0)}}^{aff})_i<0} - \frac{z_i^{(0)}}{(d_{z^{(0)}}^{aff})_i}\} = 1,$$

$$y^{(0)} = \frac{(x^{(0)})^T z^{(0)}}{4} = 5.74218750,$$

$$y_{aff}^{(0)} = \frac{(x^{(0)}+\alpha_{x^{(0)}}^{aff} d_{x^{(0)}}^{aff})^T (z^{(0)}+\alpha_{z^{(0)}}^{aff} d_{z^{(0)}}^{aff})}{4} = 1.03435111.$$

$$\text{Then, } \tau^{(0)} = \left(\frac{y_{aff}^{(0)}}{y^{(0)}}\right)^3 = 0.00584483.$$

Now solve for $\begin{bmatrix} d_x \\ d_\pi \\ d_z \end{bmatrix}$ in

$$\begin{bmatrix} 0 & A^T & I \\ A & 0 & 0 \\ Z^{(0)} & 0 & X^{(0)} \end{bmatrix} \begin{bmatrix} d_x \\ d_\pi \\ d_z \end{bmatrix} = \begin{bmatrix} -r_p^{(0)} \\ -r_d^{(0)} \\ -X^{(0)}Z^{(0)}e + D_{x^{(0)}}D_{z^{(0)}}e + \tau^{(0)}y^{(0)}e \end{bmatrix}$$

where $D_{x^{(0)}} = diag(d^{aff}_{x^{(0)}})$ and $D_{z^{(0)}} = diag(d^{aff}_{z^{(0)}})$. Then, $d_x = \begin{bmatrix} -1.26643159 \\ -1.16097290 \\ -1.66162262 \\ -1.55616392 \end{bmatrix}$, $d_\pi = \begin{bmatrix} 0.17544269 \\ -0.35648747 \end{bmatrix}$, and

$$d_z = \begin{bmatrix} -0.58746775 \\ -1.11939791 \\ -1.30044269 \\ -0.76851253 \end{bmatrix}.$$ Go to Step 3.

Step 3: Now

$\alpha_x^{\max} = \min\{1, \min_{i:(d_x)_i<0} -\frac{x_i}{(d_x)_i}\} = 1$ and $\alpha_z^{\max} = \min\{1, \min_{i:(d_z)_i<0} -\frac{z_i}{(d_z)_i}\} = 1$.

Also,

$$\alpha_x = \min\{1, 0.95\alpha_x^{\max}\} = 0.95 \text{ and } \alpha_z = \min\{1, 0.95\alpha_z^{\max}\} = 0.95.$$

Then, the new iterates are

$$x^{(1)} = x^{(0)} + \alpha_x d_x$$

$$= \begin{bmatrix} 5.3125 \\ 7.8125 \\ 4.3125 \\ 1.8125 \end{bmatrix} + 0.95 \begin{bmatrix} -1.26643159 \\ -1.16097290 \\ -1.66162262 \\ -1.55616392 \end{bmatrix}$$

$$= \begin{bmatrix} 4.10938999 \\ 6.70957575 \\ 2.73395851 \\ 0.33414427 \end{bmatrix}$$

$$\pi^{(1)} = \pi^{(0)} + \alpha_z d_\pi$$

$$= \begin{bmatrix} -0.5000 \\ -1.0000 \end{bmatrix} + 0.95 \begin{bmatrix} 0.17544269 \\ -0.35648747 \end{bmatrix}$$

$$= \begin{bmatrix} -0.33332944 \\ -1.33866310 \end{bmatrix}$$

$$z^{(1)} = z^{(0)} + \alpha_z d_z$$

$$= \begin{bmatrix} 0.6250 \\ 1.1250 \\ 1.6250 \\ 2.1250 \end{bmatrix} + 0.95 \begin{bmatrix} -0.58746775 \\ -1.11939791 \\ -1.30044269 \\ -0.76851253 \end{bmatrix}$$

$$= \begin{bmatrix} 0.06690564 \\ 0.06157198 \\ 0.38957944 \\ 1.39491310 \end{bmatrix}$$

Now

$$\left|\left|Ax^{(1)} - b\right|\right| = 0.37123106$$
$$\left|\left|A^T\pi^{(1)} + z^{(1)} - c\right|\right| = 0.11250000$$
$$(x^{(1)})^T z^{(1)} = 2.21925950.$$

All of these quantities are greater than the tolerance ε, so Iteration 2 is started. It will take six iterations to reach the tolerance. The results of the first six iterations are summarized below in Tables 6.2, 6.3, and 6.4.

Table 6.2 Primal iterates of Example 6.4

k	x_1	x_2	x_3	x_4	Primal objective $c^T x^{(k)}$
0	5.31250000	7.81250000	4.31250000	1.81250000	−31.56250000
1	4.10938999	6.70957575	2.73395851	0.33414427	−25.74732147
2	3.37518246	8.24333111	0.15128033	0.01942898	−26.61220958
3	3.33542378	8.32883506	0.00756402	0.00097530	−26.66394146
4	3.33343786	8.33310842	0.00037820	0.00004877	−26.66653041
5	3.33333856	8.33332209	0.00001891	0.00000244	−26.66665985
6	3.33333359	8.33333277	0.00000095	0.00000012	−26.66666633

Table 6.3 Dual iterates of Example 6.4

k	π_1	π_2	$b^T\pi$
0	−0.50000000	−1.00000000	−25.00000000
1	−0.33332944	−1.33866310	−26.74653532
2	−0.33183129	−1.33632428	−26.68149006
3	−0.33325784	−1.33348383	−26.66741437
4	−0.33332956	−1.33334086	−26.66670405
5	−0.33333314	−1.33333371	−26.66666854
6	−0.33333332	−1.33333335	−26.66666676

Table 6.4 Dual slack iterates for Example 6.4

k	z_1	z_2	z_3	z_4
0	0.62500000	1.12500000	1.62500000	2.12500000
1	0.06690564	0.06157198	0.38957944	1.39491310
2	0.00757159	0.00307860	0.33492303	1.33941602
3	0.00038010	0.00015411	0.33341243	1.33363842
4	0.00001901	0.00000771	0.33333729	1.33334859
5	0.00000095	0.00000039	0.33333353	1.33333410
6	0.00000005	0.00000002	0.33333334	1.33333337

Table 6.5 Residuals and τ for Example 6.4

k	τ	$Ax^{(k)} - b$	$A^T\pi^{(k)} + z^{(k)} - c$	$(x^{(k)})^T z^{(k)}$
0	–	7.424621202	2.250000000	22.968750000
1	0.005844830	0.371231060	0.112500000	2.219259500
2	0.001302445	0.018561553	0.006183470	0.127624175
3	4.03E-07	0.000930441	0.000309173	0.006373953
4	5.16E-11	4.65E-05	1.55E-05	0.000318652
5	6.46E-15	2.33E-06	7.73E-07	1.59E-05
6	8.08E-19	1.16E-07	3.86E-08	7.97E-07

6.5 Primal-Dual Interior Point Method in MATLAB®

This section contains MATLAB code for the predictor-corrector version of the primal-dual interior point method. In particular, the function

```
function [xsol, objval] = PD_InteriorPoint(c, A, b)
```

is created where it takes as arguments the parameters of a linear program that is in standard form where c is the objective coefficient vector, A, the matrix of coefficients of the equality constraints and b, the vector of right-hand sides of equality constraints.

The function will return the final primal solution in xsol, the final objective function value in objval.

Example 6.5

Consider the linear program in (6.4)

$$\begin{array}{ll} \textit{maximize} & -3x_1 - 2x_2 \\ \textit{subject to} & x_1 + 2x_2 \leq 20 \\ & 2x_2 + x_2 \leq 15 \\ & x_1 \geq 0, x_2 \geq 0. \end{array}$$

Slack variables x_3 and x_4 must be added to get the linear program to standard form. Then the following MATLAB statements create the parameters for the function SimplexMethod.

```
>> c=[-3; -2; 0; 0;];
>> A=[1 2 1 0;
        2 1 0 1];
>> b=[20;15];
```

then the function can be called by writing (with output following)

```
>>function [xsol, objval] = PD_InteriorPoint(c, A, b)
```

problem solved

xsol =
3.3333
8.3333
0.0000
0.0000

objval =

-26.6667

6.5.1 MATLAB Code

```
function [xsol, objval] = PD_InteriorPoint(c, A, b)
% PD_InteriorPoint solves a linear programming in standard form
%                          min  c'*x
%                          s.t. A*x = b
%                               x >= 0
% using the predictor corrector primal-dual path following method
% of Mehrotra.
%
% Inputs:
%  c = n*1 vector, objective coefficients
%  A = m*n matrix with m < n, A is full rank matrix
%  b = m*1 vector, RHS
%
% Outputs:
%  xsol = n*1 vector, final solution
%  objval is  scalar,  final objective value

[m n]=size(A); % number of constraint and variables
e=ones(n,1);
%% Step 0: Initialization
%%%%%%%%%%%%%%%%%%%%%%%%%%%%
% Obtain an initial interior solution [x(0), pie(0), z(0)]^T such that
% x(0)>0 and z(0)>0. Let k=0 and epsi be some small positive number
% (tolerance). Go to STEP 1.
k=0;                        %counter
epsi=1/10^6;                %tolerance
eta=.95;                    %step length dampening constant
%generate a warm start point
lambda=(A*A')\(2*b); %Lagrange multiplier
x_bar=.5*A'*lambda;  %solve min ||x||+lambda*(b-Ax)
pie_bar=(A*A')\(A*c);%solve min ||A'*pie -c||
z_bar=c-A'*pie_bar;
```

```
del_x=max([0; -1.5*min(x_bar)]);
del_z=max([0; -1.5*min(z_bar)]);
del_x_bar=del_x+.5*(x_bar+del_x*e)'*(z_bar+del_z*e)/sum(z_bar+del_z);
del_z_bar=del_z+.5*(x_bar+del_x*e)'*(z_bar+del_z*e)/sum(x_bar+del_x);
x(:,k+1)=x_bar+del_x_bar; %initial x(0), primal variable
pie(:,k+1)=pie_bar;       %initial pie(0), slack variable of dual
z(:,k+1)=z_bar+del_z_bar; %initial z(0), dual variable
obj_pd(:,k+1)=[c'*x(:,k+1); b'*pie(:,k+1)];
Norm(:,k+1)=[norm(A*x(:,k+1)-b);
norm(A'*pie(:,k+1)+z(:,k+1)-c); x(:,k+1)'*z(:,k+1);];
while k>=0
    %% Step 1: Affine Scaling Direction Generation
    %%%%%%%%%%%%%%%%%%%%%%%%%%%%%%%%%%%%%%%%%%%%%%%%%%%%%%%%%%
    % Solve KKT system for affine direction d_affine in the algorithm.
    % GO to STEP 2.
    r_p=A*x(:,k+1)-b;                %primal residuals
    r_d=A'*pie(:,k+1)+z(:,k+1)-c;    %dual residuals
    X=diag(x(:,k+1));                %diag(x(k))
    Z=diag(z(:,k+1));                %diag(z(k))
    coeffi_kkt=[zeros(size(A',1), n) A' eye(size(A',1), size(X,2)); ...
        A zeros(m, size(A',2)) zeros(m, size(X,2)); ...
        Z zeros(size(X,1), size(A',2)) X];%coefficient matrix of KKT system
    d_aff=-coeffi_kkt\[r_d; r_p; X*Z*e];  %solve the KKT system
    d_x_aff=d_aff(1:n);                   %affine direction of x(k)
    d_z_aff=d_aff(n+m+1:end);             %affine direction of z(k)
    %% Step 2: Centering Parameter Generation
    %%%%%%%%%%%%%%%%%%%%%%%%%%%%%%%%%%%%%%%%%%%%%%%%%%
    % Compute alpha_x_affine, alpha_z_affine, y(k), y_affine(k) and let
    % tau(k) = (y_affine(k)/y(k))^3. Solve KKT system for corrector
    % direction d in the algorithm. Go to STEP 3.
    x_temp=x(:,k+1);
    flag_x=find(d_x_aff<0);
    alpha_x_aff = ...
        min([1; min(-x_temp(flag_x)./d_x_aff(flag_x))]);%alpha_x_affine
    z_temp=z(:,k+1);
    flag_z=find(d_z_aff<0);
    alpha_z_aff = ...
        min([1; min(-z_temp(flag_z)./d_z_aff(flag_z))]);%alpha_z_affine
    y(k+1)=x(:,k+1)'*z(:,k+1)/n;    %y(k)
    y_aff(k+1) = ...                                       %y_affine(k)
        (x(:,k+1)+alpha_x_aff*d_x_aff)'*(z(:,k+1)+alpha_z_aff*d_z_aff)/n;
    tau(k+1)=(y_aff(k+1)/y(k+1))^3;%tau(k)
    %%%%%%%%%%%%%%%%%%%%%%%%%%%%%%%%%%%%%%%%%%%%%%%%%%
    D_x=diag(d_x_aff);               %D_x(k)
    D_z=diag(d_z_aff);               %D_z(k)
```

```
    d = ...                                              %solve the KKT system
        -coeffi_kkt\[r_d; r_p; X*Z*e+D_x*D_z*e-tau(k+1)*y(k+1)*e];
    d_x=d(1:n);                       %d_x
    d_pie=d(n+1:n+m);                 %d_pie
    d_z=d(n+m+1:end);                 %d_z
    %% Step 3: New Primal and Dual solution Generation
    %%%%%%%%%%%%%%%%%%%%%%%%%%%%%%%%%%%%%%%%%%%%%%%%%%%%%%%%%%%%%%%%%%%%%%%
    % Compute alpha_x and alpha_z and let x(k+1)=x(k)+alpha_x*d_x,
    % pie(k+1)=pie(k)+alpha_z*d_pie, and z(k+1)=z(k)+alpha_z*d_z.
    % If the stopping criteria is met i.e. ||A*x(k+1) - b||<= epsi,
    % ||A^T*pie(k+1) + z(k+1) - c||<= epsi, and (x(k+1))^T*z(k+1)<= epsi,
    % then STOP. Else k=k+1, go to STEP 1.
    flag_x=find(d_x<0);
    alpha_x_max = ...
        min([1; min(-x_temp(flag_x)./d_x(flag_x))]);% minimum ratio test for x
    alpha_x=min([1; eta*alpha_x_max]); %alpha_x
    flag_z=find(d_z<0);
    alpha_z_max = ...
        min([1; min(-z_temp(flag_z)./d_z(flag_z))]);%minimum ratio test for z
    alpha_z=min([1; eta*alpha_z_max]); %alpha_z
    k=k+1;%update the counter
    %%%%%%%%%%%%%%%%%%%%%%%%%%%%%%%%%%%%%%%%%%%%%%%%%%%%%%%%%%%%%%%%%%%%%%%%
    x(:,k+1)=x(:,k)+alpha_x*d_x;           %generate x(k+1)=x(k)+alpha_x*d_x
    pie(:,k+1)=pie(:,k)+alpha_z*d_pie;     %generate pie(k+1)=pie(k)+alpha_z*d_pie
    z(:,k+1)=z(:,k)+alpha_z*d_z;           %generate z(k+1)=z(k)+alpha_z*d_z
    obj_pd(:,k+1)=[c'*x(:,k+1); b'*pie(:,k+1)];%primal and dual objective value
    Norm(:,k+1) = ...
        [norm(A*x(:,k+1)-b); norm(A'*pie(:,k+1)+z(:,k+1)-c); x(:,k+1)'*z(:,k+1);];
    if isempty(find(Norm(:,k+1) >= epsi))%if residual <= epsi, then optimal STOP.
        disp('problem solved')
        break
    end
end
xsol=x(:,end);            %optimal solution
objval=obj_pd(end,end);%optimal objective value
```

6.6 Exercises

Exercise 6.1

(a) Use the Newton-Raphson method to compute the first two iterates (i.e., $x^{(1)}$ and $x^{(2)}$) for solving the following system of equations. Use $x^{(0)} = [1, 1, 1]^T$ as an initial point.

$$\begin{aligned} x_1^2 + x_2^2 - 4x_3 &= 0 \\ x_1^2 + \qquad x_3^2 &= 1/4 \\ x_1^2 + x_2^2 + x_3^2 &= 1. \end{aligned}$$

(b) Implement the Newton-Raphson method in MATLAB and use it to approximate the solution to the system of equations in (a) up to a tolerance of 10^{-5} using the same initial point.

Exercise 6.2

The Newton-Raphson method can be used for unconstrained problems of the form

$$\begin{array}{ll} \textit{minimize} & f(x) \\ \textit{subject to} & x \in R^n \end{array}$$

where $f(x)$ is a twice differentiable function from R^n to R.

(a) Derive the Newton-Raphson direction for this case.

(b) Apply the Newton-Raphson method to the following problem

$$\begin{array}{ll} \textit{minimize} & 4x_1^2 - 4x_1x_2 + 2x_2^2 \\ \textit{subject to} & \begin{bmatrix} x_1 \\ x_2 \end{bmatrix} \in R^2. \end{array}$$

Use as an initial point $x^{(0)} = (2,3)^T$.

(c) Let $x^{(k)}$ be a current iterate in using the Newton-Raphson method for unconstrained minimization. Prove that if the Hessian $\nabla f(x^{(k)})$ is positive semidefinite at $x^{(k)}$ and the Newton-Raphson direction is non-zero, then $f(x^{(k+1)}) \leq f(x^{(k)})$.

Exercise 6.3

Consider the following system of linear equations (6.12) that arises from using the Newton-Raphson method on the KKT conditions of the Barrier problem

$$\begin{bmatrix} 0 & A^T & I \\ A & 0 & 0 \\ Z & 0 & X \end{bmatrix} \begin{bmatrix} d_x \\ d_\pi \\ d_z \end{bmatrix} = \begin{bmatrix} 0 \\ 0 \\ -XZe + \mu e \end{bmatrix}.$$

Prove the following vectors solve the system

$d_x = (I - XZ^{-1}A^T(AXZ^{-1}A^T)^{-1}A)(-x + \mu Z^{-1}e)$

$d_\pi = (AXZ^{-1}A^T)^{-1}A(x - \mu Z^{-1}e)$

$d_z = A^T(AXZ^{-1}A^T)^{-1}A(-x + \mu Z^{-1}e)$, where x is a current primal vector for which $X = diag(x)$.

Exercise 6.4

Consider the linear program

$$\begin{array}{ll} \textit{minimize} & -55x_1 - 45x_2 \\ \textit{subject to} & 4x_1 + 6x_2 \leq 210 \\ & x_1 + 3x_2 \leq 90 \\ & 15x_1 + 8x_2 \leq 600 \\ & x_1 \geq 0, x_2 \geq 0. \end{array}$$

(a) Compute the first two iterates of the predictor-corrector primal-dual interior point method applied to the linear program above using the initial point generation method in Section 4.5.1. Use a tolerance of $\varepsilon = 10^{-6}$.

(b) Use the MATLAB function PD_InteriorPoint in Section 6.5 to solve the linear program above. Summarize the iterations in tables showing for each iteration the primal values, dual values, objective values, and residual values. See Example 6.4 for an illustration.

(c) Solve the linear program by using the MATLAB linprog function and verify the solution from (b).

Exercise 6.5

Consider the linear program

$$\begin{array}{ll} \textit{minimize} & 9x_1 + 15x_2 + 8x_3 + 12x_4 \\ \textit{subject to} & x_1 + 2x_2 + 2x_3 + x_4 = 1200 \\ & 4x_1 + 3x_2 + x_3 + 7x_4 \leq 3500 \\ & 3x_1 + 3x_2 + 5x_3 + 6x_4 \leq 4200 \\ & x_1 \geq 0, x_2 \geq 0, x_3 \geq 0, x_4 \geq 0. \end{array}$$

(a) Compute the first two iterates of the predictor-corrector primal-dual interior point method applied to the linear program above using the initial point generation method in Section 4.5.1. Use a tolerance of $\varepsilon = 10^{-6}$.

(b) Use the MATLAB function PD_InteriorPoint in Section 6.5 to solve the linear program above. Summarize the iterations in tables showing for each iteration the primal values, dual values, objective values, and residual values. See Example 6.4 for an illustration.

(c) Solve the linear program by using the MATLAB linprog function and verify the solution from (b).

Exercise 6.6

(a) Consider the linear program from Exercise 6.4. Use the predictor-corrector interior point method and repeat part (b) of Exercise 6.4, but this time start with the infeasible point $(x^{(0)}, \pi^{(0)}, z^{(0)})$ where all components are equal to 1. (Note: The MATLAB function PD_InteriorPoint must be modified to handle infeasible starting points; see Section 3.2.1)

(b) Compare the results from Exercise 6.5 (a) with the results from Exercise 6.4 (b). Which converges faster?

Exercise 6.7

Prove that if $\bar{\delta_x}$ and $\bar{\delta_z}$ are positive, then $x^{(0)} > 0$ and $z^{(0)} > 0$.

Exercise 6.8

Consider the following interior point method called, the affine scaling method, for solving linear programs in standard form, i.e., minimize $c^T x$ subject to $Ax = b$, $x \geq 0$.

Affine Scaling Method

Step 0: Let $x^{(0)}$ be an initial feasible solution such that $Ax^{(0)} = b$ and $x^{(0)} > 0$. Set $k = 0$ and let $\varepsilon > 0$ be a tolerance.

Step 1: Compute $\pi^{(k)}$ by solving $[A(D^{(k)})^2 A^T]\pi^{(k)} = A(D^{(k)})^2 c$ where $D^{(k)} = diag(x^{(k)})$.

Step 2: Compute $r^{(k)} = c - A^T \pi^{(k)}$, then let $d_x^{(k)} = -(D^{(k)})^2 r^{(k)}$.

Step 3: Let $x^{(k+1)} = x^{(k)} + \sigma \alpha d_x^{(k)}$ where $\alpha = \min\{-\frac{x_i^{(k)}}{(d_x^{(k)})_i} | \ (d_x^{(k)})_i < 0\}$ and $0 < \sigma < 1$, where $x_i^{(k)}$ is the ith component of $x^{(k)}$ and $(d_x^{(k)})_i$ is the ith component of $d_x^{(k)}$.

Step 4: If $\|Ax^{(k+1)} - b\| \leq \varepsilon$, $\|A^T \pi^{(k+1)} + r^{(k+1)} - c\| \leq \varepsilon$, and $(x^{(k+1)})^T r^{(k+1)} \leq \varepsilon$, then STOP.

Else go to Step 1.

(a) Consider the linear program

$$\begin{array}{ll} minimize & x_1 + x_2 \\ subject\ to & 3x_1 + x_2 \leq 18 \\ & x_2 \leq 6 \\ & x_1 \geq 0, x_2 \geq 0. \end{array}$$

Convert the linear program to standard form and compute the first two iterates of the affine scaling method applied to the linear program starting with the initial feasible point $x^{(0)} = \begin{bmatrix} 2 & 4 & 8 & 2 \end{bmatrix}$.

(b) Code in MATLAB the affine scaling method and using a tolerance of $\varepsilon = 10^{-6}$, solve the linear program in part (a) using the MATLAB code.

(c) Prove that $Ad_x^{(k)} = 0$. Why must this result be necessary?

(d) Prove that $c^T d_x^{(k)} \leq 0$. What are the implications of this fact?

(e) Compare and contrast the affine scaling method with predictor corrector primal-dual interior point method.

Notes and References

The simplex method once dominated the landscape for linear programming solution methodology, but the fact that it has an exponential worst case complexity drove researchers to consider other methods that would offer a better theoretical (i.e., polynomial) worst-case complexity. The ellipsoidal method of Khachian (1979) was the first method for linear programming that was shown to exhibit polynomial worst-case complexity, but paradoxically its practical performance was shown to be very poor. Research in interior point methods began in earnest with the development of the polynomial time interior point method of Karmarkar (1984). Since then, a flurry of research began culminating in many different strategies for developing polynomial time interior point methods for linear programming and extensions to quadratic programming and especially convex optimization problems, e.g., Nesterov and Nemirovski (1994). In this chapter we focused on the primal-dual path-following interior methods whose development is detailed in Monteiro and Adler (1989 a, b) and the predictor-corrector method in Mehrotra (1992). The method of Mehrotra is actually a heuristic and there is no proof of polynomial complexity of the method. Mizuno, Todd, and Ye (1993) give a polynomial predictor-corrector primal-dual path following method.

Primal-dual path-following interior point strategies heavily rely on the Newton-Raphson method. See the Dennis and Schnable (1983) and Ortega and Rheinboldt (1970) for more information about the Newton-Raphson method and its convergence properties. Renegar (1988) was the first to consider path-following strategies for interior point methods that rely on using the Newton-Raphson method. The barrier reformulation of linear programs has it origins in the work of Fiacco and McCormick (1968). The book by Wright (1997) is an excellent monograph on primal-dual interior point strategies that include not only path-following strategies, but others as well, e.g., potential reduction and infeasible start methods. The affine scaling method of Exercise 6.7 was first introduced by Dikin (1967) and more recently by Barnes (1986) and Vanderbei, Meketon, and Freedman (1986). Other references for interior point methods include Fang and Puthenpura (1993), Nocedal and Wright (1999), Roos, Terlaky, and Vial (2006), Saigal (1995), Ye (1997), and Vanderbei (2008).

7

Quadratic Programming

7.1 Introduction

In this chapter, we consider a generalization of linear programming that involves linear constraints, but the objective function is a quadratic function and so will contain terms that involve the products of pairs of variables. Such an optimization problem is called a quadratic programming (QP) problem and is a very important class of problems as many applications can be modeled in this framework. We illustrate the importance of quadratic programming through financial portfolio applications. A characterization of optimality of quadratic programs is given that will serve as the basis for algorithmic development.

7.2 QP Model Structure

The quadratic programming (QP) problem can be stated as

$$\begin{array}{ll} \textit{minimize} & c^T x + \frac{1}{2} x^T Q x \\ \textit{subject to} & Ax \leq b \\ & Ex = d \\ & x \geq 0 \end{array}$$

where $c, x \in R^n, Q$ is a $n \times n$ symmetric matrix, A is an $m_1 \times n$ matrix, and E is an $m_2 \times n$ matrix. Constraints are partitioned into linear inequality constraints and linear equality constraints. Note that if $Q = 0$, then the problem becomes a linear program.

In addition, when there are no constraints and variables are unrestricted, the quadratic programming problem is

$$\begin{array}{ll} \textit{minimize} & c^T x + \frac{1}{2} x^T Q x \\ \textit{subject to} & x \in R^n. \end{array}$$

This problem will be referred to as the unconstrained quadratic programming problem (UQP).

Example 7.1

The following problem

$$\begin{array}{ll} \textit{minimize} & 4x_1^2 + x_1x_2 + 2x_1x_3 + 3.5x_2^2 + x_2x_3 + 3x_3^2 + 2x_1 - x_2 + x_3 \\ \textit{subject to} & x_1 + 2x_3 \leq 5 \\ & 3x_2 + x_3 \leq 2 \\ & x_1 + x_3 = 2 \\ & x_1 \geq 0, x_2 \geq 0, x_3 \geq 0 \end{array}$$

is a quadratic program where

$$Q = \begin{bmatrix} 8 & 1 & 2 \\ 1 & 7 & 1 \\ 2 & 1 & 6 \end{bmatrix}, c = \begin{bmatrix} 2 \\ -1 \\ 1 \end{bmatrix}, x = \begin{bmatrix} x_1 \\ x_2 \\ x_3 \end{bmatrix},$$

$$A = \begin{bmatrix} 1 & 0 & 2 \\ 0 & 3 & 1 \end{bmatrix}, b = \begin{bmatrix} 5 \\ 2 \end{bmatrix}, E = \begin{bmatrix} 1 & 0 & 1 \end{bmatrix}, d = [2].$$

Example 7.2 (Least Squares Fit)

Quadratic problems arise naturally in statistics. Suppose that you have observed values $(t_1, u_1), (t_2, u_2), ..., (t_n, u_n)$ where t_i is the unemployment rate in year i and u_i is the rate of inflation for year i. Based on these observations, you believe that the unemployment rate and inflation rate in a year are related. In particular, you believe that t_i and u_i are related by a polynomial function

$$p(t) = x_0 + x_1 t + x_2 t^2 \cdots + x_k t^k$$

where the degree of the polynomial k is determined in advance. However, the coefficients $x_0, x_1, ..., x_k$ are not known. The goal is to choose the values for these coefficients so that the absolute difference between the observed values u_i and $p(t_i)$, i.e.,

$$|u_i - p(t_i)|$$

are as small as possible. Let $x = (x_0, x_1, ..., x_n)$, then one strategy is to minimize the function

$$\begin{aligned} \varphi(x) &= \sum_{i=1}^{n} (u_i - p(t_i))^2 \\ &= \sum_{i=1}^{n} (u_i - \sum_{j=1}^{k} x_j t_i^j)^2. \end{aligned}$$

We can express $\varphi(x)$ in terms of norms on vector quantities. Let

$$A = \begin{bmatrix} 1 & t_1 & t_1^2 & \cdots & t_1^k \\ 1 & t_2 & t_2^2 & & t_2^k \\ & & & \ddots & \vdots \\ 1 & t_n & t_n^2 & \cdots & t_n^k \end{bmatrix} \text{ and } b = \begin{bmatrix} u_1 \\ u_2 \\ \vdots \\ u_n \end{bmatrix}$$

then,

$$\begin{aligned}\varphi(x) &= \|b - Ax\|^2 \\ &= (b - Ax) \cdot (b - Ax) \\ &= b \cdot b - 2b \cdot Ax + Ax \cdot Ax \\ &= b \cdot b - 2A^T b \cdot x + xA^T Ax.\end{aligned}$$

(Note: $\cdot$ indicates the dot product of two vectors.) Now if we let $Q = A^T A$ and $c = -A^T b$, then it is clear that minimizing $\varphi(x)$ subject to $x \in R^{k+1}$ is equivalent to minimizing $\frac{1}{2}x^T Qx + cx$ subject to $x \in R^{k+1}$, which is an unconstrained quadratic program. The optimal solution provides the coefficients for $p(t)$ that represent a least squares fit of the observed data.

Note: It is generally the case that the number of observations n is greater than $k+1$, so that the system $Ax = b$ will have more rows than columns, i.e., it is an over-determined system of linear equations and so an exact solution will not exist, which makes the search for a best approximating solution meaningful.

7.3 QP Application: Financial Optimization

We consider a financial portfolio problem of Markowitz (1952) that was briefly introduced in Chapter 1. This model is perhaps the most well-known instance of a quadratic program and served as the basis of the research in portfolio selection by Markowitz that was awarded the 1992 Nobel Prize in Economic Sciences. We start the development from first principles.

Consider an investor that wishes to allocate funds into n financial securities now ($t = 0$), and will hold the investment until time $t = T$ in the future. Let $w_i =$ the dollar amount invested in security i and $w = \sum_{i=1}^{n} w_i$, then $x_i = \frac{w_i}{w}$ is the proportion of total funds invested in security i. The vector

$$x = \begin{bmatrix} x_1 \\ x_2 \\ \vdots \\ x_n \end{bmatrix}$$ represents the portfolio of investments.

Portfolio Returns

If the current ($t = 0$) price of security i is p_i^0 and the price of security i at time T is p_i^T, then the rate of return of security i, denoted by r_i, over the time period $[0, T]$ is defined as

$$r_i = \frac{p_i^T - p_i^0}{p_i^0}.$$

If x_i is the amount allocated to security i, then the return from investment in security i is $r_i x_i$. Then, the return of a portfolio x denoted by r_p is defined as

$$r_p = \sum_{i=1}^{n} r_i x_i,$$

which says that the return of the portfolio is the sum of the returns from investments in individual securities. It is clear that a rational investor desires that this quantity is higher rather than lower. The challenge is that the price p_i^T of a security i in the future at time T is essentially random, and so the return r_i will be considered random as well. This means that r_p would be a weighted sum of random quantities. In general, we assume that the rate of return of individual securities and hence portfolio returns are modeled by random variables. r_i will be a random variable with mean μ_i and variance σ_i^2.

Instead of dealing with the full complexities of randomness, it is simpler to consider the expectation of the portfolio return, which is denoted by $\bar{r_p} = E(r_p)$, where $E(\cdot)$ is the expectation operator on a random variable, and so we have by the linearity of expectation

$$\begin{aligned} E(r_p) &= E(\sum_{i=1}^{n} r_i x_i) = \sum_{i=1}^{n} E(r_i x_i) \\ &= \sum_{i=1}^{n} E(r_i) x_i = \sum_{i=1}^{n} \mu_i x_i \end{aligned}$$

where $\mu_i = E(r_i)$.

Risk in Portfolio Selection

If the investor only cared about selecting a portfolio x to maximize the expected portfolio return $E(r_p)$, then she would invest only in the security i with the highest expected return μ_i. But this strategy ignores risk, i.e., the possiblity that the expected returns will not realize, which could result in losses or lower returns. The intuitive idea behind the Markowitz approach is to diversify investments into several securities, i.e., "don't put all eggs in the same basket". The key quantities to consider are the covariances between the returns of pairs of assets. The covariance between the returns r_i and r_j of securities i and j is denoted by $\sigma_{ij} = cov(r_i, r_j)$ where

$$\sigma_{ij} = E((r_i - \mu_i)(r_j - \mu_j)) \text{ for } i \neq j$$

and

$$\sigma_{ij} = \sigma_{ii} = \sigma_i^2 \text{ for } i = j.$$

The covariance between the returns of two securities measures how the two securities move together. Positive covariance, i.e., $\sigma_{ij} > 0$ indicates that when one security goes up (down) then the other security goes up (down) as well. Negative covariance, i.e., $\sigma_{ij} < 0$ indicates that when one security goes up (down) the other goes down (up). If $\sigma_{ij} = 0$, then we say that the securities are uncorrelated, i.e., the movement of one security has no relation with the movement of the other.

Then, it is ideal for a pair of securities to exhibit negative or smaller covariance as long as the net return is sufficiently high. For example, consider one security that represents a company that makes ice cream and another security that represents a company that makes umbrellas and raincoats. These securities may exhibit negative or low correlation as business for each depends on the weather and each company is affected in opposite ways by the weather. If it is a longer winter with more rain, then the stock in the ice cream company will be lower, but the stock in the umbrella and raincoat company will be higher. If winter is less harsh and there is warmer and better weather, then the stock for the company that makes ice cream will do better than the stock for the company that makes umbrellas and raincoats. In either case, as long as the net return is good enough, it is safer to invest in both companies rather than just one. In fact, it can be shown that for sufficiently low covariance between securities, it is superior (less risk with same return) to invest in more securities rather than fewer; see Luenberger (1998).

In the Markowitz approach, an investor should consider the variance of the return of a portfolio, which is denoted by σ_p^2 where

$$\sigma_p^2 = var(r_p) = E[(r_p - E(r_p))^2]$$

$$= E[(\sum_{i=1}^{n} r_i x_i - \sum_{i=1}^{n} \mu_i x_i)^2] = E[(\sum_{i=1}^{n} x_i(r_i - \mu_i)(\sum_{j=1}^{n} x_j(r_j - \mu_j))]$$

$$E[(\sum_{i=1}^{n}\sum_{j=1}^{n} x_i x_j (r_i - \mu_i)(r_j - \mu_j))] = \sum_{i=1}^{n}\sum_{j=1}^{n} \sigma_{ij} x_i x_j$$

$$= \sum_{i=1}^{n} \rho_{ij}\sigma_i\sigma_j x_i x_j$$

where

$$\rho_{ij} = \sigma_{ij}/\sigma_i\sigma_j.$$

The quantity ρ_{ij} is the correlation coefficient between r_i and r_j and normalizes the covariance to be between 1 and -1. An investor will seek to select a portfolio x that minimizes σ_p. In doing so, the investor will seek to invest in securities that have lower or negative covariance with other securities since the covariance is the coefficient for each $x_i x_j$ term. Thus, portfolio variance is seen as a measure by which to evaluate the riskiness of a portfolio. However, the investor also needs a goal for the rate of return for the portfolio x since minimizing just the variance of a portfolio may not generate sufficient return for the investor. To this end, the investor specifies a constraint that forces her portfolio selection to achieve at least a certain amount of expected return R. That is, the investor enforces the constraint

$$E(r_p) = \sum_{i=1}^{n} \mu_i x_i \geq R.$$

Markowitz Portfolio Selection Model

Now all of the components are in place for specification of the Markowitz portfolio model, also known as mean-variance optimization (MVO). The most well-known version of the MVO model has the objective of minimizing the variance of the portfolio and is given as follows

$$\begin{array}{ll} minimize & \sum_{i=1}^{n}\sum_{j=1}^{n} \sigma_{ij} x_i x_j \\ subject\ to & \sum_{i=1}^{n} \mu_i x_i \geq R \\ & \sum_{i=1}^{n} x_i = 1 \\ & x_i \geq 0, i = 1, ..., n. \end{array}$$

The objective is to find a portfolio x that minimizes variance subject to meeting the expected return goal. There is also a budget constraint where the sum of all investments is equal to the budget. The non-negativity constraints ensure that shorting of securities is prohibited. A negative value of x_i indicates that a security is sold short, meaning that the investor borrows shares of the security and sells the shares at the current price and must later return those shares to the original owner (often a stock brokerage). The non-negativity constraints are optional. In matrix form, the MVO model is

$$\begin{array}{ll} minimize & \frac{1}{2} x^T Q x \\ subject\ to & \mu^T x \geq R \\ & e^T x = 1 \\ & x \geq 0 \end{array}$$

where $Q = [\sigma_{ij}]$ for $1 \leq i, j \leq n$, and so Q is symmetric since the $\sigma_{ij} = \sigma_{ji}$. Let $A = [\mu_1, ..., \mu_n]$ and $b = [R]$ and $E = [1, ..., 1]$ and $d = [1]$ and thus the MVO formulation is a quadratic program with $c = 0$. Note that the original objective function is multiplied by $\frac{1}{2}$ for mathematical convenience which does not affect the original formulation.

A three ($n = 3$) security MVO problem with no short selling has the form

$$\begin{array}{ll} minimize & \sigma_1^2 x_1^2 + \sigma_2^2 x_2^2 + \sigma_3^2 x_3^2 + 2\sigma_{12} x_1 x_2 + 2\sigma_{13} x_1 x_3 + 2\sigma_{23} x_2 x_3 \\ subject\ to & \mu_1 x_1 + \mu_2 x_2 + \mu_3 x_3 \geq R \\ & x_1 + x_2 + x_3 = 1 \\ & x_1 \geq 0, x_2 \geq 0, x_3 \geq 0. \end{array}$$

Example 7.3
Consider three securities with the expected returns given in Table 7.1

Table 7.1 Expected security returns

Expected return	Security 1 ($i = 1$)	Security 2 ($i = 2$)	Security 3 ($i = 3$)
μ_i	9.73%	6.57%	5.37%

with covariances given in Table 7.2.

Table 7.2 Covariance of returns

Covariance σ_{ij}	$i = 1$	$i = 2$	$i = 3$
$i = 1$	0.02553	0.00327	0.00019
$i = 2$		0.013400	-0.00027
$i = 3$			0.00125

We wish to form a portfolio with minimum variance with short selling allowed that achieves an expected return of at least 5.5%. The corresponding model is

$$\begin{aligned} \textit{minimize} \quad & (0.02553)x_1^2 + (0.013400)x_2^2 + (0.00125)x_3^2 + 2(0.00327)x_1x_2 \\ & +2(0.00019)x_1x_3 + 2(-0.00027)x_2x_3 \\ \textit{subject to} \quad & 0.0972x_1 + 0.0657x_2 + 0.0537x_3 \geq 0.055 \\ & x_1 + x_2 + x_3 = 1. \end{aligned}$$

Solving this model gives the optimal portfolio

$$x_1 = 0.0240 \qquad x_2 = 0.0928 \qquad x_3 = 0.8832$$

with risk (variance) $\sigma_P^2 = 0.033069$. The expected return of the portfolio is

$$\bar{r_p} = 0.0972(0.0240) + 0.0657(0.0928) + 0.0537(0.8832) = 0.0558.$$

Thus, we see that the optimal portfolio meets the return goal and it will not be possible to have another portfolio that achieves an expected return of 5.58% with lower variance. Observe that the optimal portfolio allocates most of the investment into the third asset whose expected return is almost enough to satisfy the return goal of 5.5% and thus there is some investment in the other two riskier (higher standard deviation) assets that possess higher expected return.

We can solve the MVO model once for each R from 5.5% to 9.5% to get the results in Table 7.3.

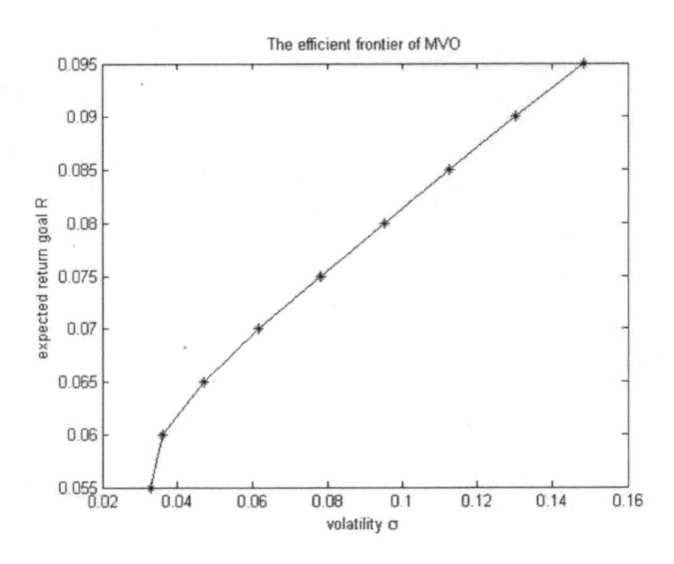

FIGURE 7.1
Risk return tradeoff of portfolios from Example 7.3.

Table 7.3 Optimal portfolios for different R

Return goal R	x_1	x_2	x_3	σ_P
5.5%	0.0240	0.0928	0.8832	0.0329
6%	0.1142	0.1101	0.7757	0.0363
6.5%	0.2232	0.1309	0.6459	0.0471
7%	0.3321	0.1518	0.5161	0.0617
7.5%	0.4410	0.1727	0.3863	0.0780
8%	0.5500	0.1935	0.2565	0.0950
8.5%	0.6589	0.2144	0.1267	0.1125
9%	0.7678	0.2352	−0.0030	0.1303
9.5%	0.8768	0.2561	−0.1329	0.1482

A graph of the portfolios in Table 7.3 with expected return of a portfolio on the vertical axis and portfolio standard deviation (volatility) on the horizontal axis is given in Figure 7.1. Such a graph is called the efficient frontier. The graph clearly indicates the tradeoff between risk (standard deviation) and reward (expected return) of portfolios. The major insight is that the only way to achieve higher expected reward is to take on more risk.

It is important to note that the matrix Q must be at least positive semi-definite (i.e., $x^TQx \geq 0$ for all $x \in R^n$) in the context of MVO since the quadratic term x^TQx represents variance of a random variable (portfolio) return, which is a non-negative quantity. For a general quadratic program it is not necessarily the case that Q must be positive semi-definite. However, there are advantages of Q being positive definite.

7.4 Solving Quadratic Programs Using MATLAB®

In this section we introduce the MATLAB function quadprog that enables solution of quadratic programming problems. Quadratic programs can be solved with MATLAB by using the function quadprog. To use quadprog, a quadratic program is specified in the following form

$$\begin{array}{ll} \textit{minimize} & c^T x + \frac{1}{2} x^T Q x \\ \textit{subject to} & Ax \leq b \\ & A_{eq} x = b_{eq} \\ & l_b \leq x \leq u_b, \end{array}$$

which assumes that constraints are grouped according to inequality constraints, equality constraints, and bounds on the decision variables. The first set of constraints $Ax \leq b$ represents inequality constraints (of the less than or equal to type). Note that any constraint that is originally an inequality constraint of the greater than or equal to type ($\leq$) must be converted to a less than or equal to equivalent. The second set of constraints $A_{eq}x = b_{eq}$ represents the equality constraints, and $l_b \leq x \leq u_b$ represents the lower and upper bounds on the decision variables. Then, A, A_{eq} are matrices and b, b_{eq}, l_b, u_b are vectors. f is a vector that represents the cost coefficients of the linear term of the objective function and Q is the matrix of the quadratic term of the objective function. These quantities are represented in MATLAB as Q, f, A, b, Aeq, beq, lb, ub and are used as arguments for the quadprog function.

For example, the statement

[x, fval] = quadprog(Q,f, A, b, Aeq, beq, lb, ub)

returns a vector x that represents the optimal solution and the optimal objective function value fval of the quadratic program specified by the data.

Example 7.4
Consider the MVO quadratic program in Example 7.3

$$\begin{array}{ll} \textit{minimize} & (0.02553)x_1^2 + (0.013400)x_2^2 + (0.00125)x_3^2 + 2(0.00327)x_1x_2 \\ & +2(0.00019)x_1x_3 + 2(-0.00027)x_2x_3 \\ \\ \textit{subject to} & 0.0972x_1 + 0.0657x_2 + 0.0537x_3 \geq 0.055 \\ & x_1 + x_2 + x_3 = 1. \end{array}$$

Then,

$$f = \begin{bmatrix} 0 \\ 0 \\ 0 \end{bmatrix}, Q = \begin{bmatrix} 0.02553 & 0.00327 & 0.00019 \\ 0.00327 & 0.013400 & -0.00027 \\ 0.00019 & -0.00027 & 0.00125 \end{bmatrix}$$

$$A = \begin{bmatrix} -0.0972 & -0.0657 & -0.0537 \end{bmatrix}, b = [-0.055]$$

$$A_{eq} = \begin{bmatrix} 1 & 1 & 1 \end{bmatrix}, b_{eq} = [1].$$

The vectors and matrices for this QP are created in MATLAB by the following statements

```
Q=[0.02553,0.00327,0.00019;0.00327,0.013400,-0.00027;
   0.00019,-0.00027,0.00125];
f=[0,0,0];
A=[-0.0972,-0.0657,-0.0537];
b=[-0.055];
Aeq=[1,1,1];
beq=[1];
lb=[ ];
ub=[ ];
```

Then, the quadprog function is called with the following statement

```
[x, fval] = quadprog(Q,f, A, b, Aeq, beq, [ ], [ ]),
```

which outputs the following values for the optimal portfolio

```
x =
0.0240
0.0928
d0.8832

fval=
5.4176e-004
```

(Note: fval needs to be doubled to equal the variance of the portfolio since the quadratic term (variance) in the objective has a coefficient of 0.5, i.e., $\frac{1}{2}x^T Qx$.)

7.4.1 Generating the Efficient Frontier Using MATLAB

The following MATLAB code generates the optimal portfolios in Example 7.3 as seen in Table 7.3 using the quadprog function, and plots the associated efficient frontier.

```
%%%%% Three asset MVO problem in Example 7.3 %%%%%
```

```
n=3;
%%%%% Data for MVO problem %%%%%
mu=[9.73 6.57 5.37]/100;   % expected returns of assets
Q=[.02553 .00327 .00019; %covariance matrix
.00327 .01340 -.00027;
.00019 -.00027 .00125];
goal_R=[5.5:.5:9.5]/100; % expected return goals range from 5.5% to 9.5%
for a=1:length(goal_R)
c=zeros(n,1);
A=-mu;
b=-goal_R(a);
Aeq=[ones(1,n);];
beq=[1;];

%%%%% quadratic optimization call %%%%%
[x(a,:), fval(a,1)] = quadprog(Q, c, A,b, Aeq,beq, [],[]);
std_devi(a,1)=(2*fval(a,1))^.5; %standard deviation = (x'*Q*x)^.5
end

%%%%% efficient frontier plot %%%%%
plot(std_devi, goal_R, '-k*')
xlabel('volatility \sigma')
ylabel('expected return goal R')
title('The efficient frontier of MVO')
```

7.5 Optimality Conditions for Quadratic Programming

We know from duality theory that if a linear program has a finite optimal solution, implies that there is a finite optimal solution for its dual with an optimal objective function value that is the same as the primal objective function value. The characterization of optimality for linear programming can be summarized as the satisfaction of (1) primal feasibility, (2) dual feasibility, and (3) complementary slackness. In fact, we have seen that these conditions are both necessary and sufficient. The complementary slackness conditions provide the key relationship between the primal and dual solutions. This allows us to effectively answer such questions as "If a vector x is feasible for a linear programming problem, is it an optimal solution?".

We seek a similar characterization for quadratic programs. However, a quadratic program is a non-linear optimization problem because the objective function has terms that are products of variables, and it is the non-linearity that prevents a nice characterization of optimality like that for linear programming. Instead, we seek necessary optimality conditions for a vector x to be a local minimum (defined below) for a quadratic program and identify conditions that are sufficient for global optimality. In other words, the best that one can do is to identify conditions that any local or global optimal solution must possess, but unfortunately these conditions are not unique to optimal solutions, and so in this sense the characterization is not as powerful as in the linear programming setting.

7.5.1 Local and Global Optima

In the linear programming setting, characterization of optimality is for global optimal solutions, i.e., for feasible vectors x that have the best objective function value among all feasible solutions. Non-linear optimization problems can exhibit curvature that makes it difficult to determine whether a feasible vector x is a global optimal solution. It is easier in non-linear optimization to characterize local minima, i.e., vectors x that are feasible and have the best objective function value within a neighborhood of x; see Figure 7.2. So in general for quadratic programs, the characterization of optimality will be for local minima, although it is possible to characterize global optimality under certain conditions.

Consider the following optimization problem (P)

$$\begin{array}{ll} \textit{minimize} & f(x) \\ \textit{subject to} & x \in S, \end{array}$$

where $f(x)$ is a non-linear function that is a twice continuously differentiable function from R^n to R and S is a polyhedron and a subset of R^n. Observe that the quadratic programming problem is an instance of problem P.

Let $B(x^*, \varepsilon) = \{x \in R^n | \; \|x - x^*\|_2 \leq \varepsilon\}$, which is a ball in R^n with radius ε. In R^1, $B(x^*, \varepsilon)$ is the line segment $[x^* - \varepsilon, x^* + \varepsilon]$, in R^2 it is the disc with center $\bar{x}$ and radius ε, in R^3 is a sphere with center x^* and radius ε, etc.

The formal definitions of the concept of local and global minimum are as follows.

Definition 7.5
A vector $x^* \in S$ is a local minimum for P if there is an $\varepsilon > 0$ such that $f(x^*) \leq f(x)$ for all $x \in B(x^*, \varepsilon) \cap S$.

A vector $x^* \in S$ is a strict local minimum for P if there is an $\varepsilon > 0$ such that $f(x^*) \leq f(x)$ for all $x \in B(x^*, \varepsilon) \cap S$ and $x^* \neq x$.

Definition 7.6
A vector $x^* \in S$ is a global minimum for P if $f(x^*) \leq f(x)$ for all $x \in S$.

A vector $x^* \in S$ is a strict global minimum for P if $f(x^*) < f(x)$ for all $x \in S$.

One can define analogous definitions for a local or global maximum.
Figure 7.2 illustrates the definition of a local minimum.

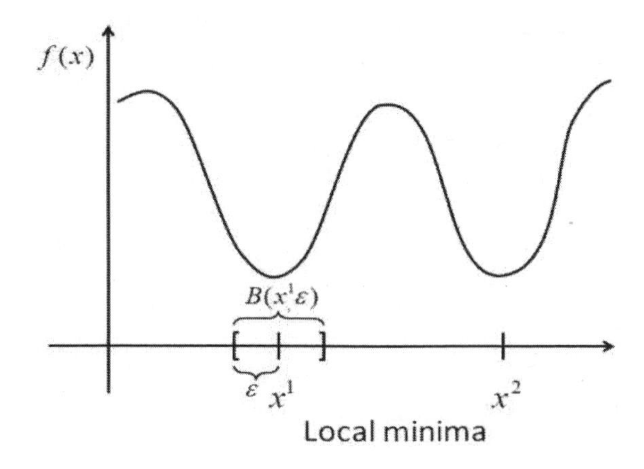

FIGURE 7.2
Illustration of local minima.

7.5.2 Unconstrained Quadratic Programs

We first consider the unconstrained quadratic programming problems

$$\begin{array}{ll} \textit{minimize} & c^T x + \frac{1}{2} x^T Q x \\ \textit{subject to} & x \in R^n. \end{array}$$

Suppose one had a vector $x \in R^n$. How can one verify that x is a local or even global optimal solution? Verifying that a solution x is a local or global solution using the definitions above is difficult in practice since it entails checking that a candidate optimal solution is better than possibly an infinite number of solutions from R^n.

We now give some constructs that will enable necessary conditions for a vector $x \in R^n$ to be a local or global optimal solution that suggest more computationally effective methods for verification of the conditions. The idea is that if a vector x is a local or global optimal solution, then it is not possible to move along any direction vector $d \in R^n$ from x that will lead to an improved vector $x^{'} \in R^n$. In other words, it will be impossible to write $x^{'} = x + \alpha d$ for some step length $\alpha > 0$ for any direction $d \in R^n$ and have $c^T x^{'} + \frac{1}{2} x^{'T} Q x^{'} < c^T x + \frac{1}{2} x^T Q x$. Such a direction d that would lead to an improvement is called a descent direction.

The development of the optimality conditions for the unconstrained quadratic programming will be done below on the more general unconstrained non-linear programming problem (UNLP)

$$\begin{array}{ll} \textit{minimize} & f(x) \\ \textit{subject to} & x \in R^n. \end{array}$$

We assume that $f(x)$ is non-linear and differentiable, and so the unconstrained quadratic programming problem is an instance of this general problem. The optimality conditions that will be derived below follow and generalize the case when $f(x)$ is a differentiable function of a single variable.

We formally define a descent direction as follows.

Definition 7.7
A vector $d \in R^n$ is a descent direction of $f(x)$ at $x = x^*$ if

$$f(x^* + \varepsilon d) < f(x^*) \text{ for all } \varepsilon > 0 \text{ and sufficiently small.}$$

The next result gives the conditions under which a vector d is a descent direction.

Theorem 7.8
Suppose that $f(x)$ is differentiable at x^* and there is a vector d such that

$$\nabla f(x^*)^T d < 0.$$

then d is a descent direction.

Proof:

Since $f(x)$ is differentiable, then by Taylor expansion we get

$$f(x) = f(x^*) + \nabla f(x^*)^T(x - x^*) + \|x - x^*\| \, o(x^*, x - x^*) \tag{7.1}$$

where $\lim_{x \to x^*} o(x^*, x - x^*) = 0$. Let $x = x^* + \alpha d$ in (7.1), then

$$f(x^* + \alpha d) = f(x^*) + \alpha \nabla f(x^*)^T(d) + \alpha \|d\| \, o(x^*, \alpha d) \tag{7.2}$$

where $\lim_{\alpha \to 0} o(x^*, \alpha d) = 0$. From (7.2), we get

$$\frac{f(x^* + \alpha d) - f(x^*)}{\alpha} = \nabla f(x^*)^T(d) + \|d\| \, o(x^*, \alpha d). \tag{7.3}$$

As $\alpha \to 0$ the second term on the right-hand side converges to 0 faster than any other term in (7.3). Now since $\nabla f(x^*)^T(d) < 0$, then for all α sufficiently small

$$f(x^* + \alpha d) - f(x^*) < 0,$$

i.e., d is a descent direction. ■

For a quadratic function $f(x) = \frac{1}{2}x^T Q x + c^T x$, the gradient is $\nabla f(x) = Qx + c$, so d is a descent direction at x^* for $f(x)$ if $(Qx^* + c)^T d < 0$.

The next corollary formalizes the idea that at a local minimum there cannot be any descent directions.

Corollary 7.9 (First-Order Necessary Condition)

Suppose $f(x)$ is differentiable at x^*. If x^* is a local minimum, then $\nabla f(x^*) = 0$.

Proof:

Suppose that $\nabla f(x^*) \neq 0$, then $d = -\nabla f(x^*)$ is a descent direction for $f(x)$ at x^*. ■

This condition generalizes the necessary condition for the case for a differentiable single variable function $f(x)$ where one sets the first derivative to 0, i.e., $f'(x) = 0$. Recall from single-variable calculus that solutions to this equation were called stationary or critical points.

For a quadratic function $f(x)$, the first-order necessary condition is then Corollary 7.10.

Corollary 7.10

If x^* is a local minimum for UQP, then $\nabla f(x^*) = Qx^* + c = 0$.

As in the single-variable case, the first-order condition suggest, that the equation $\nabla f(x^*) = 0$ should be solved for potential local minimizers. However, $\nabla f(x^*) = 0$ is in general a system of non-linear equations, but for the quadratic case we get a linear system of equations. It is important to note that solutions to $\nabla f(x) = 0$ are not automatically local minimums, but potential local minimums. However, any vector x that is a local minimum must satisfy this system. Further analysis is often required to determined which of the solutions to the system is a local minimum.

Example 7.11

Consider the problem

$$\begin{array}{ll} \textit{minimize} & f(x) = 2x_1^2 + x_2^2 + x_3^2 + x_1x_2 + x_1x_3 + x_2x_3 \\ \textit{subject to} & x \in R^n. \end{array}$$

The system $\nabla f(x) = 0$ can be written as

$$\begin{bmatrix} 4x_1 + x_2 + x_3 \\ x_1 + 2x_2 + x_3 \\ x_1 + x_2 + 2x_3 \end{bmatrix} = \begin{bmatrix} 0 \\ 0 \\ 0 \end{bmatrix},$$

which is a linear system whose only solution is $x^* = \begin{bmatrix} 0 & 0 & 0 \end{bmatrix}^T$. However, it is not known at this point, just based on satisfying the first-order necessary conditions, that x^* is a local minimum. (Note: that the function is simple enough for one to guess that the optimal solution is indeed x^*.)

The next result gives additional necessary conditions in the case that $f(x)$ is at least twice differentiable, and thus applies to quadratic functions. $H(x^*)$ is the Hessian of $f(x)$ at x^*.

Theorem 7.12 (Second-Order Necessary Condition)

Suppose that $f(x)$ is twice differentiable at x^*. If x^* is a local minimum, then $\nabla f(x^*) = 0$ and $H(x^*)$ is positive semidefinite.

Proof:

Since x^* is a local minimum, by Corollary 7.9 $\nabla f(x^*) = 0$. Now suppose that $H(x^*)$ is not positive semidefinite. This implies that there is a vector d such that $d^T H(x^*)d < 0$.

Now $f(x)$ is twice differentiable, so

$$\begin{aligned} f(x^* + \alpha d) &= f(x^*) + \alpha \nabla f(x^*)^T(d) + 1/2\alpha^2 d^T H(x^*)d + \alpha^2 \|d\|^2 o(x^*, \alpha d) \\ &= f(x^*) + \tfrac{1}{2}\alpha^2 d^T H(x^*)d + \alpha^2 \|d\|^2 o(x^*, \alpha d) \end{aligned}$$

where $\lim_{\alpha \to 0} o(x^*, \alpha d) = 0$. Then,

$$\tfrac{f(x^*+\alpha d)-f(x^*)}{\alpha^2} = \tfrac{1}{2} d^T H(x^*)d + \|d\|^2 o(x^*, \alpha d).$$

Since $d^T H(x^*)d < 0$, then for all $\alpha > 0$ sufficiently small $f(x^* + \alpha d) - f(x^*) < 0$, which implies that d is a descent direction for $f(x)$ at x^*contradicting that x^* is local minimum. ■

The second-order necessary condition for a quadratic function translates to the requirement that for any local minimum (in addition to the first-order condition), the matrix Q is positive semidefinite.

The following result is a sufficient condition for a local minimum which means that if the conditions are satisfied for a vector x^*, then it can be concluded that it is a strict local minimum.

Theorem 7.13 (Sufficiency Condition)

Suppose that $f(x)$ is twice differentiable at x^*. If $\nabla f(x^*) = 0$ and $H(x^*)$ is positive definite, then x^* is a strict local minimum.

Proof: See Bertsekas (2003).

This result is a generalization of the familiar case from calculus where $f(x)$ is a twice differentiable function of a single variable and x^* is a number such that $f'(x^*) = 0$ and $f''(x^*) > 0$, from which one can conclude that x^* is a strict local minimum. The gradient $\nabla f(x)$ and the Hessian $H(x)$ are the generalizations of the first and second derivatives for multi-variate functions.

Example 7.14

In Example 7.11, the Hessian of $f(x)$ is

$$H(x) = \begin{bmatrix} 4 & 1 & 1 \\ 1 & 2 & 1 \\ 1 & 1 & 2 \end{bmatrix}.$$

The principal minors are $\triangle_1 = 4, \triangle_2 = 7$, and $\triangle_3 = 10$ and so $H(x)$ is positive definite and is also positive semi-definite, thus we can conclude that $x^* = \begin{bmatrix} 0 & 0 & 0 \end{bmatrix}^T$ is a strict local minimum.

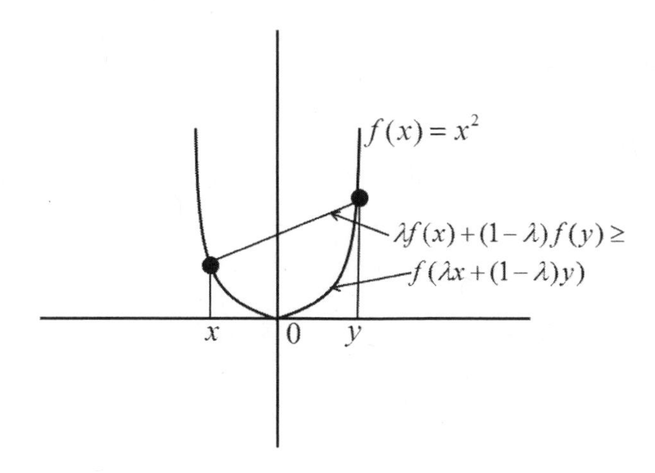

FIGURE 7.3
Convex function.

7.5.2.1 Convex Unconstrained Quadratic Programming (Global Optimality)

Consider the quadratic function $f(x) = x^2$ and the point $x^* = 0$. The first-order condition is satisfied at x^*, i.e., $\nabla f(x^*) = f^{'}(x^*) = 2x^* = 0$ and $H(x^*) = [2]$ is positive definite. Thus, we can conclude, based on Theorem 7.3 above, that $x^*$is a strict local minimum. However, it easy to see that x^* is actually a strict global minimum, yet all of the conditions derived above are unable to mathematically detect this. In general, for non-linear optimization one may obtain a vector x that is in fact a global optimal solution, but it may be very difficult to verify that it is globally optimal.

However, the quadratic function $f(x) = x^2$ is an example of what is called a convex function, and when $f(x)$ is a convex function, the case simplifies considerably. In fact, as will be shown, first-order necessary conditions are sufficient for global optimality.

The salient feature of a convex function, as can be seen from the graph of $f(x) = x^2$, is that $f(x)$ over any line segment $[x, y]$ in the domain R^1 is dominated by the line segment between the points $(x, f(x))$ and $(y, f(y))$; see Figure 7.3. This observation motivates the formal definition of a convex function, which is given next.

Definition 7.15
Let S be a convex set in R^n. A function $f(x) : S \rightarrow R$ is convex if

$$f(\lambda x + (1 - \lambda)y) \leq \lambda f(x) + (1 - \lambda)f(y)$$

for all $x, y \in S$ and for all $\lambda \in [0, 1]$. $f(x)$ is strictly convex if the inequality is strict for all $x \neq y$ and $\lambda \in (0, 1)$.

A natural question at this point is "When is a quadratic function convex?". The next result gives the answer.

Theorem 7.16

A quadratic function $f(x) = c^T x + \frac{1}{2}x^T Qx$ is convex on S if and only if the Hessian $H(x) = Q$ is positive semidefinite on S.

Proof:

=>

Suppose that $f(x)$ is convex and that Q is not positive semidefinite. Then, there exists a d such that $d^T Qd < 0$. Now let $x = td$ for any real number t, then $f(td) = tc^T d + \frac{1}{2}t^2 d^T Qd$ goes to $-\infty$ as $t \to \infty$, and so $f(x)$ is not convex, a contradiction.

<=

Suppose that Q is positive semidefinite and for any x and $y \in S$ and any $\lambda \in [0, 1]$ consider

$$
\begin{aligned}
f(\lambda x + (1-\lambda)y) &= f(y + \lambda(x-y)) \\
&= c^T(y + \lambda(x-y)) + \tfrac{1}{2}(y + \lambda(x-y))^T Q(y + \lambda(x-y)) \\
&= \lambda c^T x + (1-\lambda)c^T y + \tfrac{1}{2}y^T Qy + \lambda(x-y)^T Qy + \tfrac{1}{2}\lambda^2(x-y)^T Q(x-y) \\
&\leq \lambda c^T x + (1-\lambda)c^T y + \tfrac{1}{2}y^T Qy + \lambda(x-y)^T Qy + \tfrac{1}{2}\lambda(x-y)^T Q(x-y) \\
&= \lambda c^T x + (1-\lambda)c^T y + \tfrac{1}{2}\lambda x^T Qx + \tfrac{1}{2}(1-\lambda)y^T Qy \\
&= \lambda f(x) + (1-\lambda)f(y),
\end{aligned}
$$

thus $f(x)$ is a convex function. ■

Corollary 7.17

The function $f(x)$ is strictly convex on S if Q is positive definite.

Proof: Similar to the proof of Theorem 7.16 above. ■

We now classify problems when the objective function is convex. In particular, we call the following problem

$$
\begin{array}{ll}
\textit{minimize} & f(x) = c^T x + \frac{1}{2}x^T Qx \\
\textit{subject to} & x \in R^n
\end{array}
$$

an unconstrained convex quadratic programming (UCQP) problem when $f(x)$ is quadratic and convex or strictly convex function. Note that R^n is a convex set.

Now we present the main result concerning the UCQP problem.

Theorem 7.18

Let $f(x) = c^T x + \frac{1}{2}x^T Qx$ and suppose that Q is positive semidefinite, then the unconstrained quadratic programming problem has a global minimum x^* if and only if x^* solves the system $\nabla f(x) = Qx + c = 0$.

Proof:

=>

Suppose that x^* is a global minimum and let $d = Qx^* + c \neq 0$. We will show that it is possible to construct a new point $x^* + \alpha d$ for some step length α. To this end, let

$$f(x^* + \alpha d) = c^T(x^* + \alpha d) + \tfrac{1}{2}(x^* + \alpha d)^T Q(x^* + \alpha d)$$
$$= c^T x^* + \alpha c^T d + \tfrac{1}{2}(x^*)^T Q x^* + \alpha d^T Q x^* + \tfrac{1}{2}\alpha^2 d^T Q d$$
$$= f(x^*) + \alpha d^T(c + Qx^*) + \tfrac{1}{2}\alpha^2 d^T Q d$$
$$= f(x^*) + \alpha d^T d + \tfrac{1}{2}\alpha^2 d^T Q d. \tag{7.4}$$

Now for sufficiently small $\alpha < 0$, we have $\alpha d^T d < 0$, and so (7.4) is less than $f(x^*)$ and thus $f(x^* + \alpha d) < f(x^*)$, which contradicts that x^* is a global minimum.

<=

Proof of the converse is left to the reader. ■

Theorem 7.18 says that for UCQPs the first order necessary conditions are sufficient for global optimality.

Example 7.19

The solution $x^* = \begin{bmatrix} 0 & 0 & 0 \end{bmatrix}$ for the problem

$$\begin{array}{ll} \textit{minimize} & f(x) = 2x_1^2 + x_2^2 + x_3^2 + x_1x_2 + x_1x_3 + x_2x_3 \\ \textit{subject to} & x \in R^n \end{array}$$

can now be seen to be a strict global minimum since the Hessian $H(x)$ is positive definite on R^n.

Application to Least Squares Fitting

In Example 7.2 the problem of determining the coefficients $x_0, x_1, x_2, ..., x_k$ of the polynomial function $p(t) = x_0 + x_1 t + x_2 t^2 \cdots + x_k t^k$ of degree k that gave a least squares fit to the observed data $(t_1, u_1), (t_2, u_2), ..., (t_n, u_n)$ was formulated as an unconstrained quadratic program

$$\begin{array}{ll} \textit{minimize} & \varphi(x) = \tfrac{1}{2}x^T Q x + cx \\ \textit{subject to} & \\ & x \in R^{k+1} \end{array}$$

where

$$Q = A^T A \text{ and } c = -A^T b$$

and

$$A = \begin{bmatrix} 1 & t_1 & t_1^2 & \cdots & t_1^k \\ 1 & t_2 & t_2^2 & & t_2^k \\ & & & \ddots & \vdots \\ 1 & t_n & t_n^2 & \cdots & t_n^k \end{bmatrix} \text{ and } b = \begin{bmatrix} u_1 \\ u_2 \\ \vdots \\ u_n \end{bmatrix}.$$

Now any local minimum x has to satisfy the first-order conditions

$$\nabla\varphi(x) = 2A^T Ax - 2A^T b = 0$$

or equivalently

$$A^T Ax = A^T b$$
$$x = (A^T A)^{-1} A^T b.$$

The Hessian of $\varphi(x)$ is given by $H\varphi(x) = 2A^T A$. Observe that A has full row rank if it is assumed that the observed values $t_1, t_2, ..., t_n$ are different. Then, it is a fact from linear algebra that the matrix $A^T A$ is positive definite and so the function $\varphi(x)$ is strictly convex, which implies that the solution $x = (A^T A)^{-1} A^T b$ will be a strict global minimum.

7.5.3 Convex Optimization

UCQPs are an instance of a broader class of problems of the form

$$\begin{array}{ll} \textit{minimize} & f(x) \\ \textit{subject to} & x \in S \subseteq R^n \end{array}$$

where $f(x)$ is a convex function over the convex set S. A problem from this class is called a convex optimization problem (COP) and has the property that any local minimum for a COP is a global minimum; see Theorem 7.20 below. Furthermore, if $f(x)$ is differentiable, then a sufficient condition for a global minimum x^* for COP is that $\nabla f(x^*) = 0$; see Bertsekas(2003).

Theorem 7.20

A local minimizer of a convex function $f(x)$ over a convex set $S \subseteq R^n$ is a global minimizer. If $f(x)$ is strictly convex, then a local minimizer is a strict and unique global minimum.

Proof:

Let x^* be a local minimizer of $f(x)$ over a convex set S. Then, there exists an $\varepsilon > 0$ such that $f(x^*) \leq x$ for all $x \in B(x^*, \varepsilon)$. Let $y \in S$, but not an element of $B(x^*, \varepsilon)$. Now the line segment between y and x^*, i.e.,

$$\lambda y + (1 - \lambda)x^* \text{ for all } \lambda \in [0, 1],$$

is in S since S is convex and x^* and y are in S. Select $\lambda^* > 0$ with $0 < \lambda^* < 1$, but sufficiently small so that $\lambda^* y + (1 - \lambda^*)x^* = x^* + \lambda^*(y - x^*) \in B(x^*, \varepsilon)$ and so

$$f(x^*) \leq f(x^* + \lambda^*(y - x^*)),$$

then since $f(x)$ is a convex function over S, we have

$$\begin{aligned} f(x^*) &\leq f(x^* + \lambda^*(y - x^*)) = f(\lambda^* y + (1 - \lambda^*)x^*) \\ &\leq \lambda^* f(y) + (1 - \lambda^*) f(x^*), \end{aligned}$$

then

$$f(x^*) \leq \lambda^* f(y) + (1 - \lambda^*) f(x^*),$$

or equivalently

$$f(x^*) \leq f(y).$$

Therefore x^* is a global minimum. If $f(x)$ is a strictly convex function, then the inequalities above hold strictly for all $y \neq x^*$. ■

7.5.4 Equality-Constrained Quadratic Programs

Next we consider equality-constrained quadratic programs (EQP) of the form

$$\begin{array}{ll} \textit{minimize} & f(x) = \frac{1}{2}x^T Q x + c^T x \\ \textit{subject to} & Ax = b. \end{array}$$

The key idea is to turn an EQP into an unconstrained problem by defining a new function called the Lagrangian. This is accomplished by multiplying the vector $Ax - b$ by a vector π of multipliers, i.e., $\pi^T(Ax - b)$ and subtracting this quantity from $\frac{1}{2}x^T Q x + c^T x$. The resulting function, the Lagrangian, is $L(x, \pi) = \frac{1}{2}x^T Q x + c^T x - \pi^T(Ax - b)$.

Then, we have the unconstrained problem

$$\begin{array}{lll} \textit{minimize} & L(x, \pi) = \frac{1}{2}x^T Q x + c^T x - \pi^T(Ax - b) & \text{(UQP)} \\ \textit{subject to} & x \in R^n, \pi \in R^m. & \end{array}$$

Note that $L(x, \pi)$ is differentiable. Then, proceeding as in calculus, the critical points of $L(x, \pi)$ are obtained. Finding the critical points amounts to computing the partial derivatives of $L(x, \pi)$, setting them to zero, and solving the resulting system of equations, i.e.,

$$\frac{dL}{dx} = Qx + c - A^T \pi = 0 \quad (7.5)$$

$$\frac{dL}{d\pi} = Ax - b = 0. \quad (7.6)$$

By (7.5) we have $Qx = A^T \pi - c$ and so $x = Q^{-1}(A^T \pi - c)$ and substituting into (7.6) gives $A(Q^{-1}(A^T \pi - c)) = b$ and so

$$\pi = (AQ^{-1}A^T)^{-1}(b + AQ^{-1}c). \quad (7.7)$$

Substituting this back into (7.7) we get

$$x = Q^{-1}(A^T(AQ^{-1}A^T)^{-1}(b + AQ^{-1}c) - c). \quad (7.8)$$

Observe that the system (7.5) and (7.6) can be written in block matrix form as

$$\begin{bmatrix} Q & -A^T \\ A & 0 \end{bmatrix} \begin{bmatrix} x \\ \pi \end{bmatrix} = \begin{bmatrix} -c \\ b \end{bmatrix}. \quad (7.9)$$

In order for the solutions x and π to be meaningful, the matrices Q and A must possess the right properties so that all of the matrix inversions in (7.7) and (7.8) exist. A minimal requirement is that the matrix

$$\begin{bmatrix} Q & -A^T \\ A & 0 \end{bmatrix}$$

is invertible so that x and π can be solved for.

Assuming that Q and A have the appropriate properties, the system of equations (7.5) and (7.6) or equivalently (7.9) will define a set of necessary conditions that any local optimal solution x of EQP must satisfy. Assume that A is such that $m < n$, i.e., the number of columns are greater than the number of rows. Then summarizing the discussion above, a set of necessary conditions for x to be a local minimum for EQP is given as follows.

Theorem 7.21 (Necessary Conditions for EQP)

Assume that x is a local minimum for EQP, A has full row rank, and the matrix

$$\begin{bmatrix} Q & -A^T \\ A & 0 \end{bmatrix}$$

is invertible, then there exists a vector π so that (7.9) is satisfied.

It is important to note that these are necessary conditions, meaning that any local minimizer must satisfy condition (7.9). If a vector x does not satisfy (7.9), then it may not be a local minimizer.

Example 7.22

Consider the MVO problem from Example 7.3, but with the expected return goal as an equality constraint instead of an inequality constraint. Then, the MVO problem is an EQP with

$$Q = \begin{bmatrix} 0.02553 & 0.00327 & 0.00019 \\ 0.00327 & 0.013400 & -0.00027 \\ 0.00019 & -0.00027 & 0.00125 \end{bmatrix}$$

$$A = \begin{bmatrix} 0.0972 & 0.0657 & 0.0537 \\ 1 & 1 & 1 \end{bmatrix}, b = \begin{bmatrix} -0.055 \\ 1 \end{bmatrix}.$$

There is no linear term in this MVO problem so, $c = 0$. Thus, (7.7) and (7.8) become

$$\pi = (AQ^{-1}A^T)^{-1}b \tag{7.10}$$
$$x = Q^{-1}(A^T(AQ^{-1}A^T)^{-1}b). \tag{7.11}$$

To solve (7.10), one should find the inverse of Q through some factorization (e.g., Cholesky if Q positive definite or LU factorization) and then solve the system

$$(AQ^{-1}A^T)\pi = b \tag{7.12}$$

for π. Then, (7.11) can be solved by solving the system

$$Qx = A^T\pi. \tag{7.13}$$

Now

$$Q^{-1} = \begin{bmatrix} 40.5196 & -10.0559 & -8.3310 \\ -10.0559 & 77.4487 & 18.2574 \\ -8.3310 & 18.2574 & 805.2099 \end{bmatrix},$$

then

$$(AQ^{-1}A^T) = \begin{bmatrix} 2.9525 & 51.5513 \\ 51.5513 & 922.9192 \end{bmatrix},$$

and so solving system (7.12)

$$\begin{bmatrix} 2.9525 & 51.5513 \\ 51.5513 & 922.9192 \end{bmatrix} \begin{bmatrix} \pi_1 \\ \pi_2 \end{bmatrix} = \begin{bmatrix} -0.055 \\ 1 \end{bmatrix}$$

gives

$$\pi = \begin{bmatrix} \pi_1 \\ \pi_2 \end{bmatrix} = \begin{bmatrix} -1.5179 \\ 0.0859 \end{bmatrix}.$$

Now

$$A^T\pi = \begin{bmatrix} 0.0006 \\ 0.0010 \\ 0.0011 \end{bmatrix},$$

so system (7.13) is

$$\begin{bmatrix} 0.02553 & 0.00327 & 0.00019 \\ 0.00327 & 0.013400 & -0.00027 \\ 0.00019 & -0.00027 & 0.00125 \end{bmatrix} \begin{bmatrix} x_1 \\ x_2 \\ x_3 \end{bmatrix} = \begin{bmatrix} 0.0006 \\ 0.0010 \\ 0.0011 \end{bmatrix}$$

whose solution is

$$x = \begin{bmatrix} x_1 \\ x_2 \\ x_3 \end{bmatrix} = \begin{bmatrix} 0.0053 \\ 0.0892 \\ 0.9055 \end{bmatrix},$$

which represents a portfolio with an expected return of 5.5%. Since the matrix Q is positive definite, the quadratic objective function $f(x)$ of the MVO problem is strictly convex and the constraints $Ax = b$ define a convex set and so x is a strict global minimum variance portfolio.

We now consider the development of sufficient conditions for optimality for EQPs. That is, we identify under what conditions will a vector x that satisfies (7.9) be a local or global optimal solution to EQP. We present a few mathematical preliminaries before presenting a sufficient condition.

Definition 7.23

Let A be an $m \times n$ matrix. The set of all vectors $q \in R^n$ such that $Aq = 0$ is called the null space of A and is denoted by $N(A)$.

It is not hard to show that if q_1 and q_2 are in $\mathcal{N}(A)$, then any linear combination of q_1 and q_2 is also in $\mathcal{N}(A)$. Thus, the null space is a subspace in R^n and can be shown to have dimension $n - m$ when A has full row rank m. A major implication of $\mathcal{N}(A)$ being a subspace is that there will be a set of basis vectors for $\mathcal{N}(A)$.

Let Z be a matrix of dimension $n \times (n-m)$ that consists of the columns of a basis for the null space of A. Then, the quantity $Z^T QZ$ is called the reduced Hessian matrix of $f(x)$. Our first result sheds light on when the necessary condition becomes a sufficient condition.

Theorem 7.24 (Sufficient Condition for EQP)

Suppose that A has full row rank and that the reduced Hessian $Z^T QZ$ of $f(x)$ is positive definite. Then, the vector x^* that satisfies the first-order necessary conditions

$$\begin{bmatrix} Q & -A^T \\ A & 0 \end{bmatrix} \begin{bmatrix} x \\ \pi \end{bmatrix} = \begin{bmatrix} -c \\ b \end{bmatrix}$$

is a unique global optimal solution for EQP.

The following is a useful fact from linear algebra. Any vector x that satisfies $Ax = b$ can be written as $x = x^{'} + p$ where $Ax^{'} = b$ and $Ap = 0$, i.e., $p \in \mathcal{N}(A)$.

Proof of Theorem 7.24:

Let x^* be a vector that satisfies the first-order conditions, and thus satisfies $Ax^* = b$, then $x^* = x^{'} + p$ for some feasible $x^{'}$, i.e., $(Ax^{'} = b)$ and $p \in \mathcal{N}(A)$. Now $x^{'} = x^* - p$ and consider the objective function $f(x) = \frac{1}{2}x^T Qx + c^T x$ at $x^{'}$, i.e.,

$$\begin{aligned} f(x^{'}) &= f(x^* - p) \\ &= c^T(x^* - p) + \tfrac{1}{2}(x^* - p)^T Q(x^* - p) \end{aligned}$$

$$= c^T x^* - c^T p + \tfrac{1}{2} x^{*T} Q x^* - p^T Q x^* + \tfrac{1}{2} p^T Q p$$
$$= f(x^*) - c^T p - p^T Q x^* + \tfrac{1}{2} p^T Q p. \qquad (7.14)$$

Now, since x^* satisfies the first-order necessary conditions $Qx^* = -c + A^T \lambda^*$ and thus,

$$p^T Q x^* = p^T(-c + A^T \lambda^*)$$
$$= -p^T c$$

since $p \in \mathcal{N}(A)$. So by substitution of this term into (7.14), we get

$$f(x') = f(x^*) + \tfrac{1}{2} p^T Q p.$$

Now, since $p \in \mathcal{N}(A)$, then p can be written as a linear combination of the vectors in the basis of $\mathcal{N}(A)$, i.e., $p = Zw$ where w is the vector of weights of the basis vectors in the linear combination. Then,

$$f(x') = f(x^*) + \tfrac{1}{2} w^T Z^T Q Z w$$

and since the reduced Hessian is positive definite (i.e., $w^T Z^T Q Z w > 0$ for all $w \neq 0$), we can conclude that

$$f(x') > f(x^*) \text{ for any } x' \text{ such that } Ax' = b\ .$$

Therefore, x^* is a strict global minimum for EQP. ■

Example 7.25

For the MVO problem in Example 7.22, the matrix Q is positive definite so the reduced Hessian $Z^T Q Z$ is positive definite since $x^T Z^T Q Z x = y^T Q y > 0$ for all $y = Zx \in R^n$ where $x \neq 0$, and so the solution x is a strict global minimum, which is consistent with the earlier determination that it was a strict global minimum based on convexity.

7.5.4.1 Alternative Solution Methods for EQP

EQPs can be solved by directly using the matrix equations (7.7) and (7.8) as in Example 7.22. An alternative strategy would be to use matrix factorization methods similar to LU factorization on the matrix

$$K = \begin{bmatrix} Q & -A^T \\ A & 0 \end{bmatrix}.$$

However, this is complicated by the fact that in general this matrix is indefinite and so methods such as Cholesky Factorization, which require a matrix to be positive definite, cannot be used. Methods called symmetric indefinite factorization are appropriate here. In this method a factorization of the form

$$PMP^T = LBL^T$$

is obtained where M is a symmetric matrix (for EQPs $M = K$), P is a permutation matrix, L is a lower triangular matrix and B is a block diagonal matrix with block matrices of dimension 1 or 2; see Golub and Van Loan (1989) for more details.

7.5.5 Inequality Constrained Convex Quadratic Programming

We now consider quadratic programs of the following form

$$\begin{array}{lll} \text{(Q)} & \textit{minimize} & f(x) = \frac{1}{2}x^T Qx + c^T x \\ & \textit{subject to} & Ax = b \\ & & x \geq 0 \end{array}$$

with equality and inequality constraints, i.e., non-negativity restrictions on the variables. We refer to this class of problem as EIQP. An EIQP can model the situation where one is constructing an MVO portfolio, but with restrictions on short selling.

Our main goal here is to develop a predictor-corrector primal-dual path-following interior point method for quadratic programming. We will assume for the purpose of algorithm construction that A has full row rank and Q is positive definite, thereby making EIQP a convex optimization problem. To characterize the optimality conditions for algorithmic development of EIQP, we proceed as in the case for developing primal-dual path-following interior point methods for linear programs in standard form where we first convert the problem to an unconstrained problem via the Lagrangian. The non-negativity constraints will be handled through using a barrier function as was done for linear programming problems in standard form in Chapter 6.

In particular, we consider the following barrier problem for EIQP analogous to the barrier problem developed for linear programs in standard form

$$\begin{array}{c} \textit{minimize } c^T x + \frac{1}{2}x^T Qx - \mu \sum\limits_{i=1}^{n} \ln(x_i) \\ \textit{subject to } Ax = b \end{array} \tag{7.15}$$

where $\mu > 0$. We can take the Lagrangian of (7.15) and set the partial derivatives equal to 0, which is equivalent to obtaining the following system of non-linear equations.

$$-Qx + A^T\pi + z = c \tag{7.16}$$
$$Ax = b \tag{7.17}$$
$$XZe = \mu e \tag{7.18}$$

These are the KKT conditions for (7.15). If $\mu = 0$, this system will represent the KKT conditions for (Q) the original quadratic program. The strategy of the development of the predictor-corrector primal-dual path-following method for quadratic programming will now parallel the development for the linear programming case in Chapter 6. In particular, the Newton-Raphson method is used to successively approximate the system (7.16)–(7.17) each time using a smaller value of μ that will be expressed as in the linear programming case as the product of a centering value that is changed adaptively and a duality measure. The only major difference compared to the linear programming case is now the presence of the matrix Q in the KKT system of equations, and the fact that the step lengths will be identical for primal, dual, and dual slack iterates.

7.5.6 Predictor-Corrector Algorithm for Convex QP

We now present the details of the predictor-corrector method.

Step 0: Obtain an initial interior solution $(x^{(0)}, \pi^{(0)}, z^{(0)})$ such that $x^{(0)} > 0$ and $z^{(0)} > 0$ (see below). Let $k = 0$ and ε be some small positive number (tolerance). Go to Step 1.

Step 1: Solve

$$\begin{bmatrix} -Q & A^T & I \\ A & 0 & 0 \\ Z^{(k)} & 0 & X^{(k)} \end{bmatrix} \begin{bmatrix} d_x^{aff} \\ d_\pi^{aff} \\ d_z^{aff} \end{bmatrix} = \begin{bmatrix} -r_d^{(k)} \\ -r_p^{(k)} \\ -X^{(k)}Z^{(k)}e \end{bmatrix}$$

for $\begin{bmatrix} d_x^{aff} \\ d_\pi^{aff} \\ d_z^{aff} \end{bmatrix}$, where $X^{(k)} = diag(x^{(k)})$, $Z^{(k)} = diag(z^{(k)})$, $r_p^{(k)} = Ax^{(k)} - b$ are the primal residuals and $r_d^{(k)} = -Qx^{(k)} + A^T\pi^{(k)} + z^{(k)} - c$ are the dual residuals. Go to Step 2.

Step 2: Compute

$$\alpha^{aff} = \min\{1, \min_{i:(d_x^{aff})_i<0} \{-\frac{x_i^{(k)}}{(d_x^{aff})_i}\}, \min_{i:(d_z^{aff})_i<0} \{-\frac{z_i^{(k)}}{(d_z^{aff})_i}\}\}$$

, $y^{(k)} = \frac{(x^{(k)})^T z^{(k)}}{n}$, $y_{aff}^{(k)} = \frac{(x^{(k)}+\alpha_x^{aff}d_x^{aff})^T(z^{(k)}+\alpha_z^{aff}d_z^{aff})}{n}$ and let $\tau^{(k)} = \left(\frac{y_{aff}^{(k)}}{y^{(k)}}\right)^3$ and solve for $\begin{bmatrix} d_x \\ d_\pi \\ d_z \end{bmatrix}$ in

$$\begin{bmatrix} -Q & A^T & I \\ A & 0 & 0 \\ Z^{(k)} & 0 & X^{(k)} \end{bmatrix} \begin{bmatrix} d_x \\ d_\pi \\ d_z \end{bmatrix} = \begin{bmatrix} -r_d^{(k)} \\ -r_p^{(k)} \\ -X^{(k)}Z^{(k)}e + D_{x^{(k)}}D_{z^{(k)}}e + \tau^{(k)}y^{(k)}e \end{bmatrix}$$

where $D_{x^{(k)}} = diag(d_x^{aff})$ and $D_{z^{(k)}} = diag(d_z^{aff})$. Go to Step 3.

Step 3: Compute $\alpha = \min\{1, \eta\alpha_x^{\max}, \eta\alpha_z^{\max}\}$ where $\eta \in [0.9, 1)$ and

$$\alpha_x^{\max} = \min_{i:(d_x^{aff})_i<0} \{-\frac{x_i^{(k)}}{(d_x^{aff})_i}\} \text{ and } \alpha_z^{\max} = \min_{i:(d_z^{aff})_i<0} \{-\frac{z_i^{(k)}}{(d_z^{aff})_i}\}$$

and let

$$\begin{aligned} x^{(k+1)} &= x^{(k)} + \alpha d_x \\ \pi^{(k+1)} &= \pi^{(k)} + \alpha d_\pi \\ z^{(k+1)} &= z^{(k)} + \alpha d_z. \end{aligned}$$

If the stopping criteria is met, i.e.,

$$\begin{gathered} \left\| Ax^{(k+1)} - b \right\| \leq \varepsilon \\ \left\| A^T\pi^{(k+1)} + z^{(k+1)} - c \right\| \leq \varepsilon \\ (x^{(k+1)})^T z^{(k+1)} \leq \varepsilon, \end{gathered}$$

then STOP. Else $k = k + 1$ and go to Step 1.

Example 7.26

Consider an MVO problem of the form

$$\begin{array}{ll} \text{minimize} & x^TQx \\ \text{subject to} & Ax = b \\ & x \geq 0 \end{array}$$

where

$$Q = \begin{bmatrix} 0.02553 & 0.00327 & 0.00019 \\ 0.00327 & 0.013400 & -0.00027 \\ 0.00019 & -0.00027 & 0.00125 \end{bmatrix}$$

$$A = \begin{bmatrix} 0.0972 & 0.0657 & 0.0537 \\ 1 & 1 & 1 \end{bmatrix}, b = \begin{bmatrix} 0.055 \\ 1 \end{bmatrix}.$$

Since $x \geq 0$, short selling is explicitly prohibited. We illustrate one iteration of the predictor-corrector method in this problem.

Step 0: Let $\varepsilon = 10^{-8}$ and $\eta = 0.95$ and let

$$x^{(0)} = \begin{bmatrix} 1 \\ 1 \\ 1 \end{bmatrix}, \pi^{(0)} = \begin{bmatrix} 1 \\ 1 \end{bmatrix}, \text{ and } z^{(0)} = \begin{bmatrix} 1 \\ 1 \\ 1 \end{bmatrix}.$$

Check the stopping condition for initial point

$$\max\left\{\begin{array}{c}\left\|Ax^{(0)}-b\right\| \\ \left\|-Qx^{(0)}+A^T\pi^{(0)}+z^{(0)}\right\| \\ (x^{(0)})^T z^{(0)}\end{array}\right\} = \max\left\{\begin{array}{c}2.0065 \\ 3.5623 \\ 3.0000\end{array}\right\} = 3.5623 > \varepsilon.$$

So go to Step 1.

Iteration 1
Step 1:
Solve

$$\begin{bmatrix} -Q & A^T & I \\ A & 0 & 0 \\ Z^{(0)} & 0 & X^{(0)} \end{bmatrix}\begin{bmatrix} d_x^{aff} \\ d_\pi^{aff} \\ d_z^{aff} \end{bmatrix} = \begin{bmatrix} -r_d^{(0)} \\ -r_p^{(0)} \\ -X^{(0)}Z^{(0)}e \end{bmatrix}$$

where

$$X^{(0)} = \begin{bmatrix} 1 & 0 & 0 \\ 0 & 1 & 0 \\ 0 & 0 & 1 \end{bmatrix}, Z^{(0)} = \begin{bmatrix} 1 & 0 & 0 \\ 0 & 1 & 0 \\ 0 & 0 & 1 \end{bmatrix}$$

$$-r_p^{(0)} = \begin{bmatrix} -0.1617 \\ -2 \end{bmatrix}, -r_d^{(0)} = \begin{bmatrix} -2.0683 \\ -2.0493 \\ -2.0525 \end{bmatrix}$$

$$-X^{(0)}Z^{(0)}e = \begin{bmatrix} -1 \\ -1 \\ -1 \end{bmatrix},$$

then

$$d_x^{aff} = \begin{bmatrix} -1.0916 \\ -0.5589 \\ -0.3495 \end{bmatrix}, d_\pi^{aff} = \begin{bmatrix} -18.0533 \\ -0.4330 \end{bmatrix}$$

$$d_z^{aff} = \begin{bmatrix} 0.0916 \\ -0.4411 \\ -0.6505 \end{bmatrix}.$$

Step 2:

$$\alpha^{aff} = \min\{1, \min\{\tfrac{1}{1.0916}, \tfrac{1}{0.5589}, \tfrac{1}{0.3495}\}, \min\{\tfrac{1}{0.4411}, \tfrac{1}{.6505}\}\}$$

$$= \tfrac{1}{1.0916} = 0.9160$$

$$y^{(0)} = \frac{(x^{(0)})^T z^{(0)}}{n} = \frac{\begin{bmatrix} 1 & 1 & 1 \end{bmatrix}^T \begin{bmatrix} 1 \\ 1 \\ 1 \end{bmatrix}}{3} = \tfrac{3}{3} = 1,$$

$$y_{aff}^{(0)} = \frac{(x^{(0)}+\alpha^{aff}d_x^{aff})^T(z^{(0)}+\alpha^{aff}d_z^{aff})}{n}$$

$$= \frac{\left(\begin{bmatrix} 1 \\ 1 \\ 1 \end{bmatrix} + 0.9160 \begin{bmatrix} -1.0916 \\ -0.5589 \\ -0.3495 \end{bmatrix}\right)^T \left(\begin{bmatrix} 1 \\ 1 \\ 1 \end{bmatrix} + 0.9160 \begin{bmatrix} -1.0916 \\ -0.5589 \\ -0.3495 \end{bmatrix}\right)}{3}$$

$$= 0.1885$$

and so

$$\tau^{(0)} = \left(\frac{y_{aff}^{(0)}}{y^{(0)}}\right)^3 = \left(\frac{0.1885}{1}\right)^3 = 0.0067.$$

Now solving for $\begin{bmatrix} d_x \\ d_\pi \\ d_z \end{bmatrix}$ in

$$\begin{bmatrix} -Q & A^T & I \\ A & 0 & 0 \\ Z^{(0)} & 0 & X^{(0)} \end{bmatrix} \begin{bmatrix} d_x \\ d_\pi \\ d_z \end{bmatrix} = \begin{bmatrix} -r_d^{(0)} \\ -r_p^{(0)} \\ -X^{(0)}Z^{(0)}e + D_{x^{(0)}}D_{z^{(0)}}e + \tau^{(0)}y^{(0)}e \end{bmatrix}$$

where

$$D_{x^{(0)}} = diag(d_x^{aff}) = \begin{bmatrix} -1.0916 & 0 & 0 \\ 0 & -0.5589 & 0 \\ 0 & 0 & -0.3495 \end{bmatrix}$$

and

$$D_{z^{(0)}} = diag(d_z^{aff}) = \begin{bmatrix} 0.0916 & 0 & 0 \\ 0 & -0.4411 & 0 \\ 0 & 0 & -0.6505 \end{bmatrix},$$

then

$$d_x = \begin{bmatrix} -1.0730 \\ -0.6265 \\ -0.3005 \end{bmatrix}, d_\pi = \begin{bmatrix} -26.2540 \\ 0.2771 \end{bmatrix},$$

$$d_z = \begin{bmatrix} 0.1796 \\ -0.6133 \\ -0.9202 \end{bmatrix}.$$

Step 3:
Now

$$\alpha_x^{\max} = \min\{1, \min_{i:(d_x)_i<0}\{-\frac{x_i^{(0)}}{(d_x)_i}\}\}$$

$$= \min\{1, \tfrac{1}{1.0730}, \tfrac{1}{0.6265}, \tfrac{1}{0.3005}\} = 0.9320$$

and

$$\alpha_z^{\max} = \min\{1, \min_{i:(d_z)_i<0}\{-\frac{z_i^{(0)}}{(d_z)_i}\}\}$$
$$= \min\{1, \tfrac{1}{0.6133}, \tfrac{1}{0.9202}\} = 1,$$

and so

$$\alpha = \min\{1, 0.95\alpha_x^{\max}, 0.95\alpha_z^{\max}\} = 0.8854,$$

then

$$x^{(1)} = x^{(0)} + \alpha d_x$$

$$= \begin{bmatrix} 1 \\ 1 \\ 1 \end{bmatrix} + (0.8854)\begin{bmatrix} -1.0730 \\ -0.6265 \\ -0.3005 \end{bmatrix} = \begin{bmatrix} 0.0499 \\ 0.4453 \\ 0.7339 \end{bmatrix}$$

$$\pi^{(1)} = \pi^{(0)} + \alpha d_\pi$$

$$= \begin{bmatrix} 1 \\ 1 \end{bmatrix} + (0.8854)\begin{bmatrix} -26.2540 \\ 0.2771 \end{bmatrix} = \begin{bmatrix} -22.2453 \\ 1.2453 \end{bmatrix}$$

$$z^{(1)} = z^{(0)} + \alpha d_z$$

$$= \begin{bmatrix} 1 \\ 1 \\ 1 \end{bmatrix} + (0.8854)\begin{bmatrix} 0.1796 \\ -0.6133 \\ -0.9202 \end{bmatrix} = \begin{bmatrix} 1.1590 \\ 0.4570 \\ 0.1853 \end{bmatrix}.$$

Check the stopping condition for point $(x^{(1)}, \pi^{(1)}, z^{(0)})^T$

$$\max\left\{\begin{array}{c} \|Ax^{(1)} - b\| \\ \|-Qx^{(1)} + A^T\pi^{(1)} + z^{(1)}\| \\ (x^{(1)})^T z^{(1)} \end{array}\right\} = \max\left\{\begin{array}{c} 0.2300 \\ 0.4083 \\ 0.3974 \end{array}\right\} = 0.4083 > \varepsilon,$$

so another iteration should be performed. After 8 iterations the tolerance is satisfied and so the interior point method stops with a final point $x^{(8)} = \begin{bmatrix} 0.0053 \\ 0.0892 \\ 0.9055 \end{bmatrix}$, which is the optimal value for the primal variables x. Table 7.4 contains the values of the primal variables and corresponding objective function value (which is half of the portfolio variance) of each iteration. Table 7.5 contains the values of the dual and dual slack iterates and Table 7.6 contains the residuals and centering parameter values for each iteration.

Table 7.4 Primal iterates

k	$x_1^{(k)}$	$x_1^{(k)}$	$x_1^{(k)}$	$\frac{1}{2}(x^{(k)})^T Qx^{(k)}$
0	1	1	1	0.02328000
1	0.05000000	0.44527099	0.73394639	0.00168851
2	0.01266180	0.09415527	0.90724988	0.00055890
3	0.00990667	0.07393041	0.91686626	0.00054909
4	0.00967408	0.07326719	0.91709542	0.00054869
5	0.00841256	0.07777401	0.91381622	0.00054775
6	0.00603015	0.08642421	0.90754583	0.00054685
7	0.00532042	0.08900251	0.90567708	0.00054679
8	0.00526378	0.08920825	0.90552796	0.00054679

Table 7.5 Dual and dual slack iterates

k	$\pi_1^{(k)}$	$\pi_2^{(k)}$	$z_1^{(k)}$	$z_2^{(k)}$	$z_3^{(k)}$
0	1	1	1	1	1
1	−22.2450337	1.24530581	1.15905425	0.45699246	0.18529698
2	−24.4184074	1.3175511	1.07371087	0.3022101	0.00926485
3	−2.53175704	0.13727947	0.11045674	0.03055316	0.00052571
4	−0.15523674	0.0094748	0.0063285	0.00152766	2.75E-05
5	−0.03118035	0.00279798	0.00088148	7.64E-05	2.14E-06
6	−0.01526178	0.00193184	0.00016232	3.82E-06	1.56E-07
7	−0.01198005	0.00175241	1.22E-05	1.91E-07	8.35E-09
8	−0.01172687	0.00173856	6.14E-07	9.55E-09	4.18E-10

Table 7.6 Residuals and centering parameter

k	$\tau^{(k)}$	$\|Ax^{(k)} - b\|$	$\|-Qx^{(k)} + A^T\pi^{(k)} + z^{(k)}\|$	$(x^{(k)})^T z^{(k)}$
0	-	2.00652608	3.56236104	3
1	0.00669789	0.22996533	0.40827753	0.39743625
2	0.00484609	0.01411286	0.02505579	0.05045532
3	0.00352535	0.00070564	0.00125279	0.00383507
4	2.10E-06	3.68E-05	6.54E-05	0.00019837
5	4.93E-05	2.81E-06	4.98E-06	1.53E-05
6	0.00193041	1.87E-07	3.31E-07	1.45E-06
7	0.00060927	9.58E-09	1.70E-08	8.95E-08
8	9.61E-07	4.79E-10	8.51E-10	4.46E-09

7.6 Exercises

Exercise 7.1 Suppose we are given the points (0,0), (1,2), (3,2), and (4,5). Find the equation of the line that best fits this data set.

Exercise 7.2
Consider three securities with expected returns given as follows

Expected security returns

Expected return	Security 1 ($i = 1$)	Security 2 ($i = 2$)	Security 3 ($i = 3$)
μ_i	12.73%	7.57%	6.37%

with covariances

Covariance of returns

Covariance σ_{ij}	$i = 1$	$i = 2$	$i = 3$
$i = 1$	0.02559	0.00327	0.00019
$i = 2$		0.01640	−0.00320
$i = 3$			0.00525

Using MATLAB, find mean-variance optimal portfolios of the three assets for each expected return goal R from 6.5% to 12.5% and plot the return goal R versus the standard deviation (of portfolio return) for each optimal portfolio.

Exercise 7.3
Find any local minima for the problem

$$\begin{array}{ll} \textit{minimize} & e^{x_1 - x_2} + e^{x_1 + x_2} \\ \textit{subject to} & (x_1, x_2, x_3)^T \in R^2. \end{array}$$

Are any of the local minima global?

Exercise 7.4
Find any local minima for the problem

$$\begin{array}{ll} \textit{minimize} & e^{x_1 - x_2} + e^{x_2 - x_1} \\ \textit{subject to} & (x_1, x_2)^T \in R^2. \end{array}$$

Are any of the local minima global?

Exercise 7.5
You wish to invest in 3 securities S, B, and M. The expected returns for each security are $\mu_S = 10.73\%, \mu_B = 7.37\%, \mu_M = 6.27\%$. The standard deviations of returns of the securities are $\sigma_S = 16.67\%, \sigma_B = 10.55\%, \sigma_M = 3.40\%$. The correlations of the returns between securities are given in the following table:

ρ_{ij}	S	B	M
S	1	0.2199	0.0366
B		1	−0.0545
M			1

You wish to find an efficient (i.e., mean-variance optimal) portfolio of these 3 securities with an expected return of exactly 9%. Also, you will allow short selling of stocks. A colleague comes to you and says a portfolio with 56% of wealth in stocks (S), 20% in bonds (B), and the rest in the money market (M) is efficient or at least approximately efficient. Prove or disprove your colleague's claim.

Exercise 7.6

Suppose in Exercise 7.5 that you restrict short selling. Is your colleague's portfolio efficient or approximately efficient? Why or why not?

Exercise 7.7

Consider an investing environment where there are n financial securities that are risky (i.e., returns are modeled as random variables) that can be purchased or sold, and all investors are mean-variance optimizers (i.e., investors only form efficient portfolios when investing and do not speculate on individual stocks). Furthermore, suppose that all investors have the same estimates of the mean, variance, and covariances of the securities. However, each investor may have their own expected portfolio return goal R. Then, show that any investor only needs to invest in two efficient portfolios F_1 and F_2 consisting of the n securities and these two funds are the same for all investors. (Note: By investing in two funds, F_1 and F_2, we mean that an investor puts a proportion of wealth α in F_1 and $(1 - \alpha)$ in F_2.)

Exercise 7.8

The correlation ρ between securities A and B is .10 with expected returns and standard deviations for each security given by the table

Security	$\bar{r}$	σ
A	12%	14%
B	20%	30%

(a) Find the proportions w_A of A and w_B of B that define a portfolio of A and B having minimum standard deviation.

(b) What is the value of the standard deviation of portfolio in (a)?

(c) What is the expected return of the portfolio in (a)?

Exercise 7.9

Consider a general non-linear optimization problem (P) of the form

$$\begin{array}{ll} \textit{minimize} & f(x) \\ \textit{subject to} & g(x) \leq 0 \\ & h(x) = 0 \\ & x \in R^n \end{array}$$

where $f(x) : R^n \rightarrow R$,

$$g(x) = \begin{bmatrix} g_1(x) \\ g_2(x) \\ \vdots \\ g_l(x) \end{bmatrix} \text{ and } h(x) = \begin{bmatrix} h_1(x) \\ h_2(x) \\ \vdots \\ h_m(x) \end{bmatrix}$$

with $g_i(x) : R^n \rightarrow R$ for $i = 1,..,l$ and $h_j(x) : R^n \rightarrow R$ for $j = 1,...,m$. So $g(x) : R^n \rightarrow R^l$ and $h(x) : R^n \rightarrow R^m$. Assume all functions are differentiable.

The first-order KKT (Karush-Kuhn-Tucker) necessary conditions for P are as follows:

Let x^* be a local minimum of P and let $I = \{i | g_i(x^*) = 0\}$. Suppose that the gradients of $g_i(x^*)$ for $i \in I$ and gradients of $h_i(x)$ for $i = 1,...,m$ are linearly independent. Then, there exist vectors π and v such that

(1) $\nabla f(x^*) + \nabla g(x^*)^T \pi + \nabla h(x^*)^T v = 0$
(2) $\pi \geq 0$
(3) $\pi^T g(x^*) = 0$

(a) Consider the following quadratic programming problem

$$\begin{array}{ll} \textit{minimize} & f(x) = \frac{1}{2}x^T Qx + c^T x \\ \textit{subject to} & Ax \leq b \\ & Ex = d \\ & x \geq 0 \end{array}$$

where Q is an $n \times n$ matrix, A is an $l \times n$ matrix, and E is a $m \times n$ matrix. Write out the KKT first-order necessary conditions for this problem.

(b) Consider the problem

$$\begin{array}{ll} \textit{minimize} & x_1^2 + x_2^2 - 2x_1 - 4x_2 + 5 \\ \textit{subject to} & -x_1 - x_2 \geq -2 \\ & -x_1 + x_2 = 1 \\ & x_1 \geq 0, x_2 \geq 0. \end{array}$$

Find all points that satisfy the KKT first-order necessary conditions. Are any of these optimal solutions?

Exercise 7.10

As seen in Exercise 7.2, the Newton-Raphson method can be used to solve first-order conditions for optimization problems. For example, consider the unconstrained optimization problem

$$\begin{array}{ll} \textit{minimize} & f(x) = \frac{1}{2}x^T Qx + c^T x \\ \textit{subject to} & x \in R^n \end{array}$$

where Q is an $n \times n$ positive definite matrix. Using any starting point $x^{(0)} \in R^n$, use the Newton-Raphson method to compute the first iterate $x^{(1)}$ and show that this is the optimal solution.

Exercise 7.11
Consider a quadratic program of the form

$$\begin{array}{ll} \textit{minimize} & f(x) = \frac{1}{2}x^T Qx + c^T x \\ \textit{subject to} & Ax = b \\ & x \in R^n. \end{array}$$

Compute the first two iterates of the predictor-corrector path-following interior point method applied to the QP where

$$c = \begin{bmatrix} -6 \\ 5 \\ -11 \\ -9 \end{bmatrix}, Q = \begin{bmatrix} 21 & -4 & 10 & 0 \\ 0 & 35 & 23 & 16 \\ 2 & 3 & 15 & -27 \\ -8 & 0 & -9 & 59 \end{bmatrix},$$

$$A = \begin{bmatrix} -4 & 26 & 7 & 15 \\ 9 & 12 & -7 & 0 \\ 0 & 6 & 14 & -8 \end{bmatrix}, b = \begin{bmatrix} 90 \\ 150 \\ 100 \end{bmatrix}$$

using $\eta = 0.95$.

Exercise 7.12
Consider a quadratic program of the form

$$\begin{array}{ll} \textit{minimize} & f(x) = \frac{1}{2}x^T Qx + c^T x \\ \textit{subject to} & Ax = b \\ & x \in R^n. \end{array}$$

Compute the first two iterates of the predictor-corrector path following interior point method applied to the QP where

$$c = \begin{bmatrix} -3.5 \\ -4.5 \\ -2 \end{bmatrix}, Q = \begin{bmatrix} 1.45 & -1.20 & -0.65 \\ -0.90 & 1.05 & 0.45 \\ -0.95 & 0.25 & 0.85 \end{bmatrix},$$

$$A = \begin{bmatrix} 5 & 12 & 4 \\ -3 & 5 & 11 \end{bmatrix}, b = \begin{bmatrix} 17 \\ 35 \end{bmatrix}$$

using $\eta = 0.95$.

Notes and References

Quadratic programming is an important class of non-linear programming problems. The major application of this chapter concerned mean-variance optimization for constructing efficient portfolios; see Marc Steinbach (2001) for an extensive review and Best (2010). There are many other applications in the areas of production planning with economies of scale, chemical engineering process control; see Haverly (1978), Lasdon et al. (1979), Floudas and Aggarwal (1990), and VLSI circuit layout design; see Kedem and Watanabe (1983). The difficulty of solving quadratic programming relies on the nature of the matrix Q. When Q is positive definite, the quadratic programming problem becomes a convex optimization problem, and so finding the optimal solution can be done efficiently, in principle. General optimality conditions for non-linear programming problems are known as the Karush-Kuhn-Tucker conditions; see Bazaraa, Sherali, and Shetty (2006). The predictor-corrector interior point method in this chapter is based on Monteiro and Adler (1989). Besides interior point methods, there are methods that mimic the strategy of simplex method to solve quadratic programs; see Wolfe (1959) and active set methods; see Nocedal and Wright (2000). Other strategies solve the KKT conditions of a quadratic program called the linear complementarity problem through the use of a pivoting strategy; see Lemke (1962), Murty (1988), Cottle et al. (1992).

8

Linear Optimization under Uncertainty

8.1 Introduction

A fundamental assumption in linear programming is that the coefficients are assumed to be known. In other words, all of the linear programs up to now have been deterministic problems. However, in reality, these quantities are at best estimations, and in many instances parameters are essentially random values and thus it is often a challenge to select parameter values for a model. Sensitivity analysis, developed in Chapter 4, may shed light on the range of data perturbations for a linear program that is allowed while keeping the original optimal solution optimal. However, there are several limitations. First, the kinds of parameter perturbations for which sensitivity analysis is amenable is limited. Only simple cases were considered, e.g., perturbations of the right-hand side vector or perturbations of cost coefficients only. It is not easy to perform sensitivity analysis when different combinations of parameters are modified, e.g., cost coefficients, constraint parameters, and right-hand sides are simultaneously perturbed. In this case, it is often better to resolve the model. Second, even if sensitivity analysis is possible, it is not clear what can be done about decisions when considering a data perturbation that is outside the range for which the current optimal is still valid. In this case, the model would need to be resolved.

The most pressing concern about using a deterministic optimization model is that often we need to make decisions now before we can know the true values or have better estimations of the parameters, and thus a deterministic model does not have the ability to account for possible deviations of parameter values that may occur. In many of these situations, the linear programming solution will be misleading.

In this chapter, two major approaches for dealing with uncertainty in linear programming are considered. The first approach is called stochastic programming and is an important class of models for which uncertainty in parameters can be incorporated. The second approach is called robust optimization where the essence is to allow a range of values for parameters. What is important about both approaches is that uncertainty is proactively incorporated into models before solving.

8.2 Stochastic Programming

Stochastic programming is branch of mathematical programming that explicitly incorporates uncertainty in parameter values. A model of uncertainty about the outcomes for the parameters that are deemed random is specified. In particular, a probability distribution is assumed to be known that characterizes the random nature of the parameters. The idea of stochastic programming is illustrated first through an example.

Motivation

A supplier of personal computers produces three types of computers called Home PC, Pro PC, and Workstation PC. Each PC requires a central processing unit (CPU) and memory boards. The production requirements, demand, and unit prices are given in Table 8.1. The demands are denoted by d_H , d_P, and d_W. It is important for the supplier to meet at all demands for its products. In addition, the supplier can obtain the same PC types pre-assembled from other wholesale manufacturers, but at a premium cost (i.e., higher than the cost of producing it) which is given in the last column of Table 8.2. The supplier has a total of 1500 CPUs and 2000 memory boards. We assume that all PCs produced or purchased from other manufacturers are sold.

Table 8.1 PC information

Computer	CPU	Memory boards	Demand
Home PC	1	2	d_H
Pro PC	1	4	d_P
Workstation PC	1	1	d_W

For example, the Pro PC requires 1 central processing unit (CPU) and 2 memory boards per unit, and is sold at \$1000 per unit, the demand is d_P units, and the cost of obtaining one unit from another manufacturer is \$800, which is considered a wholesale price not accessible by the customers of the supplier. The cost per unit of each type of computer is given below from units assembled by the supplier and from units obtained at wholesale prices.

Table 8.2 PC cost

Computer	Cost (assembled)	Cost (wholesale)
Home PC	225	275
Pro PC	350	400
Workstation PC	250	300

Assuming all parameters are known, then the deterministic production model (DPM) is

$$\begin{array}{llllllll}
\textit{minimize} & 225x_1 + 350x_2 + 250x_3 + 275y_1 + 400y_2 + 300y_3 & & & & & & \\
\textit{subject to} & x_1 & +x_2 & +x_3 & & & & \leq 1500 \\
& 2x_1 & +4x_2 & +x_3 & & & & \leq 2000 \\
& x_1 & & & +y_1 & & & = d_H \\
& & x_2 & & & +y_2 & & = d_P \\
& & & x_3 & & & +y_3 & = d_W \\
& x_1 \geq 0, & x_2 \geq 0, & x_3 \geq 0, & y_1 \geq 0, & y_2 \geq 0, & y_3 \geq 0 &
\end{array}$$

where x_1 = number of Home PCs to build, x_2 = number of Pro PCs to build, x_3 = number of Workstation PCs to build, y_1 = number of Home PCs to purchase at wholesale price, y_2 = number of Pro PCs to purchase at wholesale price, and y_3 = number of Workstation PCs to purchase at wholesale price.

This model assumes that the demand is deterministic and thus decisions regarding production and purchasing of PCs will be determined simultaneously to meet demand.

Demand Uncertainty

However, it has been observed in the past that demand for each type of computer can vary over time. The central question is how much of each type of computer should be produced before the demands for each type are known?

The supplier has identified the following three equally likely possibilities (or scenarios) for demand in Table 8.3

Table 8.3 Demand scenarios

Demand	Scenario 1	Scenario 2	Scenario 3
Home PC	300	400	500
Pro PC	500	650	800
Workstation PC	200	300	400

For example, in Scenario 2 there is a demand of 300 units for Home PCs, 650 units for Pro PCs, and 300 units for Workstation PCs. Scenario 2 can be seen as the average demand case, Scenario 1 as the low demand case, and Scenario 3 the high demand case.

Average Case Strategy

One popular and often used method to deal with uncertainty in data is to use average values for parameters in DPM. In this case, the average demands are given by Scenario 2 where $\bar{d}_H = 400, \bar{d}_P = 650$, and $\bar{d}_W = 300$. Solving the model using the average case gives us the following optimal solution.

Optimal decisions	Produced	Purchased
Home PC	400	
Pro PC	225	4250
Workstation PC	300	
Total Cost	$413,750	

The supplier could be content with this solution if the realization of demands were going to be the expected demands. However, as a check, the supplier computes the optimal production/acquisition plans for the high, and low, demand cases which are given below.

High-Demand Strategy
Solving the DPM model for Scenario 3 only gives the following solution:

Optimal decisions	Produced	Purchased
Home PC	500	
Pro PC	150	650
Workstation PC	400	
Total Cost	$525,000	

Low-Demand Strategy
Solving the model for Scenario 1 only gives the following solution:

Optimal decisions	Produced	Purchased
Home PC	300	
Pro PC	300	200
Workstation PC	200	
Total Cost	$302,500	

Sensitivity to Demand
It is clear that a strategy that assumes demand values to be at their average values would not perform well if demands actual follow scenario 1 or 3 since the optimal solutions for the latter cases are much different than the average case optimal solution. If production plans are executed according to the optimal scenario 2 (average demand) plan and then scenario 3 occurs, then the supplier will have underproduced/purchased all PCs and will have missed out on additional demand. If scenario 1 occurs, then there will be over production of all PC types. Clearly the demand values of scenarios 1 and 3 fall outside the range for which the average case optimal solution remains optimal (why?). In general, it is a challenge to assume only a single scenario for demand since the model is very sensitive to demand values. What kind of solution would perform well under any of the three scenarios given that it is in general impossible to have a perfect solution under all possible outcomes. This bring us to the motivation of stochastic programming.

8.2.1 Stochastic Programming with Recourse

The idea in stochastic programming is to incorporate uncertainty of the parameters into a model to generate a solution that performs well under the random outcomes. The premise is that a "here and now" solution has to be made before uncertainty is resolved, and the solution should be reasonable under any realization of uncertainty that might occur. In the current example, the production of the three types of computers have to be made now before uncertainty is resolved and these decisions are referred to as the first-stage decisions and will affect how many computers of each type will be purchased from other manufacturers once demand has been realized. The purchasing decisions are corrective or recourse actions in the sense that they are made to correct the imperfect first-stage decisions made earlier due to the impact of uncertainty. The uncertainty in the model will be incorporated in the stochastic model via the three scenarios.

The production decision variables x_1, x_2, and x_3 are the "here and now" or first-stage decisions. The purchasing decisions occur after demand realization and after the first-stage decisions, and so will be indexed in addition by scenarios $s = 1, 2, 3$. Therefore, we denote the purchasing (recourse) decisions as $y^-_{i,s}$ for $i = 1, 2, 3$ and $s = 1, 2, 3$. For example, $y^-_{2,3}$ represents the amount of Pro PCs purchased after realization of high demand. We also have surplus variables $y^+_{i,s}$ for $i = 1, 2, 3$ and $s = 1, 2, 3$ to account for the case when the first-stage production decision x_i exceeds the realized demand for PC type i. Then recourse variables will be non-negative.

Since the costs of production and purchasing should relate to uncertain demand levels, a reasonable objective of the supplier is to minimize expected costs where in addition to costs of production at the first stage, there is cost due to purchasing in a scenario if there is a shortfall of production from the first stage and a cost of excess production if the first-stage decision exceeds the realized demand for a PC type in a scenario. Then, the stochastic programming model with recourse for the PC supplier is

$$
\begin{aligned}
\textit{minimize}\ & 225x_1 + 350x_2 + 250x_3 && (8.1)\\
& +\tfrac{1}{3}(275y^-_{1,1} + 50y^+_{1,1} + 400y^-_{2,1} + 50y^+_{2,1} + 300y^-_{3,1} + 50y^+_{3,1}) && (8.2)\\
& +\tfrac{1}{3}(275y^-_{1,2} + 50y^+_{2,2} + 400y^-_{2,2} + 50y^+_{2,2} + 300y^-_{3,2} + 50y^+_{3,2}) && (8.3)\\
& +\tfrac{1}{3}(275y^-_{1,3} + 50y^+_{1,3} + 400y^-_{2,3} + 50y^+_{2,3} + 300y^-_{3,3} + 50y^+_{3,3}) && (8.4)
\end{aligned}
$$

subject to

$$
\begin{aligned}
& x_1 + x_2 + x_3 \le 1500 && (8.5)\\
& 2x_1 + 4x_2 + x_3 \le 2000 && (8.6)\\
& x_1 \ge 0,\quad x_2 \ge 0,\ x_3 \ge 0 && (8.7)\\
& x_1 + y^-_{1,1} - y^+_{1,1} = 300 && (8.8)\\
& x_2 + y^-_{2,1} - y^+_{2,1} = 500 && (8.9)\\
& x_3 + y^{-1}_{3,1} - y^+_{3,1} = 200 && (8.10)\\
& y^-_{1,1}, y^+_{1,1}, y^-_{2,1}, y^+_{2,1}, y^-_{3,1}, y^+_{3,1} \ge 0 && (8.11)\\
& x_1 + y^-_{1,2} - y^+_{1,2} = 400 && (8.12)
\end{aligned}
$$

$$x_2 \qquad + y_{2,2}^- - y_{2,2}^+ = 650 \tag{8.13}$$

$$x_3 \qquad + y_{3,2}^{-1} - y_{3,2}^+ = 300 \tag{8.14}$$

$$y_{1,2}^-, y_{1,2}^+, y_{2,2}^-, y_{2,2}^+, y_{3,2}^-, y_{3,2}^+ \geq 0 \tag{8.15}$$

$$x_1 \qquad + y_{1,3}^- - y_{1,3}^+ = 500 \tag{8.16}$$

$$x_2 \qquad + y_{2,3}^- - y_{2,3}^+ = 800 \tag{8.17}$$

$$x_3 \qquad + y_{3,3}^{-1} - y_{3,3}^+ = 400 \tag{8.18}$$

$$y_{1,3}^-, y_{1,3}^+, y_{2,3}^-, y_{2,3}^+, y_{3,3}^-, y_{3,3}^+ \geq 0. \tag{8.19}$$

The objective function (8.1) minimizes the profit from each type of computer produced plus the expected cost (8.2)-(8.4) from purchasing PCs in scenarios of shortfall and penalties in scenarios of overproducing PCs. The first set of constraints (8.5)-(8.7) are the first-stage constraints and ensure that the amount of computers produced don't exceed the CPU and memory board supply. The next three sets of constraints determine the amount of each type of computer to purchase from other manufacturers and determines the amount of excess production under scenarios 1, 2, and 3, respectively given the first-stage production decisions x_1, x_2, and x_3. These are the second-stage constraints, and each set of constraints corresponds to a scenario realization of demands. The first scenario constraints are given by (8.8)-(8.11), the second scenario constraints are given by (8.12)-(8.15), and the third scenario constraints are given by (8.16)-(8.19). This form of the stochastic program is called the *extensive form* since the constraints corresponding to all three scenarios are completely specified. The extensive form remains as a linear program.

Solving the stochastic programming model we get the following:

The optimal first-stage production decisions are

Home PC	Pro PC	Workstation PC
300	300	200

The recourse purchasing decisions are

Scenario	Home PC	Pro PC	Workstation PC
low (s=1)		purchase 200	
average (s=2)	purchase 100	purchase 350	purchase 100
high (s=3)	purchase 200	purchase 500	purchase 200

The optimal objective function value of the stochastic program is $420,000, which represents the expected cost of first-stage and recourse decisions. The stochastic solution can be interpreted as follows: produce 300 units of Home PCs, 300 units of Pro PCs, and 200 units of Workstation PCs now, before demand has been resolved. Once demand is realized, i.e., some scenario s unfolds, then the purchasing decisions are given by the corresponding scenario dependent recourse decisions, e.g., if scenario 3 unfolds, then the supplier purchases 200 units of Home PCs, 500 units of Pro PCs, and 200 units of

Workstation PCs. Note that for this problem instance, it is never optimal to overproduce PCs in the first stage. An empty entry indicates that the first-stage production decision for that case meets demand exactly.

Observe that for any of the single scenario deterministic problems it is never optimal to purchase Home or Workstation PCs from other manufacturers. The stochastic solution, on the other hand, has the opposite property where it is optimal to purchase across all PC types. The stochastic solution as a hedge against uncertainty considers all possibilities of demand and therefore generates a balanced production and purchasing strategy across the PC types. This highlights the differences between deterministic and stochastic solutions.

8.2.1.1 Two-Stage Stochastic Programming with Recourse

The supplier stochastic programming model above is an instance of what is known as a two-stage stochastic program with recourse. This class of models represents the case where a set of decisions x must be taken without full knowledge of information regarding some of the parameters, e.g., the demand for different types of PCs, and then a second-stage set of recourse decisions y, e.g., the purchasing of PCs are made upon realization of uncertainty. Each stage has its own set of corresponding constraints, and the constraints in the second stage are further classified according to various random outcomes, i.e., scenarios. Uncertainty will be represented by the random vector ξ, which in the case for the supplier problem represents the uncertainty in demand.

Then, the general form of a two-stage stochastic program with recourse is

$$\begin{array}{ll} \textit{minimize} & c^T x + E_\xi Q(x,\xi) \\ \textit{subject to} & Ax = b \\ & x \geq 0 \end{array}$$

where

$$Q(x,\xi) = \min_y \{q^T y | Wy = h - Tx, y \geq 0\}.$$

Observe that the objective of the general problem is a minimization and in this form the terms of the objective function are considered costs (a maximization problem will have cost terms that are negative, representing benefits, e.g., profits or revenue). The first term of the objective represents any immediate cost incurred by the first-stage decision x and the second term is the expected cost of the recourse decisions obtained by solving $Q(x,\xi)$ and taking the expectation with respect to ξ, which is denoted by E_ξ. The parameters that are random come from elements in q, h, or T, which collectively form the vector ξ. It will be assumed that the number of realizations of ξ is finite. The constraints $Ax = b, x \geq 0$ represent the first-stage constraints, and without loss of generality are expressed as equality constraints except for the non-negativity restriction. The recourse constraints are represented by $Wy = h - Tx, y \geq 0$ for each realization of ξ.

Example 8.1

In the PC supplier problem, we have the following correspondence for the first stage

$$c = \begin{bmatrix} -225 \\ -350 \\ -250 \end{bmatrix}, A = \begin{bmatrix} 1 & 1 & 1 \\ 2 & 4 & 1 \end{bmatrix} \text{ and } b = \begin{bmatrix} 1500 \\ 2000 \end{bmatrix}.$$

The random vector ξ consists of only the three demand components for the 3 types of PCs. We let $\xi(s)$ represent the realization of ξ corresponding to scenario s ($s = 1, 2, 3$) and thus,

$$\xi(1) = \begin{bmatrix} \xi_1(1) \\ \xi_2(1) \\ \xi_3(1) \end{bmatrix} = \begin{bmatrix} 300 \\ 500 \\ 200 \end{bmatrix}, \xi(2) = \begin{bmatrix} \xi_1(2) \\ \xi_2(2) \\ \xi_3(2) \end{bmatrix} = \begin{bmatrix} 400 \\ 650 \\ 300 \end{bmatrix}$$

and

$$\xi(3) = \begin{bmatrix} \xi_1(3) \\ \xi_2(3) \\ \xi_3(3) \end{bmatrix} = \begin{bmatrix} 500 \\ 800 \\ 400 \end{bmatrix}.$$

Then, the recourse function $Q(x, \xi(s)) =$

$$\begin{array}{ll} \textit{minimize} & (275y^-_{1,s} + 50y^+_{1,s} + 400y^-_{2,s} + 50y^+_{2,s} + 300y^-_{3,s} + 50y^+_{3,s}) \\ \textit{subject to} & y^-_{1,s} - y^+_{1,s} = \xi_1(s) - x_1 \\ & \qquad y^-_{2,s} - y^+_{2,s} = \xi_2(s) - x_2 \\ & \qquad\qquad y^-_{3,s} - y^+_{3,s} = \xi_3(s) - x_3 \\ & y^-_{1,s}, y^+_{1,s}, y^-_{2,s}, y^+_{2,s}, y^-_{3,s}, y^+_{3,s} \geq 0 \end{array}$$

where $q^T = \begin{bmatrix} 275 & 50 & 400 & 50 & 300 & 50 \end{bmatrix}$ and $y^T = \begin{bmatrix} y^-_{1,s} & y^+_{1,s} & y^-_{2,s} & y^+_{2,s} & y^-_{3,s} & y^+_{3,s} \end{bmatrix}$.

The matrix $W = \begin{bmatrix} 1 & -1 & 0 & 0 & 0 & 0 \\ 0 & 0 & 1 & -1 & 0 & 0 \\ 0 & 0 & 0 & 0 & 1 & -1 \end{bmatrix}$ and matrix T in this case is equal to the 3×3 identity matrix. $h^T = \begin{bmatrix} \xi_1(s) & \xi_2(s) & \xi_3(s) \end{bmatrix}$ where the random parameters are the components of vector h, which are uncertain according to the random vector ξ. In this case, all three components of h are random and all components of T are fixed, i.e., not random. Thus,

$$h - Tx = \begin{bmatrix} \xi_1(s) \\ \xi_2(s) \\ \xi_3(s) \end{bmatrix} - \begin{bmatrix} 1 & 0 & 0 \\ 0 & 1 & 0 \\ 0 & 0 & 1 \end{bmatrix} \begin{bmatrix} x_1 \\ x_2 \\ x_3 \end{bmatrix} = \begin{bmatrix} \xi_1(s) - x_1 \\ \xi_2(s) - x_2 \\ \xi_3(s) - x_3 \end{bmatrix}.$$

The expected recourse function can be written as

$$E_\xi Q(x, \xi(s)) = \sum_{s=1}^{3} p(s) Q(x, \xi(s))$$

where $p(s) = \frac{1}{3}$ is the probability of scenario s. In general, for the case where there is a finite number scenarios with a discrete probability density function $p(s)$, we have $E_\xi Q(x, \xi(s)) = \sum_{s=1}^{S} p_s Q(x, \xi(s))$ where S is the number scenarios.

Let $Q(x) = E_\xi Q(x, \xi(s))$, then the general two-stage stochastic program with recourse can be written as

$$\begin{array}{ll} \textit{minimize} & c^T x + Q(x) \\ \textit{subject to} & Ax = b \\ & x \geq 0. \end{array}$$

We can write a more compact model by representing the complete objective function as $f(x, \xi) = c^T x + Q(x, \xi)$ and writing the first-stage constraints as $X = \{x \mid Ax = b, x \geq 0\}$, then the stochastic program with recourse model can be written as

$$\min_{x \in X} E_\xi f(x, \xi).$$

Expected Value of Perfect Information (EVPI)

If it were possible for the PC supplier to know, i.e., have perfect information about which demand scenario will occur before production, then the supplier will solve the corresponding single-scenario problem. Suppose that for each production cycle the supplier will know the demand scenario and that each of the three possible scenarios, will in the long run, occur one third of the time. Thus, the long run average cost will be the average of the costs from each scenario problem, which is 1/3($302,500 + $413,750 + $525,000) = $413,750. This is the expected cost under perfect information.

Now the difference between the expected cost of the optimal solution of the stochastic program and the expected cost under perfect information is called the Expected Value of Perfect Information or EVPI, and in the case of the computer supplier, the EVPI = $420,000 − $413,750 = $6,500. This can be interpreted as the loss of profit due to uncertainty in the demand or the amount that the supplier would have to pay for perfect information.

More formally, let a single-scenario (deterministic) problem corresponding to scenario s be denoted as

$$\min_{x \in X(s)} \quad f(x, \xi(s))$$

where $\xi(s)$ is the realization of uncertainty (demands) under scenario s, $X(s)$ is the constraint set of the single-scenario problem under scenario s, and $f(x, \xi(s)) = c^T x + Q(x, \xi(s))$. Let $x^*(\xi(s))$ denote the optimal solution of the single-scenario problem under s. Then, the expected performance under perfect information is $E_\xi[f(x^*(\xi(s)), \xi(s))]$ where the expectation is taken over all scenarios. The notation $\xi(s)$ is used to emphasize the dependence on a scenario s.

Also, let x^* be the optimal solution of the stochastic program with recourse $\min_{x \in X} E_\xi f(x, \xi)$. Then,

$$EVPI = E_\xi f(x^*, \xi) - E_\xi[f(x^*(\xi(s)), \xi(s))].$$

Value of the Stochastic Solution

An important measure of the value of using stochastic programming with recourse is found by comparing the performance of the optimal solution obtained from the deterministic model where the average values are used for the parameters against the optimal solution obtained from the stochastic programming model. For a minimization problem, the idea is that if the latter is less than the former, there is value in considering the optimal stochastic programming solution.

For example, in the PC supplier problem, one can solve the deterministic model with the average values for the demand, which is equivalent to solving the DPM model under scenario 2. The optimal production decisions of the deterministic problem are then used as the values of the first-stage variables in the stochastic program. Then the corresponding optimal recourse variables are obtained for each scenario and the objective function value can be computed. The cost of this strategy can be compared with the cost of the optimal stochastic solution. If the cost of the stochastic solution is lower than the cost of the objective function value of the stochastic problem using the first-stage values from the optimal solution of the deterministic average case problem and implied recourse values, then there is value in the stochastic solution.

More formally, let the average-case scenario deterministic problem be denoted as

$$\min_{x \in \bar{X}} \quad f(x, \bar{\xi})$$

where $\bar{\xi}$ is the average value of the random vector ξ, $\bar{X}$ is the constraint set where the components of ξ in the constraints are set to $\bar{\xi}$, and $f(x, \bar{\xi}) = c^T x + Q(x, \bar{\xi})$. Let $x^*(\bar{\xi})$ be the optimal solution for the average case problem and let x^* be the optimal solution for the corresponding stochastic programming with recourse problem $\min_{x \in X} E_\xi f(x, \xi)$, then the value of the stochastic solution (VSS) is defined as

$$VSS = E_\xi f(x^*(\bar{\xi}), \xi) - E_\xi f(x^*, \xi).$$

A larger VSS indicates that the optimal solution from the stochastic program is more robust to uncertainty than using the first-stage values generated from the corresponding deterministic problem using average values for parameters. A smaller VSS, i.e., a value close to 0, indicates that the deterministic solution is as good as the stochastic solution.

Example 8.2

Solving DPM using scenario 2 data for the demand gives the optimal production decisions $x_1 = 400, x_2 = 225$, and $x_3 = 300$. These values represent $x^*(\bar{\xi})$ and then are used as the first-stage values in the stochastic program (8.1)-(8.19). The recourse values are then given as follows

Scenario	Home PC	Pro PC	Workstation PC
low (s=1)	surplus 100	purchase 275	surplus 100
average (s=2)		purchase 425	
high (s=3)	purchase 100	purchase 575	purchase 100

The expected recourse cost is \$192,500. Thus, $E_\xi f(x^*(\bar{\xi}), \xi)$ is the cost of the first stage 225(400)+350(225)+250(300) + \$192,500 = \$436,250. Then,

$$VSS = \$436,250 - \$420,000 = \$16,250,$$

which represents a 3.7% expected cost reduction over the average case strategy by using the optimal stochastic solution.

8.3 More Stochastic Programming Examples

Example 8.3 (Asset Liability Matching)

An insurance company has a liability that is due in 6 months that represents the claims from its policyholders. The company will invest now in 3 securities, a stock index, a bond index, and a money market fund so that the value of the portfolio will be enough to offset the liability in 6 months. The returns of each security and the value of the liability, however, are random and the uncertainty is represented by the following scenarios each occuring with probability $\frac{1}{3}$

Scenario (s)	Stock	Bond	Money market	Liability (\$)
1	$\mu_S^1 = 17\%$	$\mu_B^1 = 12\%$	$\mu_M^1 = 13\%$	$L^1 = 1,000,000$
2	$\mu_S^2 = 15\%$	$\mu_B^2 = 9\%$	$\mu_M^2 = 10\%$	$L^2 = 1,030,000$
3	$\mu_S^3 = 7\%$	$\mu_B^3 = 17\%$	$\mu_M^3 = 10\%$	$L^3 = 1,500,000$

For example, under scenario 1 the realized returns after 6 months for stocks, bonds, and money market are 8%, 15%, and 13%, respectively. The company wishes to find a lowest-cost portfolio consisting of the three securities that will offset the liability. The company has a budget of \$950,000 to invest now in the securities. Since returns and liability values are uncertain, the value of the portfolio may be over or under the value of the liability in 6 months. In addition, the company wishes to maximize the utility for the value of the portfolio in 6 months net of satisfying the liability by creating a utility for money by assigning a penalty for being short of the liability and a reward for being over the value of the liability.

Let

$$\begin{aligned}
x_S &= \text{amount of wealth invested in the stock fund}\\
x_B &= \text{amount of wealth invested in the bond fund}\\
x_M &= \text{amount of wealth invested in the money market fund}\\
surplus^s &= \text{surplus wealth after liability is met in scenario } s\\
shortfall^s &= \text{amount short of meeting liability in scenario } s.
\end{aligned}$$

The penalty per dollar short of meeting the liability q^- is \$12 dollars and the reward of every dollar over the liability is $q^+ = -\$2$ (negative since recourse is a minimization). q^+ represents preference in achieving a surplus and q^- aversion toward a shortfall in the liability. Then, the extensive form of the stochastic program is

$$\begin{aligned}
\textit{minimize} \quad & x_S + x_B + x_M\\
& + \tfrac{1}{3}(q^+(surplus^1) + q^-(shortfall^1))\\
& + \tfrac{1}{3}(q^+(surplus^2) + q^-(shortfall^2))\\
& + \tfrac{1}{3}(q^+(surplus^3) + q^-(shortfall^3))
\end{aligned}$$

$$\begin{aligned}
\textit{subject to} \quad & \text{(first stage constraints)}\\
& x_S + x_B + x_M \leq 950,000\\
& x_S \geq 0, \quad x_B \geq 0, \ x_M \geq 0
\end{aligned}$$

$$\begin{aligned}
&\text{(Scenario 1 recourse constraints)}\\
&(1+\mu_S^1)x_S + (1+\mu_B^1)x_B + (1+\mu_M^1)x_M - surplus^1 + shortfall^1 = L^1\\
&surplus^1 \geq 0, shortfall^1 \geq 0
\end{aligned}$$

$$\begin{aligned}
&\text{(Scenario 2 recourse constraints)}\\
&(1+\mu_S^2)x_S + (1+\mu_B^2)x_B + (1+\mu_M^2)x_M - surplus^2 + shortfall^2 = L^2\\
&surplus^2 \geq 0, shortfall^2 \geq 0
\end{aligned}$$

$$\begin{aligned}
&\text{(Scenario 3 recourse constraints)}\\
&(1+\mu_S^3)x_S + (1+\mu_B^3)x_B + (1+\mu_M^3)x_M - surplus^3 + shortfall^3 = L^3\\
&surplus^3 \geq 0, shortfall^3 \geq 0.
\end{aligned}$$

Observe that this model incorporates uncertainty not only in the h vector, but the matrix T as well. Solving the stochastic program gives the first-stage solution:

Stock	Bond	Money market
$x_1 = \$0$	$x_2 = \$950000$	$x_3 = \$0$

i.e., all available money is used to invest in the bond only. Then, the recourse decisions are

	Scenario 1	Scenario 2	Scenario 3
Surplus	$64,000	$5500	$0
Shortfall	$0	$0	$388500

and the optimal objective function value is $2,457,666.67. This quantity includes the investment amount of $950,000 and expected net (penalized) amount of shortfall. If the corresponding deterministic formulation (see Exercise 8.1) uses as the return for an asset, the average over the returns of the asset over all scenarios, and for the liability, the average of the liabilities over all scenarios: then such a deterministic model will exclusively invest in the security with the highest average return, which in this case is the stock. Exercise 8.1 asks the reader to compute the value of the stochastic solution of this problem.

Example 8.4 (Transportation Problem)

Consider the problem where there is a single product type that is shipped from n plants to m warehouses. Plant i has a supply of u_i units of the product. Warehouse j has a demand of d_j units. The cost of shipping one unit from plant i to warehouse j is c_{ij}. The decision variables x_{ij} represent the number of units of product shipped from plant i to warehouse j. Recall that the objective is to minimize the cost of shipping over all possible plant-warehouse pairs. The LP is

$$\begin{array}{lll} \textit{minimize} & \sum_{i=1}^{n}\sum_{j=1}^{m} c_{ij}x_{ij} & \\ \textit{subject to} & \sum_{j=1}^{m} x_{ij} \leq u_i & i=1,...,n \\ & \sum_{i=1}^{n} x_{ij} \geq d_j & j=1,...,m \\ & x_{ij} \geq 0 & i=1,...,n,\ j=1,...,m. \end{array}$$

Now consider the situation that demands at the warehouses are random and that due to the time requirements of transporting from plants to warehouse that the shipping decisions have to be made before demand is known. We would like to formulate a stochastic program so that the shipping decisions can perform well under uncertain demand. There are two cases to consider under uncertainty.

(1) If the amount shipped from a plant i to a warehouse j is less than the realized demand at the warehouse, then there will be a shortage and we assume that the extra supply needed to fully meet demand can be procured at extra cost, e.g., through express delivery from another source.

(2) If the amount shipped from a plant i to a warehouse j is more than the realized demand at the warehouse, then there will be a surplus of products and we assume that the extra supply will be kept at the warehouse at extra holding cost.

Let $d^s = \begin{bmatrix} d_1^s \\ \vdots \\ d_m^s \end{bmatrix}$ be a vector of realization of demands at warehouses for scenario s ($s = 1, ..., S$) where d_j^s is the demand at warehouse j under scenario s. The probability of occurrence of scenario s is $p(s)$.

The stochastic program will minimize the cost of the first-stage shipping decisions x_{ij} and the expected cost of being under or over demand at warehouses after realization of uncertainty.

The first-stage constraints are to ensure that any amount of the product shipped from a plant to warehouses does not exceed its supply.

The second-stage recourse decisions are defined as follows:

$$y_{j,s}^{+} = \text{surplus at warehouse } j \text{ under scenario } s$$
$$y_{j,s}^{-} = \text{shortage at warehouse } j \text{ under scenario } s.$$

A penalty of $k_j^+ (k_j^-)$ will be incurred for every unit that is above (below) demand. Let x denote the first-stage shipping decisions, i.e., the components of x are x_{ij} for $i = 1, ..., n$ and $j = 1, ..., m$. Then, the scenario s recourse problem $Q(x, s) =$

$$\begin{array}{lll} \textit{minimize} & \sum\limits_{j=1}^{m} k_j^+ y_{j,s}^+ + \sum\limits_{j=1}^{m} k_j^- y_{j,s}^- & \\ \textit{subject to} & \sum\limits_{i=1}^{n} x_{ij} + y_{j,s}^- - y_{j,s}^+ = d_j^s & j = 1, ..., m \\ & y_{j,s}^+ \geq 0,\ y_{j,s}^- \geq 0 & j = 1, ..., m. \end{array}$$

The two-stage stochastic program transportation problem can now be written as

$$\begin{array}{lll} \textit{minimize} & \sum\limits_{i=1}^{n} \sum\limits_{j=1}^{m} c_{ij} x_{ij} + \sum\limits_{s=1}^{S} p(s) Q(x, s) & \\ \textit{subject to} & \sum_{j=1}^{m} x_{ij} \leq u_i & i = 1, ..., n \\ & x_{ij} \geq 0 & i = 1, ..., n,\ j = 1, ..., m. \end{array}$$

Note that the recourse decisions $y_{j,s}^+$ and $y_{j,s}^-$ are determined once demand is known. In this case, the model is called a stochastic program with simple recourse. Observe that the PC production stochastic programming model is also a simple recourse model.

8.3.1 Solving Two-Stage Stochastic Programs with Recourse

We now consider solution strategies for two-stage stochastic programs with recourse. Assume that the random vector ξ has a finite number of outcomes and the possible outcomes are indexed by $s = 1, ..., S$. Let p_s be the probability that outcome s occurs. The parameters q, T, W, and h and decision variables y in the recourse problem $Q(x, s)$ will be indexed by s. Then, the two-stage stochastic program with recourse can be written as

$$
\begin{array}{lll}
\textit{minimize} & c^T x + \sum_{s=1}^{S} p_s q_s^T y_s & \\
\textit{subject to} & Ax = b & \\
& T_s x + W y_s = h_s & s = 1, ..., S \\
& x \geq 0, y_s \geq 0 & s = 1, ..., S.
\end{array}
$$

This model is equivalent to the extensive form of the stochastic program. When the number of scenarios is very large, the model becomes a large-scale linear program. However, the constraints exhibit block structure, as can be seen from the following expansion of the extensive form:

$$
\begin{array}{lllllll}
\textit{minimize} & c^T x & + p_1 q_1 y_1 & + p_2 q_2 y_2 & + \cdots + & p_S q_S y_S & \\
\textit{subject to} & Ax & & & & & = b \\
& T_1 x & + \ W_1 y_1 & & & & = h_1 \\
& T_2 x & & + \ W_2 y_2 & & & = h_2 \\
& \vdots & & & \ddots & & \vdots \\
& T_S x & & & & + W_S y_S & = h_S \\
& x \geq 0 & y_1 \geq 0 & y_2 \geq 0 & \cdots & y_S \geq 0. &
\end{array}
$$

If one takes the dual of the problem above where π_0 is the dual variable associated the first-stage constraints and π_s is the dual variable associated with the set of constraints in the recourse problem $Q(x, s)$, then the dual is

$$
\begin{array}{lllllll}
\textit{maximize} & b^T \pi_0 & + h_1^T \pi_1 & + h_2^T \pi_2 & + \cdots + & h_S^T \pi_3 & \\
\textit{subject to} & A^T \pi_0 & + T_1^T \pi_1 & + T_2^T \pi_2 & + \cdots + & T_S^T \pi_S & \leq c \\
& & + W_1^T \pi_1 & & & & \leq p_1 q_1 \\
& & & + W_2^T \pi_2 & & & \leq p_2 q_2 \\
& & & & \ddots & & \vdots \\
& & & & & + W_S^T \pi_S & \leq p_S q_S.
\end{array}
$$

8.3.1.1 Dantzig-Wolfe Decomposition for Stochastic Programming

Observe that the dual has block angular structure that will be readily amenable to Dantzig-Wolfe decomposition. Thus, one strategy to solve two-stage stochastic programs with recourse is to take the dual of the extensive form and apply the Dantzig-Wolfe decomposition. However, if the number of columns in the extensive form is much larger than the number of rows, then the dual problem will have a much larger basis than the basis of the primal problem. This will result in increased computational effort due to having to factor and store larger basis matrices in the revised simplex method.

8.3.1.2 L-Shaped Method (Constraint Generation)

An alternative is to solve the primal problem (extensive form) by using a constraint generation as opposed to a column generation (Dantzig-Wolfe) approach. Analogous to Dantzig-Wolfe decomposition, the L-Shaped method of

Van Slyke and Wets (1969) provides a decomposition of the primal problem into a master problem and subproblems, but generates constraints. The motivation for the master problem is to approximate the expected recourse function $Q(x)$ by creating an outer approximation. The master problem will generate a first-stage decision x. The subproblems are the recourse problems and will take the x from the master problem and then find the best recourse decisions. For any subproblem that is infeasible at x (i.e., it is impossible to generate recourse decisions using x), a feasibility constraint is added to the master problem. If all subproblems have finite optimal recourse at x, then a constraint is added to the master problem if the resulting recourse decisions and x do not constitute an optimal solution to the problem. If decisions are optimal, the decomposition stops, else the process iterates. The steps of the L-Shaped method are detailed below and the presentation is based on Birge and Louveaux (2011).

L-Shaped Method

Step 0: (Initialization) Let $k = t = v = 0$.

Step 1: (Solve Master Problem) Let $v = v + 1$ and solve the master problem MP_v

$$\begin{array}{lll} \textit{minimize} & z = c^T x + \theta & \\ \textit{subject to} & Ax = b & \\ & D_l x \geq d_l,\ l = 1, ..., k & \text{(feasibility constraints)} \\ & E_l x + \theta \geq e_l,\ l = 1, ..., t & \text{(optimality constraints)} \\ & x \geq 0, \theta \text{ unrestricted.} & \end{array}$$

Let x^v and θ^v be an optimal solution to the master problem. If there are no optimality constraints, i.e., $t = 0$, then $\theta^v = -\infty$ and only x^v is solved for. Go to Step 2.

Step 2: (Recourse Feasibility Test) For $s = 1, ..., S$, solve the recourse feasibility linear program RF_s

$$\begin{array}{ll} \textit{minimize} & f_s = e^T y^+ + e^T y^- \\ \textit{subject to} & Wy + Iy^+ - Iy^- = h_s - T_s x^v \\ & y \geq 0, y^+ \geq 0, y^- \geq 0 \end{array}$$

until there is an index s such that the optimal solution for RF_s gives $f_s > 0$, then let σ^v be the dual multipliers associated the constraints of RF_s. Let $D_{k+1} = \sigma^v T_s$ and $d_{k+1} = \sigma^v h_s$ and set $k = k + 1$ and add the constraint

$$D_k x \geq d_k$$

to the master problem and go to Step 1.

Else, if $f_s = 0$ at the optimal solution for RF_s for all $s = 1, ..., S$, then go to Step 3.

Step 3: (Solve Subproblems) For each $s = 1, ..., S$ solve the recourse sub-problem RS_s

$$\begin{array}{ll} \textit{minimize} & w^s = q_s^T y \\ \textit{subject to} & Wy = h_s - T_s x^v \\ & y \geq 0. \end{array}$$

Let π_s^v be the simplex multipliers corresponding to the optimal solution of RS_s and

$E_{t+1} = \sum\limits_{s=1}^{S} p_s \pi_s^v T_s$ and $e_{t+1} = \sum\limits_{s=1}^{S} p_s \pi_s^v h_s$. Now let $w^v = e_{t+1} - E_{t+1} x^v$.

If $\theta^v \geq w^v$, then STOP with x^v as an optimal solution, else let $t = t + 1$ and add the constraint $E_t x + \theta \geq e_t$ to the master problem and go to Step 1.

Example 8.5

Consider the two-stage stochastic program with recourse below:

$$\begin{array}{lllll} \textit{minimize} & 2x_1 + 3x_2 + E_\xi(q^T y) & & & \\ \textit{subject to} & x_1 + x_2 & & & \leq 120 \\ & x_1 & & & \geq 40 \\ & \quad x_2 & & & \geq 25 \\ & & 3y_1 + 5y_2 & & \leq 30x_1 \\ & & y_1 + 0.625y_2 & & \leq 10x_2 \\ & & y_1 & & \leq \mathbf{r}_1 \\ & & & y_2 & \leq \mathbf{r}_2 \\ & y_1 \geq 0, y_2 \geq 0 & & & \end{array}$$

where $q^T = \begin{bmatrix} q_1 & q_2 \end{bmatrix}$ and $y^T = \begin{bmatrix} y_1 & y_2 \end{bmatrix}$. The random vector is $\xi = \begin{bmatrix} \mathbf{q}_1 \\ \mathbf{q}_2 \\ \mathbf{r}_1 \\ \mathbf{r}_2 \end{bmatrix}$ where scenario $s = 1$ is such that $\xi(1) = \begin{bmatrix} -12 \\ -14 \\ 500 \\ 100 \end{bmatrix}$ and scenario $s = 2$ is such that $\xi(2) = \begin{bmatrix} -14 \\ -16 \\ 300 \\ 300 \end{bmatrix}$. Set $p(1) = 0.4$ and $p(2) = 0.6$.

The recourse function $Q(x, \xi) =$

$$\begin{array}{llll} \textit{minimize} & \mathbf{q}_1 y_1 + \mathbf{q}_2 y_2 & & \\ \textit{subject to} & 3y_1 + 5y_2 & & \leq 30x_1 \\ & y_1 + 0.625y_2 & & \leq 10x_2 \\ & y_1 & & \leq \mathbf{r}_1 \\ & & y_2 & \leq \mathbf{r}_2 \\ & y_1 \geq 0, y_2 \geq 0. & & \end{array}$$

The random quantities are in bold. Note that for any vector x the recourse problems for each scenario (realization of uncertainty) are feasible since setting $y = 0$ will always be feasible for the recourse problems.

Finally, we have

$$W = \begin{bmatrix} 3 & 5 \\ 1 & 0.6250 \\ 1 & 0 \\ 0 & 1 \end{bmatrix} \text{ and } T = \begin{bmatrix} -30 & 0 \\ 0 & -10 \\ 0 & 0 \\ 0 & 0 \end{bmatrix}.$$

$$h = \begin{bmatrix} 0 \\ 0 \\ \mathbf{r}_1 \\ \mathbf{r}_2 \end{bmatrix} \text{ so } h_1 = \begin{bmatrix} 0 \\ 0 \\ 500 \\ 100 \end{bmatrix} \text{ and } h_2 = \begin{bmatrix} 0 \\ 0 \\ 300 \\ 300 \end{bmatrix}.$$

Step 0: Set $k = t = v = 0$.

Iteration 1

Step 1: Set $v = 1$. There are no feasibility or optimality constraints, so $\theta^1 = -\infty$ and the master problem is

$$\begin{array}{lll} \textit{minimize} & 2x_1 + 3x_2 & \\ \textit{subject to} & x_1 + x_2 & \leq 120 \\ & x_1 & \geq 40 \\ & \quad x_2 & \geq 25. \end{array}$$

The optimal solution is $x^1 = \begin{bmatrix} x_1 \\ x_2 \end{bmatrix} = \begin{bmatrix} 40 \\ 25 \end{bmatrix}$.

Step 2: Not necessary.

Step 3: Solve recourse sub-problems given x^1.

For $\xi(1)$, the recourse problem is

$$\begin{array}{lll} \textit{minimize} & -12y_1 - 14y_2 & \\ \textit{subject to} & 3y_1 + 5y_2 & \leq 1200 \\ & y_1 + 0.625y_2 & \leq 250 \\ & y_1 & \leq 500 \\ & \quad y_2 & \leq 100 \\ & y_1 \geq 0, y_2 \geq 0. & \end{array}$$

The optimal objective value is $w_1 = -3650$ and solution $y = \begin{bmatrix} 187.5 \\ 100 \end{bmatrix}$ and $\pi_1^T = \begin{bmatrix} 0 & -12 & 0 & -6.5 \end{bmatrix}$.

For $\xi(2)$, the recourse problem is

$$\begin{array}{lll} \textit{minimize} & -14y_1 - 16y_2 & \\ \textit{subject to} & 3y_1 + 5y_2 & \leq 1200 \\ & y_1 + 0.625y_2 & \leq 250 \\ & y_1 & \leq 300 \\ & \quad y_2 & \leq 300 \\ & y_1 \geq 0, y_2 \geq 0. & \end{array}$$

The optimal objective value is $w_2 = -4544$ and solution $y = \begin{bmatrix} 160 \\ 144 \end{bmatrix}$ and $\pi_2^T = \begin{bmatrix} -2.32 & -7.04 & 0 & 0 \end{bmatrix}$.

Now $e_1 = 0.4\pi_1^T h_1 + 0.6\pi_2^T h_2 = 0.4(-605) + 0.6(0) = -260$ and

$$E_1 = 0.4\pi_1^T T + 0.6\pi_2^T T$$

$$= 0.4 \begin{bmatrix} 0 & -12 & 0 & -6.5 \end{bmatrix} \begin{bmatrix} -30 & 0 \\ 0 & -10 \\ 0 & 0 \\ 0 & 0 \end{bmatrix} +$$

$$0.6 \begin{bmatrix} -2.32 & -7.04 & 0 & 0 \end{bmatrix} \begin{bmatrix} -30 & 0 \\ 0 & -10 \\ 0 & 0 \\ 0 & 0 \end{bmatrix}$$

$$= \begin{bmatrix} 41.76 & 90.24 \end{bmatrix}.$$

Then, $w^1 = e_1 - E_1 x^1 = -260 - \begin{bmatrix} 41.76 & 90.24 \end{bmatrix} \begin{bmatrix} 40 \\ 25 \end{bmatrix} = -4186.4 \geq \theta^1 = -\infty$.

So add constraint $E_1 x + \theta \geq e_1$ i.e., $41.76x_1 + 90.24x_2 + \theta \geq -260$ to the master problem and go to Step 1.

Iteration 2

Step 1: Set $v = 2$. The master problem is

$$\begin{array}{lll} \textit{minimize} & 2x_1 + 3x_2 + \theta & \\ \textit{subject to} & x_1 + x_2 & \leq 120 \\ & x_1 & \geq 40 \\ & \quad x_2 & \geq 25 \\ & 41.76x_1 + 90.24x_2 + \theta & \geq -260. \end{array}$$

The optimal solution is $x^2 = \begin{bmatrix} x_1 \\ x_2 \end{bmatrix} = \begin{bmatrix} 40 \\ 80 \end{bmatrix}$ and $\theta^2 = -9149.6$.

Step 2: Not necessary.

Step 3: Solve recourse sub-problems given x^2.

For $\xi(1)$, the recourse problem is

$$\begin{array}{lll} \textit{minimize} & -12y_1 - 14y_2 & \\ \textit{subject to} & 3y_1 + 5y_2 & \leq 1200 \\ & y_1 + 0.625y_2 & \leq 800 \\ & y_1 & \leq 500 \\ & \quad y_2 & \leq 100 \\ & y_1 \geq 0, y_2 \geq 0. & \end{array}$$

The optimal objective value is $w_1 = -4800$ and solution $y^1 = \begin{bmatrix} 400 \\ 0 \end{bmatrix}$ and $\pi_1^T = \begin{bmatrix} -4 & 0 & 0 & 0 \end{bmatrix}$.

For $\xi(2)$, the recourse problem is

$$\begin{array}{lll} minimize & -14y_1 - 16y_2 & \\ subject\ to & 3y_1 + 5y_2 & \leq 1200 \\ & y_1 + 0.625y_2 & \leq 800 \\ & y_1 & \leq 300 \\ & \qquad\qquad y_2 & \leq 300 \\ & y_1 \geq 0, y_2 \geq 0. & \end{array}$$

The optimal objective value is $w_2 = -5160$ and solution $y^2 = \begin{bmatrix} 300 \\ 60 \end{bmatrix}$ and $\pi_2^T = \begin{bmatrix} -3.2 & 0 & -4.4 & 0 \end{bmatrix}$.

Now $e_2 = 0.4\pi_1^T h_1 + 0.6\pi_2^T h_2 = 0.4(0) + 0.6(-1320) = -792$ and

$$E_2 = 0.4\pi_1^T T + 0.6\pi_2^T T$$

$$= 0.4\begin{bmatrix} -4 & 0 & 0 & 0 \end{bmatrix}\begin{bmatrix} -30 & 0 \\ 0 & -10 \\ 0 & 0 \\ 0 & 0 \end{bmatrix} +$$

$$0.6\begin{bmatrix} -3.2 & 0 & -4.4 & 0 \end{bmatrix}\begin{bmatrix} -30 & 0 \\ 0 & -10 \\ 0 & 0 \\ 0 & 0 \end{bmatrix}$$

$$= \begin{bmatrix} 105.6 & 0 \end{bmatrix}.$$

Then, $w^2 = e_2 - E_2 x^2 = -792 - \begin{bmatrix} 105.6 & 0 \end{bmatrix}\begin{bmatrix} 40 \\ 80 \end{bmatrix} = -5016$ so $w^2 \geq \theta^2 = -9149.6$.

So add constraint $E_2 x + \theta \geq e_2$ i.e., $105.6x_1 + \theta \geq -792$ to the master problem and go to Step 1.

Iteration 3

Step 1: Set $v = 3$. The master problem is

$$\begin{array}{lll} minimize & 2x_1 + 3x_2 + \theta & \\ subject\ to & x_1 + x_2 & \leq 120 \\ & x_1 & \geq 40 \\ & \qquad x_2 & \geq 25 \\ & 41.76x_1 + 90.24x_2 + \theta & \geq -260 \\ & 105.6x_1 \qquad\qquad +\theta & \geq -792. \end{array}$$

The optimal solution is $x^3 = \begin{bmatrix} x_1 \\ x_2 \end{bmatrix} = \begin{bmatrix} 66.8276 \\ 53.1724 \end{bmatrix}$ and $\theta^3 = -7849$.

Step 2: Not necessary.
Step 3: Solve recourse sub-problems given x^3.
For $\xi(1)$, the recourse problem is

$$\begin{array}{lll} \textit{minimize} & -12y_1 - 14y_2 & \\ \textit{subject to} & 3y_1 + 5y_2 & \leq 2004.8000 \\ & y_1 + 0.625y_2 & \leq 531.7328 \\ & y_1 & \leq 500 \\ & \qquad\qquad y_2 & \leq 100 \\ & y_1 \geq 0, y_2 \geq 0. & \end{array}$$

The optimal objective value is $w_1 = -7030.7000$ and solution $y^1 = \begin{bmatrix} 469.2238 \\ 100 \end{bmatrix}$ and $\pi_1^T = \begin{bmatrix} 0 & -12 & 0 & -6.5 \end{bmatrix}$.
For $\xi(2)$, the recourse problem is

$$\begin{array}{lll} \textit{minimize} & -14y_1 - 16y_2 & \\ \textit{subject to} & 3y_1 + 5y_2 & \leq 2004.8000 \\ & y_1 + 0.625y_2 & \leq 531.7328 \\ & y_1 & \leq 300 \\ & \qquad\qquad y_2 & \leq 300 \\ & y_1 \geq 0, y_2 \geq 0. & \end{array}$$

The optimal objective value is $w_2 = -7735.5$ and solution $y^2 = \begin{bmatrix} 300 \\ 220.9657 \end{bmatrix}$ and $\pi_2^T = \begin{bmatrix} -3.2 & 0 & -4.4 & 0 \end{bmatrix}$.

Now $e_3 = 0.4\pi_1^T h_1 + 0.6\pi_2^T h_2 = 0.4(-650) + 0.6(-1320) = -1052$ and

$$E_3 = 0.4\pi_1^T T + 0.6\pi_2^T T = \begin{bmatrix} 57.6 & 48 \end{bmatrix}.$$

Then, $w^2 = e_3 - E_3 x^3 = -1052 - \begin{bmatrix} 57.6 & 48 \end{bmatrix} \begin{bmatrix} 66.8276 \\ 53.1724 \end{bmatrix} = -7453.5 \geq \theta^3 = -7849$.

So add constraint $E_3 x + \theta \geq e_3$ i.e., $57.6x_1 + 48x_2 + \theta \geq -1052$ to the master problem and go to Step 1.

Iteration 4
Step 1: Set $v = 4$. The master problem is

$$
\begin{array}{lllll}
\textit{minimize} & 2x_1 + 3x_2 + \theta & & & \\
\textit{subject to} & x_1 + x_2 & & & \leq 120 \\
& x_1 & & & \geq 40 \\
& & x_2 & & \geq 25 \\
& 41.76x_1 + 90.24x_2 + \theta & & & \geq -242 \\
& 105.6x_1 & & + \theta & \geq -792 \\
& 57.6x_1 + 48x_2 & & +\theta & \geq -1052.
\end{array}
$$

The optimal solution is $x^4 = \begin{bmatrix} x_1 \\ x_2 \end{bmatrix} = \begin{bmatrix} 73.6364 \\ 46.3636 \end{bmatrix}$ and $\theta^4 = -7518.9$.

Step 2: Not necessary.
Step 3: Solve recourse sub-problems given x^4.
For $\xi(1)$, the recourse problem is

$$
\begin{array}{llll}
\textit{minimize} & -12y_1 - 14y_2 & & \\
\textit{subject to} & 3y_1 + 5y_2 & & \leq 2209.1000 \\
& y_1 + 0.625y_2 & & \leq 463.6364 \\
& y_1 & & \leq 500 \\
& & y_2 & \leq 100 \\
& y_1 \geq 0, y_2 \geq 0. & &
\end{array}
$$

The optimal objective value is $w_1 = -6213.6$ and solution $y^1 = \begin{bmatrix} 401.1364 \\ 100 \end{bmatrix}$ and $\pi_1^T = \begin{bmatrix} 0 & -12 & 0 & -6.5 \end{bmatrix}$.
For $\xi(2)$, the recourse problem is

$$
\begin{array}{llll}
\textit{minimize} & -14y_1 - 16y_2 & & \\
\textit{subject to} & 3y_1 + 5y_2 & & \leq 2209.1000 \\
& y_1 + 0.625y_2 & & \leq 463.6364 \\
& y_1 & & \leq 500 \\
& & y_2 & \leq 100 \\
& y_1 \geq 0, y_2 \geq 0. & &
\end{array}
$$

The optimal objective value is $w_2 = -8389.1$ and solution $y^2 = \begin{bmatrix} 300 \\ 261.8182 \end{bmatrix}$ and $\pi_2^T = \begin{bmatrix} -2.7908 & -3.2735 & -2.3540 & 0 \end{bmatrix}$.

Now $e_4 = 0.4\pi_1^T h_1 + 0.6\pi_2^T h_2 = 0.4(-650) + 0.6(-706.2) = -683.7200$ and

$$E_4 = 0.4\pi_1^T T + 0.6\pi_2^T T = \begin{bmatrix} 50.2345 & 67.6412 \end{bmatrix}.$$

Then, $w^4 = e_4 - E_4 x^4 = -683.7200 - \begin{bmatrix} 41.76 & 90.24 \end{bmatrix} \begin{bmatrix} 73.6364 \\ 46.3636 \end{bmatrix} = -7518.9$, so now $w^4 \leq \theta^4 = -7518.9$ so STOP.

Therefore, $x^4 = \begin{bmatrix} 73.6364 \\ 46.3636 \end{bmatrix}$, $y^1 = \begin{bmatrix} 401.1364 \\ 100 \end{bmatrix}$, and $y^2 = \begin{bmatrix} 300 \\ 261.8182 \end{bmatrix}$ is an optimal solution.

8.3.1.3 Convergence of the L-Shaped Method

The convergence of the L-Shaped method follows from the fact that it is equivalent to applying the Dantzig-Wolfe decomposition on the dual of the primal extensive form of the stochastic program. To see this, consider the dual of the master problem MP_v in the L-Shaped method.

$$\begin{array}{lll} maximize & \rho^T b + \sum_{l=1}^{k} \sigma_l d_l + \sum_{l=1}^{t} \pi_l e_l & \\ subject\ to & A^T \rho + \sum_{l=1}^{k} D_l \sigma_l + \sum_{l=1}^{t} \pi_l E_l & \leq c \\ & \sum_{l=1}^{t} \pi_l & = 1 \\ & \sigma_l \geq 0, \quad l = 1, ..., k & \\ & \pi_l \geq 0, \quad l = 1, ..., t. & \end{array}$$

The dual of subproblem RS_s is

$$\begin{array}{lll} maximize & (h_s - T_s x^v)^T \pi & \\ subject\ to & W^T \pi & \leq q. \end{array}$$

The L-Shaped method on the dual of the stochastic program is described in the following steps. We assume that the stochastic program is feasible, so the original Step 2 is omitted.

Step 1 (Master Problem) Consists of solving the dual of the master problem MP_v to obtain the solution ρ^v, σ^v, π^v and the solution x^v and θ^v for MP_v.

Step 2 (Subproblems) Consists of solving the duals of RS_s.

If a dual of a subproblem RS_j is unbounded, then there is a vector (extreme direction) σ^v such that $W^T \sigma^v \leq 0$ and $(h_j - T_j x^v)^T \sigma^v > 0$. Then, construct the columns $D_{k+1} = (\sigma^v)^T T_j$ and $d_{k+1} = (\sigma^v)^T h_j$ and add to the dual of MP_v. Observe that adding the constraint $D_{k+1} x \geq d_{k+1}$ to MP_v is equivalent to adding the columns D_{k+1} and d_{k+1} to the dual of MP_v.

If all duals of subproblems RS_s have finite optimal solutions, then construct the columns $E_{t+1} = \sum_{s=1}^{S} p_s \pi_s^v T_s$ and $e_{t+1} = \sum_{s=1}^{S} p_s \pi_s^v h_s$. Now let $w^v = e_{t+1} - E_{t+1} x^v$.

If $\theta^v \geq w^v$, then STOP with x^v, θ^v as an optimal solution to MP_v and ρ^v, σ^v, π^v as an optimal solution to the dual, else let $t = t + 1$ and add the columns E_{t+1} and e_{t+1} to the dual of MP_v (which is equivalent to adding $E_t x + \theta \geq e_t$ to MP_v) and go to Step 1.

In summary, the steps above are equivalent to Dantzig-Wolfe decomposition where the master problem is the dual of the master problem MP_v in the L-Shaped method, and columns are generated by the dual of the subproblems RS_s and added to the dual of MP_v, which is equivalent to generating constraints by the subproblems RS_s and adding to MP_v. Dantzig-Wolfe decomposition converges since it is the revised simplex method (assuming appropriate anti-cycling rules are in place), so thus the L-Shaped method converges.

8.4 Robust Optimization

We now consider an alternative approach to handle uncertainty in parameter values in linear programming called robust linear optimization. Consider the linear program

$$\begin{array}{ll} \textit{minimize} & c_1x_1 + c_2x_2 \\ \textit{subject to} & a_1x_1 + a_2x_2 \geq b_1. \end{array}$$

Let the instance $c_1 = 2$, $c_2 = 3$, $a_1 = 2, a_2 = 1$, and $b_1 = 1$ be called the nominal problem. The optimal solution is $x_1 = 0.5$ and $x_2 = 0$ with an objective function value of 1. Now suppose that the values for a_1 and a_2 are only estimates and can be inaccurate, and that the actual values that realize are $a_1 = 1.99$ and $a_2 = 1.01$. Then the optimal solution of the nominal problem is no longer feasible for this realization. This is problematic in situations where a decision based on a model has to be taken here and now and the constraints are hard in the sense that they must be satisfied after realization of actual values of parameters.

The motivation for robust optimization, like for stochastic programming, is to protect against inaccuracies that arise in the estimation of the parameters for a linear program in the context where a here and now solution must be taken before the inaccuracies are resolved, but unlike in stochastic programming, a probability distribution on possible outcomes is not necessary. Instead, an uncertainty set $\mathcal{U}$ is specified that captures the possible realizations of the parameters that are deemed to be uncertain. A robust counterpart *RC* of a linear program is the original problem but with the additional requirement that any feasible vector x must satisfy all constraints including each set of constraints corresponding to a possible realization of the uncertain parameters from the set $\mathcal{U}$.

For example, the robust counterpart of the linear program above, where the constraint coefficients a_1 and a_2 are considered uncertain, is

$$\begin{array}{ll} \textit{minimize} & 2x_1 + 3x_2 \\ \textit{subject to} & a_1x_1 + a_2x_2 \geq b_1 \\ & \forall (a_1, a_2) \in \mathcal{U} \end{array}$$

where $\mathcal{U}$ is a subset of $R^{1\times 2}$, i.e., a subset of the space consisting of 1×2 matrices where entries of matrix are real numbers. It is important to note that a vector $x^T = \begin{bmatrix} x_1 & x_2 \end{bmatrix}$ must satisy the constraint $a_1x_1 + a_2x_2 \geq b_1$ constraint for all (a_1, a_2) in $\mathcal{U}$. Suppose that $\mathcal{U} = \{ \begin{bmatrix} 1.99 \\ 0.99 \end{bmatrix}^T, \begin{bmatrix} 2.00 \\ 1.00 \end{bmatrix}^T, \begin{bmatrix} 2.01 \\ 1.01 \end{bmatrix}^T \}$, then the robust counterpart of the nominal linear program is the following linear program

$$\begin{array}{ll} \textit{minimize} & 2x_1 + 3x_2 \\ \textit{subject to} & 1.99x_1 + 0.99x_2 \geq 1 \\ & 2x_1 + 1x_2 \geq 1 \\ & 2.01x_1 + 1.01x_2 \geq 1. \end{array}$$

The optimal solution to the robust counterpart is $\bar{x}_1 = 0.5025$ and $\bar{x}_2 = 0.0000$ with a corresponding objective function value of 1.0050. The solution has the advantage of satisfying all of the constraints without increasing the objective function too much. This is a desirable property of a solution when it is important to satisfy constraints that involve inaccurate or uncertain parameters. In this case, we say the solution is robust or immune to uncertainty.

In general, for a linear program of the form

$$\begin{array}{ll} \textit{minimize} & c^T x \\ \textit{subject to} & Ax \geq b \end{array}$$

where $c, x \in R^n, b \in R^m$, and $A \in R^{m \times n}$, then the robust counterpart RC is the following optimization problem

$$\begin{array}{ll} \textit{minimize} & c^T x \\ \textit{subject to} & Ax \geq b, \forall (c, A, b) \in \mathcal{U} \end{array}$$

for a given uncertainty set $\mathcal{U} \subset R^n \times R^{m \times n} \times R^m$. Here we assume that all data can be considered uncertain. A vector x that satisfies the constraints of RC is said to be a robust feasible solution.

It is not hard to show that the robust counterpart RC can be written as

$$\underset{t,x}{\textit{minimize}} \quad \{t \mid t \geq c^T x, Ax \geq b, \forall (c, A, b) \in \mathcal{U}\}.$$

8.4.1 Constraint-wise Construction of *RC*.

One can proceed by replacing each constraint in the original linear program

$$a_i^T x \geq b_i \ (ith \text{ constraint of } Ax \geq b)$$

by its corresponding robust counterpart

$$a_i^T x \geq b_i \ \forall (a_i, b_i) \in \mathcal{U}_i$$

where $\mathcal{U}_i$ is the restriction of $\mathcal{U}$ to elements only relevant to realizations of coefficients a_i and b_i of the ith constraint.

Example 8.6
Consider the linear program

$$\begin{array}{ll} \textit{minimize} & 2x_1 + 3x_2 \\ \textit{subject to} & a_{11}x_1 + a_{12}x_2 \geq b_1 \\ & a_{21}x_1 + a_{22}x_2 \geq b_2. \end{array}$$

If the constraint and the right-hand side coefficients are uncertain, then the robust counterpart RC is

$$\begin{array}{ll} \textit{minimize} & 2x_1 + 3x_2 \\ \textit{subject to} & a_{11}x_1 + a_{12}x_2 \geq b_1 \\ & a_{21}x_1 + a_{22}x_2 \geq b_2 \\ & \forall \left(\begin{bmatrix} a_{11} & a_{12} \\ a_{21} & a_{22} \end{bmatrix} \begin{bmatrix} b_1 \\ b_2 \end{bmatrix} \right) \in \mathcal{U} \subset R^{2\times 2} \times R^{2\times 1}. \end{array}$$

Note: An element of $\mathcal{U}$ is the direct product of a 2×2 matrix and a 2×1 vector, and so two sets of matrix brackets were used to highlight the dimension of the direct product. From here on such an element will be represented using only one bracket for ease of exposition, i.e.,.

$$\begin{bmatrix} a_{11} & a_{12} \\ a_{21} & a_{22} \end{bmatrix} \begin{bmatrix} b_1 \\ b_2 \end{bmatrix} = \begin{bmatrix} a_{11} & a_{12} & b_1 \\ a_{21} & a_{22} & b_2 \end{bmatrix}$$

and

$$(a_i, b_i) = (a_{i1}, a_{i2}, b_i).$$

If $\mathcal{U} =$

$$\left[\begin{bmatrix} 0.95 & 1.95 & 0.95 \\ 2.95 & 1.95 & 1.95 \end{bmatrix}, \begin{bmatrix} 1 & 2 & 1 \\ 3 & 2 & 2 \end{bmatrix}, \begin{bmatrix} 1.05 & 2.05 & 1.05 \\ 3.05 & 2.05 & 2.05 \end{bmatrix} \right],$$

then

$$\mathcal{U}_1 = \{(0.95, 1.95, 0.95), (1, 2, 1), (1.05, 2.05, 1.05)\}$$

and

$$\mathcal{U}_2 = \{(2.95, 1.95, 1.95), (3, 2, 2), (3.05, 2.05, 2.05)\}.$$

So the robust counterpart can be written as

$$\begin{array}{lll} \textit{minimize} & t & \\ \textit{subject to} & t \geq 2x_1 + 3x_2 & \\ & a_{11}x_1 + a_{12}x_2 \geq b_1 & \forall (a_{11}, a_{12}, b_1) \in \mathcal{U}_1 \\ & a_{21}x_1 + a_{22}x_2 \geq b_2 & \forall (a_{21}, a_{22}, b_2) \in \mathcal{U}_2. \end{array}$$

As we have seen, when $\mathcal{U}$ is finite the robust counterpart will remain a linear program and hence computationally tractable. The robust counterpart will be larger than the nominal problem in proportion to the number of elements in $\mathcal{U}$.

However, if the set $\mathcal{U}$ has an infinite number of elements, the number of constraints become infinite and we have what is called a semi-infinite linear program, i.e., a linear program with an infinity of constraints which is generally an intractable class of problems.

We now give an example of an uncertainty set with an uncountably infinite number of elements. Consider that the elements of $\mathcal{U}$ above can be seen to be perturbations of the element

$$\begin{bmatrix} 1 & 2 & 1 \\ 3 & 2 & 2 \end{bmatrix}$$

where the other elements of $\mathcal{U}$ can be obtained from this nominal element by uniformly adding 0.5 to each component or by subtracting 0.5 from each component. We now consider the general case where any component can be perturbed by adding or subtracting any amount from 0 up to and including 0.5. Using the nominal element above we can construct an uncertainty $\mathcal{U}$ set with an infinite number of elements as follows

$$\mathcal{U} = \left\{ \begin{bmatrix} a_{11} & a_{12} & b_1 \\ a_{21} & a_{22} & b_2 \end{bmatrix} = \begin{bmatrix} 1 & 2 & 1 \\ 3 & 2 & 2 \end{bmatrix} + \sum_{j=1}^{6} \zeta_j P_j \right\}$$

where

$$P_1 = \begin{bmatrix} 0.5 & 0 & 0 \\ 0 & 0 & 0 \end{bmatrix}, P_2 = \begin{bmatrix} 0 & 0.5 & 0 \\ 0 & 0 & 0 \end{bmatrix}, P_3 = \begin{bmatrix} 0 & 0 & 0.5 \\ 0 & 0 & 0 \end{bmatrix}$$

$$P_4 = \begin{bmatrix} 0 & 0 & 0 \\ 0.5 & 0 & 0 \end{bmatrix}, P_5 = \begin{bmatrix} 0 & 0 & 0 \\ 0 & 0.5 & 0 \end{bmatrix}, P_6 = \begin{bmatrix} 0 & 0 & 0 \\ 0 & 0 & 0.5 \end{bmatrix}$$

where $\zeta \in B = \{\zeta \in R^6 | -1 \leq \zeta_j \leq 1, j = 1, ..., 6\}$. B is called the perturbation set.

This construction results in an uncertainty set with an infinite number of elements since there will be a element corresponding to each realization of ζ_j in the interval $[-1, 1]$, which is an uncountably infinite set.

The matrices P_j indicate the jth parameter that is to be considered uncertain where the corresponding entry contains the perturbation value, which in the example above is 0.5. For example, P_1 indicates that parameter a_{11} can be perturbed by an amount between -0.5 and 0.5. The corresponding sets $\mathcal{U}_i$ can be written as

$$\mathcal{U}_1 = \{(a_1, b_1) = (1, 2, 1) + \sum_{j=1}^{3} \zeta_1^j P_1^j | \zeta_1 \in B_1\}$$

and

$$\mathcal{U}_2 = \{(a_2, b_2) = (3, 2, 2) + \sum_{j=1}^{3} \zeta_2^j P_2^j | \zeta_2 \in B_2\}$$

where

$$P_1^1 = (0.5, 0, 0), P_1^2 = (0, 0.5, 0), P_1^3 = (0, 0, 0.5),$$
$$B_1 = \{\zeta \in R^3 | -1 \leq \zeta_j \leq 1, j = 1, ..., 3\},$$
$$P_2^1 = (0.5, 0, 0), P_2^2 = (0, 0.5, 0), P_2^3 = (0, 0, 0.5),$$

and

$$B_2 = \{\zeta \in R^3 | -1 \leq \zeta_j \leq 1, j = 1, ..., 3\}.$$

In general, for the constraint $a_i^T x \geq b_i \ \forall (a_i, b_i) \in \mathcal{U}_i$ in the robust counterpart, the uncertainty set $\mathcal{U}_i$ can be written as

$$\mathcal{U}_i = \{(a_i, b_i) = (a_i^0, b_i^0) + \sum_{j=1}^{J_i} \zeta_i^j P_i^j | \zeta \in B_i\},$$

so the constraint is can be written as

$$a_i^T x \geq b_i \ \forall \{(a_i, b_i) = (a_i^0, b_i^0) + \sum_{j=1}^{J_i} \zeta_i^j P_i^j | \zeta \in B_i\}$$

where a_i^0 (b_i^0) is the nominal value of a_i (b_i), e.g., $(a_1^0, b_1^0) = (1, 2, 1)$, B_i is the perturbation set, and J_i is the number of elements in (a_i, b_i) that are to be uncertain. Observe that P_j^i can be represented in terms of the partition into entries that correspond to a_i and b_i, which we denote as a_i^j and b_i^j, respectively. So we can write $P_i^j = (a_i^j, b_i^j)$, e.g., $P_1^2 = (0, 0.5, 0)$ where $a_1^2 = (0, 0.5)$ and $b_1^2 = (0)$. From here on, the partitioned representation will be used in place of P_i^j.

When $\mathcal{U}_i$ has an infinite number of elements, the tractability of the robust counterpart will depend on the structure of $\mathcal{U}_i$. Tractability of the robust counterpart means that it can be solved to optimality efficiently. In particular, the structure of the corresponding perturbation set B_i will characterize the tractability.

For example, let $B_i = \{\zeta \in R^{J_i} | \ \|\zeta\|_\infty \leq 1\}$, then the constraint

$$a_i^T x \geq b_i \ \forall \{(a_i, b_i) = (a_i^0, b_i^0) + \sum_{j=1}^{J_i} \zeta_i^j (a_i^j, b_i^j) | \zeta \in B_i\}$$

can be written as

$$(a_i^0 + \sum_{j=1}^{J_i} \zeta_i^j a_i^j)^T x \geq (b_i^0 + \sum_{j=1}^{J_i} \zeta_i^j b_i^j) \qquad \forall \zeta \in B_i$$

or

$$\sum_{j=1}^{J_i} \zeta_i^j [(a_i^j)^T x - b_i^j] \geq b_i^0 - (a_i^0)^T x \qquad \forall (\zeta | \ |\zeta_i^j| \leq 1, \ j = 1, ..., J_i).$$

Now the inequality will hold if and only if the minimum value on the left hand side of the inequality is greater than the right-hand side for any possible realization of ζ. In other words, we need

$$\min_{-1\leq\zeta_i^j\leq 1}\left[\sum_{j=1}^{J_i}\zeta_i^j[(a_i^j)^T x - b_i^j]\right] \geq b_i^0 - (a_i^0)^T x.$$

The minimum value on the left-hand side is $-\sum_{j=1}^{J_i}|(a_i^j)^T x - b_i^j|$ and so the representation of the constraint becomes

$$-\sum_{j=1}^{J_i}|(a_i^j)^T x - b_i^j| + (a_i^0)^T x \geq b_i^0.$$

The absolute value term $|(a_i^j)^T x - b_i^j|$ can be removed by replacing, with the following inequalities

$$\begin{aligned} d_j &\geq (a_i^j)^T x - b_i^j \\ d_j &\geq -(a_i^j)^T x + b_i^j. \end{aligned}$$

So the constraint becomes the following system of linear equations:

$$\begin{aligned} -d_j \leq (a_i^j)^T x - b_i^j &\leq d_j \quad \text{for } j = 1, ..., J_i \\ -\sum_{j=1}^{J_i} d_j + (a_i^0)^T x &\geq b_i^0. \end{aligned}$$

It is important to observe that this is a finite system of equations that represents the incorporation of the uncountably infinitely many elements in the uncertainty set $\mathcal{U}_i$. Thus, the robust counterpart *RC* will remain as a linear program and is computationally tractable. We summarize the discussion as follows.

Theorem 8.7 (Interval Uncertainty)
Consider the robust counterpart RC to the linear program min $c^T x$ *subject to* $Ax \geq b$. *If* $B_i = \{\zeta \in R^{J_i} \mid \|\zeta\|_\infty \leq 1\}$ *for* $i = 1, ..., m$, *then RC is a linear program where RC* is

$$\begin{aligned} &\textit{minimize} \quad t \\ &\textit{subject to} \quad t \geq c^T x \\ &\qquad a_i^T x \geq b_i \ \forall\{(a_i, b_i) = (a_i^0, b_i^0) + \sum_{j=1}^{J_i}\zeta_i^j(a_i^j, b_i^j) | \zeta \in B_i\} \quad i = 1, ..., m \end{aligned}$$

and J_i is the number of elements in (a_i, b_i) that are uncertain.

Example 8.8

We consider robust optimization for the following portfolio optimization problem adapted from Ben Tal et al. (2009). There are n assets including a risk-free asset. We consider the last $(i = n)$ asset to be the risk-free asset with a certain return $r_n = 5\%$ and the remaining $n-1$ assets to have returns r_i which are modeled by independent random variables with mean μ_i and standard deviation σ_i given by

$$\mu_i = 1.05 + 0.3\tfrac{(n-i)}{n-1} \text{ and } \sigma_i = 1.05 + 0.6\tfrac{(n-i)}{n-1} \text{ for } i = 1, ..., n-1,$$

and each return r_i will take on values in the interval $[\mu_i - \sigma_i, \mu_i + \sigma_i]$, i.e., values within one standard deviation away from the mean. The objective is to invest among the assets, i.e., find proportion of wealth to invest in each asset so that the return of the portfolio is maximized.

The nominal portfolio optimization problem, assuming that there is no uncertainty (variance) in returns, is

$$\begin{array}{lll} maximize & r_n y_n + \sum\limits_{i=1}^{n-1} r_i y_i & \\ subject\ to & \sum\limits_{i=1}^{n} y_i = 1 & \\ & y_i \geq 0 & i = 1, ..., n. \end{array}$$

Now consider that the uncertainty in the parameters are in the returns, which can be represented as $r_i = \mu_i + \zeta_i \sigma_i$, where ζ_i can be considered an independent random perturbation such that $-1 \leq \zeta_i \leq 1$. Then, the robust counterpart RC of the nominal problem can written as

$$\underset{t,y}{\text{minimize}} \ \{-t \mid -t \geq -r_n y_n - \sum_{i=1}^{n-1} r_i y_i, \sum_{i=1}^{n} y_i = 1, y_i \geq 0 \ \forall i \text{ and } \forall r \in \mathcal{U}\}$$

for some uncertainty set $\mathcal{U}$.

Now let $x = \begin{bmatrix} y \\ -t \end{bmatrix} \in R^n$, then RC is the semi-infinite program

$$\begin{array}{lll} minimize & x_n & \\ subject\ to & (a^0 + \sum\limits_{i=1}^{n-1} \zeta_i a^i)^T x \ \leq (b^0 + \sum\limits_{i=1}^{n-1} \zeta_i b^i) \ \forall \zeta \in B & \\ & \sum\limits_{i=1}^{n} x_i = 1 & \\ & x_i \geq 0 & i = 1, ..., n \end{array}$$

where

$$B = \{\zeta \in R^{n-1} | -1 \leq \zeta_i \leq 1, \ i = 1, ..., n-1\}$$

$$b^i = 0 \text{ for } i = 0, ..., n-1$$

$$a^0 = (-\mu_1, -\mu_2 ..., -\mu_n, -1)$$

$$a^i = \sigma_i e_i^T \text{ for } i = 1, ..., n-1$$

where e_i is the vector of n components with a 1 in the ith position and 0 elsewhere.

Now suppose that $B = \{\zeta | \; \|\zeta\|_\infty \leq 1\}$, then the uncertain constraint becomes, by Theorem 8.7,

$$-t \geq -1.05y_1 - \sum_{i=2}^{n}(\mu_i - \sigma_i)y_i.$$

The robust counterpart can now be written as

$$\underset{t,y}{\text{minimize}} \quad \{-t \mid -t \geq -1.05y_n - \sum_{i=1}^{n-1}(\mu_i - \sigma_i)y_i, \sum_{i=1}^{n} y_i = 1, y_i \geq 0 \; \forall i \; \}.$$

which is a linear program.

Solving the cases when $n = 10$, 25, 50, 100, 150, and 200 each results in a optimal solution that gives a guaranteed return of 5%. The optimal non-zero decision variables are $t = 1.05$ and $y_n = 1$ over all of the instances. This should not be surprising as the robust counterpart of the uncertainty contraint based on the set $B = \{\zeta | \; \|\zeta\|_\infty \leq 1\}$ ensures that the optimal solution to the robust counterpart is immune or protected against all possible perturbations in the constraint. One many consider this case to represent the most conservative case, i.e., ensuring 100% protection from uncertainty.

If one desires less, but reasonable levels of protection against uncertainty in an uncertain constraint where ζ_i can be considered an independent random perturbation such that $-1 \leq \zeta_i \leq 1$, one can consider the perturbation set

$$B = \{\zeta \in R^{J_i} | \; \|\zeta\|_\infty \leq 1, \|\zeta\|_2 \leq \Omega\},$$

which can be interpreted as the intersection of a unit box (length of 1 on all sides) and a ball of radius Ω. It can be shown (see Ben-Tal et al. 2009) that the robust counterpart of an uncertain constraint using B is equivalent to the following finite set of constraints

$$z_j + w_j = b^j - (a^j)^T x, \; j = 1, ..., J_i$$
$$\sum_{j=1}^{J_i} |z_j| + \Omega \sqrt{\sum_{j=1}^{J_i} w_j^2} \leq b^0 - (a^0)^T x.$$

Here, the vector (z, w, x) are the variables and the subvector x will satisfy the uncertain constraint $a_i^T x \geq b_i \; \forall \{(a_i, b_i) = (a_i^0, b_i^0) + \sum_{j=1}^{J_i} \zeta_i^j (a_i^j, b_i^j) | \zeta \in B_i\}$ with probability at least $1 - \exp\{-\Omega^2/2\}$. Not having to satisfy these constraints with probability 1 has the effect of producing solutions that are less conservative but with a better objective value. The robust counterpart of the portfolio problem using the box - ball perturbation set is a second-order conic program, which is a convex optimization problem that is equivalent to a quadratically constrained quadratic program and can be solved efficiently by using SeDuMi; see Sturm (1999).

8.5 Exercises

Exercise 8.1

Consider the stochastic program in Example 8.5. Formulate the extensive form of the problem and solve using the MATLAB® linprog function.

Exercise 8.2

Consider the stochastic programming Asset Liability Model (ALM) model in Example 8.3, which had the following return and liability scenarios:

Scenario (s)	Stock	Bond	Money market	Liability ($)
1	$\mu_S^1 = 17\%$	$\mu_B^1 = 12\%$	$\mu_M^1 = 13\%$	$L^1 = 1,000,000$
2	$\mu_S^2 = 15\%$	$\mu_B^2 = 9\%$	$\mu_M^2 = 10\%$	$L^2 = 1,030,000$
3	$\mu_S^3 = 7\%$	$\mu_B^3 = 17\%$	$\mu_M^3 = 10\%$	$L^3 = 1,500,000$

The deterministic version of the problem can be written as

$$\begin{array}{lll} \textit{minimize} & x_S + x_B + x_M + shortfall & \\ \textit{subject to} & x_S \quad +x_B \quad +x_M & \leq 950,000 \\ & (1+\mu_S)x_S + (1+\mu_B)x_B + (1+\mu_M)x_M + shortfall & = L \\ & x_S \geq 0, \quad x_B \geq 0, x_M \geq 0, shortfall \geq 0 & \end{array}$$

where μ_S, μ_B, μ_M is a single set of returns for the three assets, respectively, and L is a liability.

(a) Using the deterministic problem above, find the value of the stochastic solution (VSS) of the optimal solution of the stochastic programming ALM model found in Example 8.3.

(b) Find the EVPI of the ALM problem.

(c) Solve the stochastic ALM model in Example 8.3 using the L-Shaped Method.

(d) Now solve (using the MATLAB linprog function) the stochastic ALM model, but using the following scenarios:

Scenario (s)	Stock	Bond	Money market	Liability ($)
1	$\mu_S^1 = 8\%$	$\mu_B^1 = 15\%$	$\mu_M^1 = 13\%$	$L^1 = 1,075,000$
2	$\mu_S^2 = 15\%$	$\mu_B^2 = 8\%$	$\mu_M^2 = 10\%$	$L^2 = 1,025,000$
3	$\mu_S^3 = 20\%$	$\mu_B^3 = 8\%$	$\mu_M^3 = 11\%$	$L^3 = 1,100,000$

(e) Find the VSS and EVPI based on the optimal solution of (d).

Exercise 8.3

Suppose that a product is shipped from 2 plants to 3 warehouses where the shipping cost from plant i to warehouse j is c_{ij} and the capacity of the plants are as follows:

	Warehouse 1	Warehouse 2	Warehouse 3	Capacity
Plant 1	22	13	9	65
Plant 2	17	8	23	80

The demand at a warehouse is uncertain and the possible realizations are given by the following low-, medium-, and high-demand scenarios for each warehouse.

Scenario (s)	Warehouse 1	Warehouse 2	Warehouse 3
1	19	27	39
2	44	50	62
3	72	68	76

A penalty or reward will be incurred for any shortage or surplus in a warehouse as follows:

	Warehouse 1	Warehouse 2	Warehouse 3
Shortage penalty	16	14	21
Surplus penalty	19	28	33

(a) Formulate the problem of finding the minimum expected cost shipping flows from plants to warehouses under the three scenarios as a stochastic program.

(b) Solve the model in (a) using the MATLAB function linprog.

Exercise 8.4

Consider the following transportation contract procurement problem. A truckload (TL) carrier transports shipments for a single shipper from origin to destination in dedicated trucks, i.e., these trucks will only be used to deliver for the particular shipper and will not deliver for any other shipper on route to the destination. Typically, a shipper will contract with a TL carrier for a certain volume of trucks with a fixed price per truckload of capacity based on estimated demand. However, the actual realized demand can vary from the estimated volume. If a shipper overestimates capacity, then the TL carrier will incur an opportunity or repositioning cost (i.e., a carrier may have to drive an empty truck since the shipper may not need the capacity or have trucks idle), and thus the shipper is usually penalized in this situation. If a shipper underestimates capacity, then a carrier may not have spare capacity and so the shipper must go to the spot market (i.e., market for immediate truckload capacity), which is more expensive.

One remedy is to design a flexible contract that allows a shipper to decide on how much (fixed) volume to commit to now before demand realization and such that the commitment is binding no matter the actual volume that realizes and to decide how much variable volume the shipper can take on but at normal prices (i.e., lower than the spot market but more expensive than the rate for committed volume). Once the variable volume is determined, then the shipper can access up to this amount after demand realization at normal prices. If needed, the shipper can still access the spot market after demand realization. Assume that you have several scenarios for demand (volume) and probabilities estimated for each scenario. Formulate the problem of designing a flexible contract for a shipper as a stochastic program.

Exercise 8.5

Consider the linear program

$$\begin{array}{ll} \textit{minimize} & 4x_1 + 3x_2 \\ \textit{subject to} & a_1x_1 + a_2x_2 \geq b_1. \end{array}$$

Let the instance $c_1 = 4$, $c_2 = 2$, $a_1 = 2, a_2 = 1$, and $b_1 = 1$ be called the nominal problem. Suppose that constraint coefficients a_1 and b_1 are considered uncertain where all possible $(a_1, b_1) \in \mathcal{U}$ where

$$\mathcal{U} = \{\begin{bmatrix}1.99\\1.00\end{bmatrix}^T, \begin{bmatrix}1.99\\0.99\end{bmatrix}^T, \begin{bmatrix}2.00\\1.00\end{bmatrix}^T, \begin{bmatrix}2.01\\1.01\end{bmatrix}^T, \begin{bmatrix}2.01\\0.99\end{bmatrix}^T\}.$$

(a) Solve the nominal problem.

(b) Formulate the robust counterpart RC of the nominal problem and solve and compare with the optimal solution of the nominal problem.

Exercise 8.6

Solve the following robust portfolio optimization problem from Example 8.8 for the case $n = 10$.

$$\begin{array}{lll} \textit{minimize} & x_n & \\ \textit{subject to} & (a^0 + \sum_{i=1}^{n-1} \zeta_i a^i)^T x & \leq (b^0 + \sum_{i=1}^{n-1} \zeta_i b^i) \quad \forall \zeta \in B \\ & \sum_{i=1}^{n} x_i = 1 & \\ & x_i \geq 0 & i = 1, ..., n \end{array}$$

where

$$B = \{\zeta \in R^{n-1} | -1 \leq \zeta_i \leq 1, \, i = 1, ..., n-1\}$$

$$b^i = 0 \text{ for } i = 0, ..., n-1$$

$$a^0 = (-\mu_1, -\mu_2 ..., -\mu_n, -1)$$

$$a^i = \sigma_i e_i^T \text{ for } i = 1, ..., n-1$$

where e_i is the vector of n components with a 1 in the *ith* position and 0 elsewhere. (Hint: Use the most tractable formulation.)

Exercise 8.7

Formulate the robust portfolio optimization problem in Exercise 8.6 in the most tractable form when $B = \{\zeta \in R^{J_i} | \; \|\zeta\|_\infty \leq 1, \|\zeta\|_2 \leq \Omega\}$, i.e., the Ball-Box perturbation set.

Exercise 8.8
Consider the uncertainty constraint $a^T x \leq b \ \forall (a,b) \in \mathcal{U}$ and its representation $a^T x \geq b \ \forall \{(a,b) = (a^0, b^0) + \sum_{j=1}^{J} \zeta_i^j (a^j, b^j) | \zeta \in B\}$. Derive a tractable form of the uncertainty constraint when $B = \{\zeta \in R^J | \, \|\zeta\|_2 \leq \Omega\}$ where J is the number of uncertain parameters in (a,b).

Notes and References

Stochastic programs were considered in Dantzig (1955) and some early applications can be found in Ferguson and Dantzig (1956). In this chapter, only two-stage stochastic programming with recourse was considered. The L-Shaped method was developed for two-stage convex stochastic programming by Van Slyke and Wets (1969) and later extended to multi-stage stochastic programming with recourse see Birge (1985). An alternative form of stochastic programming considers probabilistic constraints and these models are referred to as chance-constrained programs; see Charnes and Cooper (1959). Today the field of stochastic programming has become an important framework for modeling problems and many extensions have been considered, e.g., stochastic integer programming by Haneveld and van der Vlerk (1999). Stochastic programs have found applications in many areas such as financial planning and engineering, see Mulvey and Vladimirou (1992), Ziemba and Vickson (1975), and Zenios (2008), production and supply chain management, see Santoso et al. (2005), logistics, see Laporte et al. (2002), and energy and power systems, see Takriti et al. (1996). Major references for stochastic programming include Birge and Louveaux (2011), Kall and Wallace (1994), and Kall and Myer (2011).

Robust optimization is a relatively newer field of study that has experienced dramatic growth and research interest during the last decade and like stochastic programming has found extensive application in many areas. A major impetus for robust optimization is the ability to incorporate uncertainty without necessarily having a probability distribution of random outcomes while retaining computational tractability; see Ben-Tal et al. (2009) and Bertsimas and Sim (2006). The definitive reference for robust optimization is by Ben-Tal et al. (2009). Recent work involves integrating both stochastic programming and robust optimization ideas in what is known as distributionally robust optimization; see Delage and Ye (2010) and Li and Kwon (2013).

A

Linear Algebra Review

Definition A.1 *A matrix A is a rectangular array of numbers. The size of A is denoted by $m \times n$ where m is the number of rows and n is the number of columns of A. If $m = n$, then A is called a square matrix.*

Example A.2
The following matrices

$$A = \begin{bmatrix} 1 & 4 & -3 \\ -12 & 9 & 2 \end{bmatrix}, \quad B = \begin{bmatrix} 2 \\ -4 \\ 5 \end{bmatrix}, C = \begin{bmatrix} 3 & 1 & 2 \\ 5 & 2 & 8 \\ 1 & 7 & 4 \end{bmatrix}, D = \begin{bmatrix} 1 & 0 & -5 \end{bmatrix}$$

have dimension $2 \times 3, 3 \times 1, 3 \times 3$, and 1×3, respectively. The matrix C is a square matrix with $m = n = 2$.

Let A be a $m \times n$ matrix. We denote a_{ij} as the element of A that is in the ith row and jth column. Then, the matrix A can be specified as $A = [a_{ij}]$ for $1 \leq i \leq m$ and $1 \leq j \leq n$.

Definition A.3
A vector v of dimension k is a matrix of size $k \times 1$ (a column vector) or size $1 \times k$ (row vector).

Example A.4

$$q = \begin{bmatrix} 1 & -9 & 3 \end{bmatrix} \text{ is a row vector of dimension } 1 \times 3$$

$$w = \begin{bmatrix} 9 \\ 2 \end{bmatrix} \text{ is a column vector of dimension } 2 \times 1.$$

Definition A.5
Given a matrix A, one can generate another matrix by taking the ith row of A and making it the ith column of a new matrix and so on. The resulting matrix is called the transpose of A and is denoted by A^T.

Example A.6

$$\text{If } A = \begin{bmatrix} 1 & 4 & -3 \\ -12 & 9 & 2 \end{bmatrix}, \text{ then } A^T = \begin{bmatrix} 1 & -12 \\ 4 & 9 \\ -3 & 2 \end{bmatrix}.$$

Definition A.7
A matrix A with the property that $A = A^T$ is called a symmetric matrix.

Example A.8

If $A = \begin{bmatrix} 1 & -1 & -7 \\ -1 & 2 & 5 \\ -7 & 5 & 3 \end{bmatrix}$, then $A^T = A$ so A is symmetric.

Definition A.9

A set of vectors $V = \{ v_1, v_2, ..., v_l\}$ each with the same dimension are said to be linearly independent if

$$\alpha_1 v_1 + \alpha_2 v_2 + \cdots \alpha_l v_l = 0 \text{ implies that } \alpha_1 = \alpha_2 = \cdots = \alpha_l = 0,$$

i.e., all scalars are 0. Otherwise the set of vectors V are said to be linearly dependent.

Definition A.10

A square $m \times m$ matrix A is said to be invertible if there exists a square $m \times m$ matrix B such that $AB = I = BA$ where I is the $m \times m$ identity matrix. B is called the inverse of A and is denoted as $B = A^{-1}$. A matrix A that has an inverse is said to be invertible of non-singular.

Theorem A.11

A square $m \times m$ matrix A is invertible if and only if the m columns (rows) of A form a linearly independent set of vectors.

The inverse of a square matrix A plays an important role in the solution of linear systems of equations of the form

$$Ax = b$$

since the solution can be represented mathematically as $x = A^{-1}b$.

For instance, the linear system

$$\begin{aligned} 3x_1 + 5x_2 &= 11 \\ 5x_1 - 2x_2 &= 8 \end{aligned}$$

can be represented in matrix form by letting

$$A = \begin{bmatrix} 3 & 5 \\ 5 & -2 \end{bmatrix}, \; b = \begin{bmatrix} 11 \\ 8 \end{bmatrix}, \text{ and } x = \begin{bmatrix} x_1 \\ x_2 \end{bmatrix},$$

then the solution is

$$x = A^{-1}b = \begin{bmatrix} 3 & 5 \\ 5 & -2 \end{bmatrix}^{-1} \begin{bmatrix} 11 \\ 8 \end{bmatrix} = \begin{bmatrix} 2 \\ 1 \end{bmatrix}.$$

The inverse of A is $A^{-1} = \begin{bmatrix} 0.0645 & 0.1613 \\ 0.1613 & -0.0968 \end{bmatrix}$.

The inverse of square matrices will be obtained in this book via MATLAB through the solving of the corresponding system of equations. The details for generating inverses using methods such as Gaussian Elimination or equivalent matrix factorizations can be found in Golub and van Loan (1996).

Special Matrices

Definition A.11

(1) An $n \times n$ symmetric matrix A is said to be positive semi-definite (PSD) if $x^T Ax \geq 0$ for all vectors x with dimension n.

(2) An $n \times n$ symmetric matrix A is said to be positive definite (PD) if $x^T Ax > 0$ for all vectors x with dimension n and $x \neq 0$.

Theorem A.12

(a) A positive definite matrix A is invertible.

(b) If the determinant of a matrix A is non-zero, then A is invertible.

Knowing whether a matrix is PSD or PD is useful in linear and quadratic programming, but the definitions above can be challenging to use to show that a matrix is PSD or PD. An alternative test that is helpful, at least in the context of this book, is the following test for determining whether a matrix is PD. First, we recall the definition of a determinant of a matrix.

Definition A.13

Let $A = [a_{ij}]$ be an $m \times m$ matrix. Then, the determinant of A (denoted by det A) is

$$\det A = \sum_{i=1}^{m} a_{i1} A_{i1}$$

where A_{i1} is the ith cofactor of a_{i1}, which is equal to $(-1)^{i+1}$ times the determinant of the matrix obtained by removing the ith row of A and the first column. The determinant of a 1×1 matrix is equal to the single element of the matrix and then the determinant of a 2×2 matrix

$$M = \begin{bmatrix} a & b \\ c & d \end{bmatrix} \text{ is } ad - bc.$$

Thus, to compute a determinant of a matrix A, one successively applies the definition.

Example A.14

Let

$$A = \begin{bmatrix} 3 & 1 & 2 \\ 5 & 2 & 8 \\ 1 & 7 & 4 \end{bmatrix}$$

then, $\det A = 3A_{11} + 5A_{21} + 1A_{31}$

$$= 3 \det \begin{bmatrix} 2 & 8 \\ 7 & 4 \end{bmatrix} - 5 \det \begin{bmatrix} 1 & 2 \\ 7 & 4 \end{bmatrix} + 1 \det \begin{bmatrix} 1 & 2 \\ 2 & 8 \end{bmatrix}$$

$$= 3(8 - 56) - 5(4 - 14) + 1(8 - 4)$$

$$= -144 + 50 + 4 = -90.$$

Principal Minor Test for Determining Positive Definiteness

Theorem A.15

Let A be an $m \times m$ symmetric matrix. Let Δ_l be the determinant of the upper $l \times l$ submatrix of A for $1 \leq l \leq m$. Δ_l is called the lth principal minor of A. If $\Delta_l > 0$ for $l = 1, ..., m$, then A is positive definite.

Example A.16

Let

$$A = \begin{bmatrix} 2 & -1 & -1 \\ -1 & 2 & 1 \\ -1 & 1 & 2 \end{bmatrix}.$$

Then, $\Delta_1 = \det[2] = 2 > 0$, $\Delta_2 = \det \begin{bmatrix} 2 & -1 \\ -1 & 2 \end{bmatrix} = 3 > 0$, and $\Delta_3 = \det \begin{bmatrix} 2 & -1 & -1 \\ -1 & 2 & 1 \\ -1 & 1 & 2 \end{bmatrix}$
$= 2 > 0$ so A is positive definite.

Eigenvalue Test for Positive Definiteness

Another test consists of checking the eigenvalues of A. Recall that λ is an eigenvalue for A (assume A is $m \times m$ symmetric) if it satisfies the $Ax = \lambda x$ for some non-zero vector x. In particular, the eigenvalues of A are the roots of the (characteristic) equation $\det(A - \lambda I) = 0$.

Theorem A.17

Suppose A is symmetric. If all eigenvalues of A are positive, then A is positive definite.

Example A.18

Let $A = \begin{bmatrix} 1 & -2 & 0 \\ -2 & 1 & 0 \\ 0 & 0 & 1 \end{bmatrix}$, then the eigenvalues are $\lambda = -1, 1$, and 3, and so A is not positive definite. On the other hand, if $A = \begin{bmatrix} 2 & -1 & -1 \\ -1 & 2 & 1 \\ -1 & 1 & 2 \end{bmatrix}$, then the eigenvalues are $\lambda = 1, 1$, and 4, which indicates that this matrix is positive definite.

Some Fundamental Spaces of Linear Algebra

Let A be an $m \times n$ matrix.

(1) The set $R(A) = \{y | y = Ax$ for some vector x of dimension $n\}$ is called the column space of A.

(2) The set $R(A^T) = \{y | y = A^T z$ for some vector z of dimension $m\}$ is called the row space of A.

(3)The set $N(A) = \{p | Ap = 0\}$ is called the null space of A.

Notes and References

See Golub and van Loan (1996) for further details regarding linear algebra.

Bibliography

Ahuja, R. K., Magnanti, T. L., and J.B. Orlin. 1993. *Network Flows.* Upper Saddle River: Prentice Hall.

Avriel, M. 2003. *Non-linear Programming: Analysis and Methods.* Mineola: Courier Dover Publications.

Barnes, E.R. 1986. A variation of Karmarkar's algorithm for solving linear programming problems. *Mathematical Programming*, **36**, 174–182.

Barnhart, C., Johnson, E.L., Nemhauser, G.L., and M.W.P. Savelsbergh. 1998. Branch-and-price: Column generation for solving huge integer programs. *Operations Research*, **46**(3), 316–329.

Bartels, R.H., and G.H. Golub. 1969. The simplex method of linear programming using the LU decomposition. *Communications of the ACM*, **12**, 266–268.

Bazaraa, M.S., Jarvis, J.J., and H. D. Sherali. 1977. *Linear Programming and Network Flows.* New York: John Wiley & Sons, Inc.

Bazaraa, M.S., Sherali, H.D., and C.M. Shetty. 2006. *Nonlinear Programming: Theory and Algorithms.* New York: John Wiley & Sons, Inc.

Beale, E.M.L. 1954. An alternative method for linear programming. *Proceedings of the Cambridge Philosophical Society*, **50**, 513–523.

Beale, E.M.L. 1955. Cycling in the dual simplex method. *Naval Research Logistics Quarterly*, **2**, 269–275.

Benders, J.F. 1962. Partitioning procedures for solving mixed-variables programming problems. *Numerische Mathematik*, **4**, 238–252.

Ben-Tal, A., El Ghaoui, L., and A. Nemirovski. 2009. *Robust Optimization.* Princeton: Princeton University Press.

Bertsekas, D.P. 2003. *Nonlinear Programming.* Belmont: Athena Scientific.

Bertsimas, D., and J. Tsitsiklis. 1997. *Introduction to Linear Optimization.* Belmont: Athena Scientific.

Bertsimas, D., and M. Sim. 2006. Tractable approximation to robust conic optimization problems. *Mathematical Programming Series B*, **107**(1), 5–36.

Best, M. 2010. *Portfolio Optimization.* Boca Raton: Chapman Hall/CRC Press.

Birge, J.R. 1985. Decomposition and partitioning methods for multi-stage stochastic linear programs. *Operations Research*, **33**, 989–1007.

Birge, J.R., and F. Louveaux. 2011. *Introduction to Stochastic Programming,* Second Edition. New York: Springer Science.

Bland, R.G. 1977. New finite pivoting rules for the simplex method. *Mathematics of Operations Research*, **2**, 103–107.

Borgwardt K.H. 1982. The average number of pivot steps required by the simplex method is polynomial. *Zeitschrift für Operations Research*, **26**, 157–177.

Boyd, S., and L. Vandenberghe. 2004. *Convex Optimization.* Cambridge: Cambridge University Press.

Bradley, S.P., Hax, A.C., and T.L. Magnanti. 1977. *Applied Mathematical Programming.* Boston: Addison-Wesley Publishing Company, Inc.

Charnes, A. 1952. Optimality and degeneracy in linear programming. *Econometrica*, **20**, 160–170.

Charnes, A., and W.W. Cooper. 1959. Chance-constrained programming. *Management Science*, **5**, 73–79.

Chvátal, V. 1983. *Linear Programming.* New York: W.H. Freeman and Company.

Conejo, A.J., Castillo, E., Mínguez, R., and R. García-Bertrand. 2006. *Decomposition Techniques in Mathematical Programming: Engineering and Science Applications.* Heidelberg: Springer-Verlag.

Cottle, R. W., Pang, J.S., and R.E. Stone. 1992. *The Linear Complementarity Problem.* Boston: Academic Press, Inc.

Dantzig, G.B. 1951. Maximization of a linear function of variables subject to linear inequalities. In *Activity Analysis of Production and Allocation*, ed. T.C. Koopmans, 339–347. New York-London: Wiley & Chapman-Hall.

Dantzig, G.B. 1953. *Computational Algorithm of the Revised Simplex Method.* Report RM 1266. Santa Monica: The Rand Corporation.

Dantzig, G.B. 1955. Linear programming under uncertainty. *Management Science*, **1**, 197–206.

Dantzig, G.B. 1963. *Linear Programming and Extensions.* Princeton: Princeton University Press.

Dantzig, G.B., Orden, A., and P. Wolfe. 1955. The generalized simplex method for minimizing a linear form under linear inequality constraints. *Pacific Journal of Mathematics*, **5**, 183–195.

Dantzig, G.B., and P. Wolfe. 1960. Decomposition principle for linear programs. *Operations Research*, **8**, 101–111.

Delage E., and Y. Ye. 2010. Distributionally robust optimization under moment uncertainty with application to data-driven problems. *Operations Research*, **58**(3), 596–612.

Dennis, J.E., and R.B. Schnable. 1983. *Numerical Methods for Unconstrained Optimization.* Englewood Cliffs: Prentice Hall.

Desrochers, M., Desrosiers, J., and M. Solomon. 1992. A new optimization algorithm for the vehicle routing problem with time windows. *Operations Research*, **40**, 342–354.

Dikin, I.I. 1967. Iterative solution of problems of linear and quadratic programming (Russian). *Doklady Akademii Nauk SSSR*, **174**, 747–748.

Fang, S-C., and S. Puthenpura. 1993. *Linear Optimization and Extensions: Theory and Algorithms.* Englewood Cliffs: Prentice Hall.

Farkas, J. 1901. Theorie der einfachen Ungleichungen. *Journal für die reine und angewandte Mathematik*, **124**,1–27.

Ferguson, A., and G.B. Dantzig. 1956. The allocation of aircraft to routes: An example of linear programming under uncertain demands. *Management Science*, **3**, 45–73.

Fiacco, A.V., and G.P. McCormick. 1968. *Nonlinear Programming: Sequential Unconstrained Minimization Techniques.* New York: Wiley and Sons, Inc.

Floudas, C.A., and A. Aggarwal. 1990. A decomposition strategy for global optimal search in the pooling problem. *ORSA Journal on Computing*, **2**(3), 225–235.

Ford, L.R., Jr. and D.R. Fulkerson. 1958. A suggested computation for maximal multi-commodity network flows. *Management Science*, **5**, 97–101.

Gale, D., Kuhn, H., and A. Tucker. 1951. Linear programming and the theory of games - Chapter XII, ed. Koopmans. In *Activity Analysis of Production and Allocation.* New York: John Wiley and Sons, Inc.

Gass, S.I. 1975. *Linear Programming: Methods and Applications, 4th Edition.* New York: McGraw-Hill Publishing Company, Inc.

Gilmore P.C., and R. E. Gomory. 1961. A linear programming approach to the cutting-stock problem. *Operations Research*, **9**, 849–859.

Golub, G.H., and C.F. van Loan. 1996. *Matrix Computations*, 3rd edition, Baltimore: Johns Hopkins University Press.

Haneveld, W.K., and M.H. van der Vlerk. 1999. Stochastic integer programming: General models and algorithms. *Annals of Operations Research*, **85**, 39–57.

Haverly, C.A. 1978. Studies of the behaviour of recursion for the pooling problem. *ACM SIGMAP Bulletin*, **25**(19).

Hoffman, A.J. 1953. *Cycling in the Simplex Method.* Report 294, National Bureau of Standards. Gaithersburg, MD.

Kall, P., and J. Mayer. 2011. *Stochastic Linear Programming: Models, Theory, and Computation, Second Edition.* New York: Springer.

Kall, P., and S. Wallace. 1994. *Stochastic Programming.* Chichester: John Wiley and Sons.

Karmarkar, N. 1984. A new polynomial-time algorithm for linear programming. *Combinatorica*, **4**, 373–395.

Karush, W. 1939. *Minima of functions of several variables with inequalities as side constraints.* M.Sc. dissertation. Dept. of Mathematics, Univ. of Chicago, Chicago, Illinois.

Kedem, G., and H. Watanabe. 1983. Optimization techniques for IC layout and compaction. In *Proc. IEEE International Conference Computer Design :VLSI in Computers,* 709–713.

Khachian, L.G. 1979. A polynomial algorithm in linear programming (Russian). *Doklady Akademii Nauk SSSR*, **244**, 1093–1096.

Klee, V., and G.J. Minty. 1972. How good is the simplex algorithm?, In *Inequalities III*, ed. O. Shisha, 159–175. New York: Academic Press.

Konno, H., and H. Yamazaki. 1991. Mean absolute deviation portfolio optimization model and its application to Tokyo Stock Market. *Management Science*, **37**, 519–531.

Kuhn, H., and A. Tucker. 1951. Nonlinear programming. *Proceedings of 2nd Berkeley Symposium.* Berkeley: University of California Press, 481–492.

Laporte, G., Louveaux, F.V., and L. van Hamme. 2002. An integer L-Shaped algorithm for the capacitated vehicle routing problem with stochastic demands. *Operations Research*, **50**(3), 415–423.

Lasdon, L.S. 1970. *Optimization Theory for Large Systems.* New York: Macmillan Company.

Lasdon, L.S., Waren, A.D., Sarkar, S., and F. Palacios-Gomez. 1979. Solving the pooling problem using the generalized reduced gradient and succesive linear programming algorithms. *ACM SIGMAP Bulleting*, **27**(9).

Lemke, C.E. 1954. The dual method of solving the linear programming problem. *Naval Research Logistics Quarterly*, **1**, 36–47.

Lemke, C.E. 1962. A method of solution for quadratic programs. *Management Science*, **8**, 442–455.

Li, J.Y., and R. Kwon, 2013. Portfolio selection under model uncertainty: A penalized moment-based optimization approach. *Journal of Global Optimization*, **56**(1), 131–164.

Luenberger, D. 1998. *Investment Science.* Oxford: Oxford University Press.

Mangasarian, O. 1969. *Non-linear Programming.* New York: McGraw Hill.

Markowitz, H.M. 1952. Portfolio Selection. *Journal of Finance*, **7**, 77–91.

Mehrotra, S. 1992. On the implementation of a primal-dual interior point method. *SIAM Journal on Optimization*, **2**, 575–601.

Minkowski, Hermann 1896. *Geometrie der Zahlen.* Leipzig: Teubner.

Mizuno, S., Todd, M., and Y. Ye. 1993. On adaptive step primal-dual interior point algorithms for linear programming. *Mathematics of Operations Research,* **18**, 964–981.

Monteiro, R., and I. Adler. 1989. Interior path following primal-dual algorithms, Part I: Linear Programming. *Mathematical Programming*, **44**, 27–41.

Monteiro, R., and I. Adler. 1989. Interior path following primal-dual algorithms, Part II: Convex Quadratic Programming. *Mathematical Programming*, **44**, 43–46.

Mulvey, J.M., and H. Vladimirou. 1992. Stochastic network planning for financial planning problems. *Management Science*, **38**, 1642–1664.

Murtagh, B.A. 1981. *Advanced Linear Programming: Computation and Practice.* New York: McGraw Hill.

Murty, K.G. 1983. *Linear Programming.* New York: John Wiley & Sons, Inc.

Murty, K.G. 1988. *Linear Complementarity, Linear and Non-linear Programming.* Berlin: Heldermann Verlag.

Murty, K.G. 1992. *Network Programming.* Englewood Cliffs: Prentice Hall.

Nesterov, Y.E., and A. Nemirovskii. 1992. *Interior Point Polynomial Methods in Convex Programming.* Philadelphia: SIAM Publications.

Nocedal, J., and S.J. Wright. 2006. *Numerical Optimization.* New York: Springer.

Orchard-Hayes, W. 1954. *Background, development, and extensions of revised simplex method.* Report RM 1433. Santa Monica: The Rand Corporation.

Orden, A. 1956. The trannshipment problem. *Management Science*, **2**, 276-85.

Ortega, J.M., and W.C. Rheinboldt. 1970. *Iterative Solution of Nonlinear Equations in Several Variables.* New York: Academic Press.

Renegar, J. 1988. A polynomial-time algorithm, based on Newton's method for linear programming. *Mathematical Programming*, **40**, 59–93.

Rockafellar, R.T. 1970. *Convex Analysis*, Vol. 28 of Princeton Mathematics Series. Princeton: Princeton University Press.

Roos, C., Terlaky, T., and J. Vial. 2006. *Interior Point Methods for Linear Optimization.* Heidelberg/Boston: Springer Science.

Saigal, R. 1995. *Linear Programming: A Modern Integrated Analysis.* Boston: Kluwer Academic Publishers.

Santoso,T., Ahmed, S., Goetschalckx, M., and A. Shapiro. 2005. A stochastic programming approach for supply chain network design under uncertainty. *European Journal of Operational Research*, **167**, 96–115.

Simchi-Levi, D., Chen, X., and J. Bramel. 2004. *The Logic of Logistics: Theory, Algorithms and Applications for Logistics and Supply Chain Management, 2nd Edition.* New York: Springer.

Simmonard, M. 1966. *Linear Programming.* Englewood Cliffs: Prentice Hall.

Smale, S. 1983. On the average number of steps of the simplex method. *Mathematical Programming*, **27**, 241–262.

Steinbach, M. 2001. Markowitz Revisited: Mean-Variance Models in Financial Portfolio Analysis. *SIAM Review*, **43**(1), 31–85.

Stigler, G.J. 1945. The cost of subsistence. *The Journal of Farm Economics*, **27**(2), 303–314.

Sturm, J.F. 1999. Using SeDumi 1.02, a MATLAB toolbox for optimization over symmetric cones. *Optimization Methods and Software*, **11–12**, 625–653.

Takriti, S. Birge, J.R., and E. Long. 1996. A stochastic model for the unit commitment problem. *IEEE Transactions on Power Systems*, **11**(3), 1497–1508.

Tebboth, J.R. 2001. A computational study of Dantzig-Wolfe decomposition, Ph.D. thesis, University of Buckingham.

Van Slyke, R., and R. J-B Wets. 1969. L-shaped linear programs with applications to control and stochastic programming. *SIAM Journal on Applied Mathematics,* **17**, 638–663.

Vanderbei, R.J., Meketon, M., and B. Freedman. 1986. A modification of Karmarkar's linear programming algorithm. *Algorithmica*, **1**, 395–407.

Vanderbei, R.J. 2008. *Linear Programming: Foundations and Extensions*, third edition. New York: Springer.

von Neumann, J. 1945. A model of general economic equilibrium. *Review of Economic Studies*, **13**, 1–9.

Williams, H.P. 1999. *Model Building in Mathematical Programming, 4th Edition.* New York: John Wiley and Sons, Inc.

Winston, W., and M. Venkataramanan. 2003. *Introduction to Mathematical Programming.* Pacific Grove: Brooks/Cole - Thompson Learning.

Wolfe, P. 1959. The Simplex Method for Quadratic Programming. *Econometrica*, **27**(3), 382–398.

Wright, S.J. 1997. *Primal-Dual Interior Point Methods.* Philadelphia: SIAM Publications.

Ye, Y. 1997. *Interior Point Algorithms: Theory and Analysis.* New York: John Wiley and Sons, Inc.

Zenios, S. 2008. *Practical Financial Optimization: Decision Making for Financial Engineers.* New York: Wiley-Blackwell.

Ziemba, W.T., and R.G. Vickson. 1975. *Stochastic Optimization Models in Finance.* New York: Academic Press.

Index